BIOCHEMISTRY OF CATALYSTS

BIOCHEMISTRY OF CATALYSTS

By

Dr. G.P. Garg

DISCOVERY PUBLISHING HOUSE PVT. LTD.

NEW DELHI-110 002

Published by:
Tilak Wasan
DISCOVERY PUBLISHING HOUSE PVT. LTD.
4383/4B, Ansari Road, Darya Ganj
New Delhi-110 002 (India)
Phone : +91-11-23279245; 23253475; 43596065
E-mail : discoverybooksindia@gmail.com
discoverypublishinghouse@gmail.com
namitwasan9@gmail.com
web : www.discoverypublishinggroup.com

Edition: **2019**

ISBN: 978-81-8356-598-1

Biochemistry of Catalysts

Printed at:
Infinity Imaging Systems
Delhi

Preface

Biochemistry today has made spectacular progress in unraveling the mysteries of animate nature. This progress has allowed us to gain deeper insight into the principles of vital activity and has to a very significant extent stimulated the development of applied disciplines, especially medicine. The present title *"Biochemistry of Catalysts"* is intended for those who wish to understand living organisms, especially man. Biochemistry is essential for this purpose, but it would be almost impossible for a student to survey on his own the massive body of existing knowledge, constantly augmented by a remarkable torrent of brilliant discoveries. The purpose of the book, then is to organize our knowledge into something that can be comprehended in a relatively short time and still convey a reasonable complete picture of the chemical structure and function of man. Readability without sacrifice of coverage has been a prime goal. An important device in gaining that goal is to keep attention constantly focused on function, with repeated use of rationalization to show that the chemical facts are not isolated, but part of a whole.

Biochemistry has two major goals as a fundamental science, it treats the vital functions from the standpoint of physical chemistry, and as an applied discipline, it points out practical applications for the wealth of scientific knowledge it has acquired. The dual purpose has been, as far as possible, take into account in the writing of this book. We have also tried to summarize our pedagogical experience in teaching biochemistry to students specializing in medicine and pharmacy. The material of this book has been organized according to the principle of functionally in order to trace the close relationship between the functions of the living organism and the structures of its constituents molecules as well as the chemical and physico-chemical process in which they are involved.

The aim of this book is to present a core of biochemical knowledge that is desirable for undergraduate and postgraduate students and also those involved in the field of medical, microbiology, biotechnology and pharmaceutical. Every attempt has been made to keep abrest of the advances in the subject and at the same time to include the fundamentals.

—Author

CONTENTS

Enzyme Kinetics

A catalyst is a substance that accelerates a chemical reaction without itself undergoing any net change. The rate enhancements achieved by many enzymes are extraordinarily high. For example, carbonic anhydrase, an enzyme found in red blood cells, catalyzes the reaction

$$CO_2 + H_2O \rightarrow H_2CO_3 \quad \text{... (1.1)}$$

In the presence of the enzyme, this reaction occurs about 10^7 times as rapidly as it does in the absence of the enzyme. One molecule of carbonic anhydrase can hydrate about 10^6 molecules of CO_2 a second. Kinetic analysis is one of the most basic topics of enzymology.

Such studies reveal not only how fast an enzyme can function, but also its preferences for various reactants (or substrates as they usually are called), the effect of substrate concentration on the reaction rate, and the sensitivity of the enzyme to specific inhibitors or activators. By studying the alterations in rate under different conditions we often gain clues to the mechanism of the reaction.

HISTORY

The existence of catalysts in biological materials was recognized as early as 1835 by Jöns Jakob Berzelius, the Swedish chemist who discovered several elements, introduced the way of writing chemical symbols, and coined the term "catalysis." Berzelius noted that potatoes contained something that catalyzed the breakdown of starch, and he suggested that all natural products are formed under the influence of such catalysts. But the chemical nature of biological catalysts was unknown, and it remained a mystery for many years. In the period between 1850 and 1860, Louis Pasteur demonstrated that fermentation, the anaerobic breakdown of sugar to CO_2 and ethanol, occurred in the presence of living cells and did not occur in a flask that was capped after any cells that it contained had been killed by heat. Then in 1897, Eduard Buchner discovered by accident that fermentation was catalyzed by a clear juice that he had prepared by grinding yeast with sand and filtering out the unbroken cells. Looking for a way to preserve the juice, Buchner had added sugar. It probably was a disappointment to him that the sugar was broken down rapidly and the mixture frothed with CO_2. But Buchner's discovery made it possible to explore metabolic processes such as fermentation in a greatly simplified system, without having to deal with the complexities of cell growth and multiplication, and without the barriers imposed by cell walls or membranes.

Arthur Harden and William Young soon showed that yeast extracts contained two different types of molecules, both of which were necessary for fermentation to occur. Some were small, dialyzable, heat-stable molecules such as inorganic phosphate; others were much larger, nondialyzable molecules—the enzymes— that were destroyed easily by heat. Although early investigators surmised that enzymes might be proteins, this remained in dispute until 1927, when James Sumner succeeded in purifying and crystalizing the enzyme urease from beans.

In the 1930s, John Northrop isolated and characterized a series of digestive enzymes, generalizing Sumner's conclusion that enzymes are proteins. Since then thousands of different enzymes have been purified, and the structures of many of them have been solved to atomic resolution; almost all of these molecules have proved to be proteins. Surprisingly, however, recent work has shown that some RNA molecules also have enzymatic activity.

Oxidoreductases: $A^- + B \rightleftharpoons A + B^-$

Transferases: $A - B + C \rightleftharpoons A + B - C$

Hydrolases: $A - B + H_2O \rightleftharpoons A - H + B - OH$

Lyases: $\underset{A}{\overset{X}{|}} - \underset{B}{\overset{Y}{|}} \rightleftharpoons A = B + X - Y$

Isomerases: $\underset{A}{\overset{X}{|}} - \underset{B}{\overset{Y}{|}} \rightleftharpoons \underset{A}{\overset{Y}{|}} - \underset{B}{\overset{X}{|}}$

Ligases (synthases): $A + B \rightleftharpoons A - B$

Fig. 1.1. The six main classes of enzymes and the reactions they catalyze.

NOMENCLATURE

Enzymes often are known by common names obtained by adding the suffix "-ase" to the name of the substrate or to the reaction that they catalyze. Thus, glucose oxidase is an enzyme that catalyzes the oxidation of glucose; glucose-6-phosphatase catalyzes the hydrolysis of phosphate from glucose-6-phosphate; and urease catalyzes the hydrolysis of urea. Common names also are used for some groups of enzymes. For example, an enzyme that transfers a phosphate group from ATP to another molecule is usually called a "kinase," instead of the more formal "phosphotransferase."

A systematic scheme for classifying enzymes was adopted in 1972 by the International Union of Biochemistry. In this scheme, each enzyme is designated by four numbers that indicate the main class, subclass, subsubclass, and the serial number of the enzyme in its subsubclass.

The six main classes are (1) *oxidoreductases,* (2) *transferases,* (3) *hydrolases*, (4) *lyases*, (5) *isomerases*, and (6) *ligases*, or *synthases.* Oxidoreductases catalyze oxidation–reduction reactions. Transferases catalyze the transfer of a functional group from one molecule to another. Hydrolases catalyze bond cleavage by the introduction of water. Lyases catalyze the removal of a group to form a double bond or the addition of a group to a double bond. Isomerases catalyze intramolecular rearrangements, and ligases catalyze reactions that join two molecules. Many enzymes require additional small molecules called cofactors for their activity.

Cofactors can be simple inorganic ions such as Mg^{2+} or complex organic molecules known as *coenzymes*. The cofactor usually binds tightly to a special site on the enzyme. An enzyme lacking an essential cofactor is called an *apoenzyme*, and the intact enzyme with the bound cofactor is called the *holoenzyme*.

BASIC ASPECTS OF CHEMICAL KINETICS

Before we delve into enzyme kinetics, we need to discuss some of the basic principles that apply to the kinetics of both enzymatic and nonenzymatic reactions. Let's first consider a nonenzymatic reaction that converts a single reactant (R) into a product (P):

$$R \rightarrow P \qquad ...(1.2)$$

To measure the velocity of the reaction (v), we plot the concentration of R as a function of time. The rate at any particular time *(t)* is

$$v = -\frac{d[R]}{dt} \qquad ...(1.3)$$

where $d[R]/dt$ is the slope of the plot at that time. The minus sign is needed because v is defined by convention to be a positive number, whereas $d[R]/dt$ is always negative ([R] decreases with time). However, we might equally well choose to measure the increase in the concentration of the product as a function of time, in which case we express the rate as $v = d[P]/dt$.

For a simple reaction of this type, the rate at any given time usually is found to be proportional to the remaining concentration of the reactant:

$$v = k[R] \qquad ...(1.4)$$

The proportionality constant k is the *rate constant*. The rate constant is independent of the concentration of the reactant, but it can depend on other parameters, such as temperature or pH, and, as we shall see, it may be altered by a catalyst. It has dimensions of reciprocal seconds (s^{-1}). Combining equations we have

$$\frac{d[R]}{dt} = -k[R] \qquad ...(1.5)$$

A reaction of this type is said to follow *first-order kinetics* because the rate is proportional to the concentration of a single species raised to the first power (Fig. 1.2). An example is the decay of a radioactive isotope such as ^{14}C. The rate of decay at any time (the number of radioactive disintegrations per second) is simply proportional to the amount of ^{14}C present. The rate constant for this extremely slow nuclear reaction is 8×10^{-12} s^{-1}. Another example is the initial electron-transfer reaction that occurs when photosynthetic organisms are excited with light. In this case there actually are two reactants, the electron donor and the acceptor, but they are held close together on a protein so that they react as a unit; the excited complex simply decays spontaneously to a more stable state. This process occurs with a rate constant of 3×10^{11} s^{-1}, which makes it one of the fastest reactions known.

$$v = -\frac{d[R]}{dt} = \frac{d[P]}{dt}$$

A slightly more complicated situation arises if the reaction is reversible:

$$R \underset{k_{-1}}{\overset{k_1}{\rightleftharpoons}} P \qquad ...(1.6)$$

Here k_1 is the rate constant for the forward reaction, and k_{-1}, is that for the back reaction. In a reversible reaction, the rate equation becomes

$$\frac{d[R]}{dt} = -k_1[R] + k_{-1}[P] \qquad ...(1.7)$$

The term — k_1[R] is the same as the term k_{-1} [P] describes the formation of R from P. In this case, [R] and [P] proceed from their initial values to their final, equilibrium values ($[R]_{eq}$ and $[P]_{eq}$) which generally are not zero. At equilibrium, $d[R]/dt$ must go to zero, which requires that

$$k_1[R]_{eq} = K_{-1}[P]_{eq} \qquad ...(1.8)$$

Rearranging this expression gives

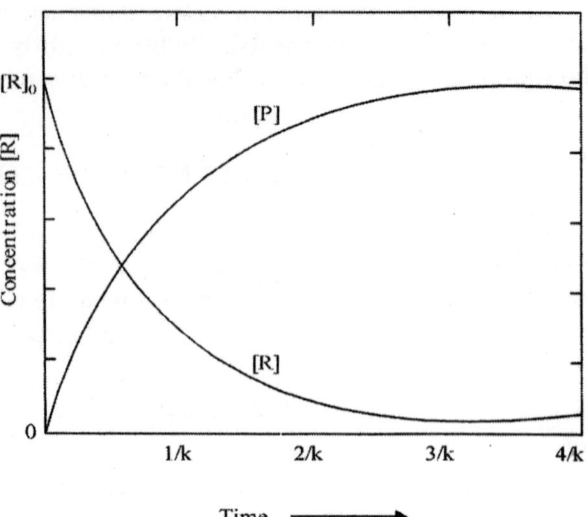

Fig. 1.2. The kinetics of a first-order reaction in which a single reactant (R) is converted irreversibly to a product (P). The concentrations of R and P are plotted as functions of time. The rate (v) at any given time can be obtained from the slope of either curve:

$$\frac{[P]_{eq}}{[R]_{eq}} = \frac{k_1}{k_{-1}} = K_{eq} \qquad ...(1.9)$$

where K_{eq} is the equilibrium constant.

Now consider a reaction involving two reactants, A and B :

$$A + B \xrightarrow{k} P \qquad ...(1.10)$$

The rate of such a bimolecular reaction usually depends on the concentrations of both reactants :

$$\frac{d[P]}{dt} = k[A][B] \qquad ...(1.11)$$

The reaction is said to follow *second-order kinetics* because its rate is proportional to a product of two concentrations. The kinetics are first order in either [A] or [B] alone, but second order overall. The rate constant for a second-order reaction has the dimensions of $M^{-1}s^{-1}$.

Thermodynamic Factors

The equilibrium constant K_{eq} for a reaction is strictly a function of thermodynamic factors. The most pertinent thermodynamic quantity here is the change in free energy that occurs in the reaction ΔG. ΔG and K_{eq} are related by the expression $\Delta G° = -RT \ln K_{eq}$. If the free energy of the products is less than the free energy of the reactants ($\Delta G < 0$), then the equilibrium constant favors the formation of products; if the free energy of the products exceeds that of the reactants ($\Delta G > 0$), the reverse reaction is favored.

We noted above that the equilibrium constant for a one-step reaction is equal to the ratio of the rate constants in the forward and reverse directions. However, the value of K_{eq} does not tell us how long it takes the reaction to reach equilibrium because it says nothing about the magnitudes of the individual rate constants. These may be very large or very small. The overall ΔG for a reaction therefore indicates only whether the reaction is possible, not how rapidly the reaction occurs.

To understand the kinetics, we must look into the machanism of the reaction. The rate at which two molecules react depends partly on how frequently the molecules collide. Collisions occur as a result of random diffusion of the reactants in the solution. The number of collisions per second is proportional to the product of the two concentrations, and the second-order rate constant k for the reaction includes a proportionality factor for this relationship.

The rate constant also includes a factor that gives the fraction of the collisions that are effective. If every collision were to result in a reaction, this second factor would be 1 and the rate constant for the reaction of two small molecules in aqueous solution would be about 10^{11} $M^{-1}S^{-1}$. Because of their large masses, proteins diffuse relatively slowly, so the frequency at which a protein and a small molecule collide is lower than the collision frequency for two small molecules. As a consequence, a reaction between a protein and a small molecule has a maximum rate constant on the order of 10^8 to 10^9 $M^{-1}S^{-1}$. But not every cullision results in a reaction. What determines the fraction of the collisions that are effective? A partial answer is that the colliding species must have a certain critical energy in order to surmount a barrier that separates the reactants fromJhe products. This is illustrated schematically. The surface represents the energy of a system in which a proton can be bound to either of two molecules.

Suppose the proton is initially on molecule A, and we are interested in how rapidly it moves to molecule B. As the proton moves from one place to the other, its electrostatic interactions with molecule A become less favorable, and the interactions with B improve. The energy of the system goes through a maximum when the proton is at an intermediate position. At this point, the system is said to be in the *transition state*. The probability that a collision leads to a reaction depends, in part, on the probability that the molecules collide with enough energy to reach this state.

The amount of free energy required to reach the transition state is called the *activation free energy*, $\Delta G^\ddagger$. We can equate $\Delta G^\ddagger$ to —RT ln $K^\ddagger$, where $K^\ddagger$ is an equilibrium constant for the formation of the transition state from the reactants. The fraction of the reactants that are in the transition state at any given moment is given approximately by $K^\ddagger$, or $e^{-\Delta G^\ddagger/RT}$. We therefore can write the overall rate constant for a reaction as

$$k = k_0 e^{-\Delta G^\ddagger / RT} \qquad \text{...(1.12)}$$

Here k_0 is an intrinsic rate constant for conversion of the transition state into the products and typically depends on the nuclear vibration frequency of a bond that is being formed or broken.

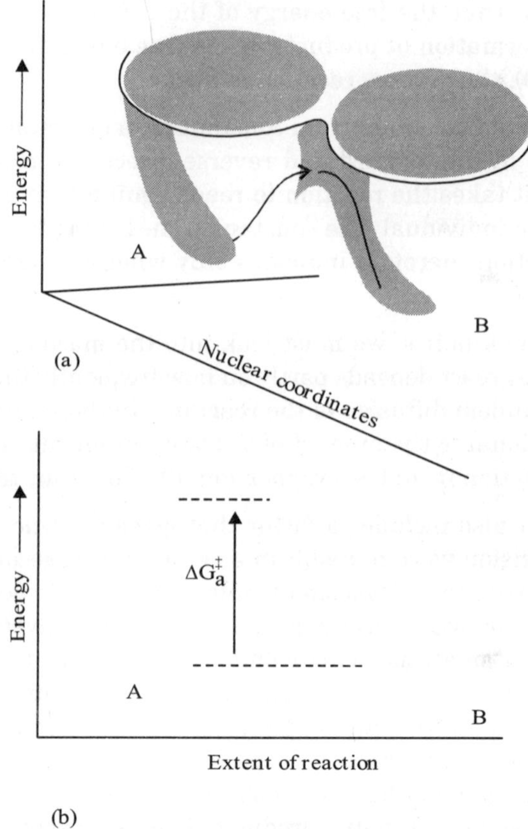

Fig. 1.3. Schematic energy diagrams for a reaction in which a proton moves from one molecule to another. In drawing *(a)*, coordinates in the plane at the bottom represent the location of the proton. The free energy of the system for a particular set of coordinates is represented by the distance of the cuplike surface above the plane. The two minima in the energy surface indicate the positions of the proton when it is bound optimally to one molecule or the other. The best route along the surface from one of these minima to the other goes through a pass, or saddle point. Drawing *(b)* shows a plot of the free energy as a function of distance along the optimal route over this pass. The activation free energy of the reaction ($\Delta G^{\ddagger}_{a}$) is the difference between the energies at the pass and at the starting point.

Energy of Activation

Indicates that, if $\Delta G^{\ddagger}$ is greater than zero as is most offer the case, a reaction can be sped up either by decreasing $\Delta G^{\ddagger}$ or by raising the temperature. Little can usually be done to change k_0. For living organisms, it generally is impractical to raise the temperature because many organisms have little or no control over the ambient temperature and can survive only within a narrow range of temperatures. Also, changing the temperature is not a very selective way to control reaction rates because most reactions speed up when the temperature is raised.

The remaining alternative is to decrease $\Delta G^{\ddagger}$. Because rate constants depend exponentially on $-\Delta G^{*}/RT$, small changes in $\Delta G^{\ddagger}$ will have a large effect on the reaction rate. At physiological temperatures, it takes a decrease of only 1.36 kcal/mole to speed up a reaction by a factor of

10, and a decrease by 8.16 kcal/ mole increases the rate by a factor of 10^6. These are relatively modest free energy changes because forming a single H bond can release anywhere from 4 to 10 kcal/mole.

In addition, because $\Delta G^{\ddagger}$ depends on the structures of the reactants and products and on the detailed nature of the reaction, it affords an opportunity to control reaction rates with a great deal of specificity. Enzymes, then, must work by decreasing the activation free energies for the specific reactions that they catalyze. They do this in a variety of ways. Our main concern here is to describe in more general terms how the kinetic properties of enzymatic reactions are measured and characterized.

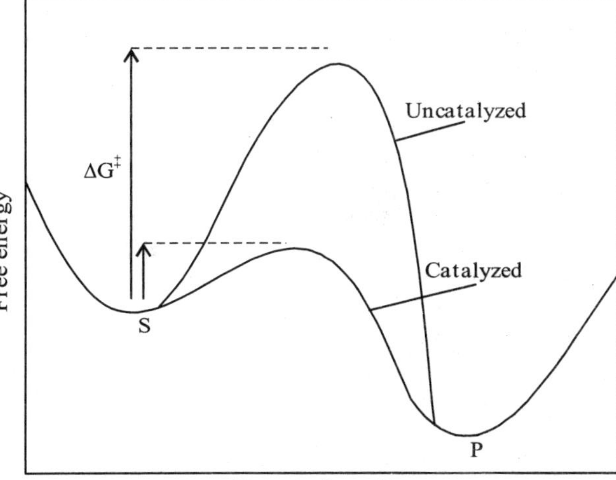

Fig. 1.4. An enzyme speeds up a reaction by decreasing $\Delta G^{\ddagger}$. The enzyme does not change the free energy of the substrate (S) or product (P); it lowers the free energy of the transition state. The two vertical arrows indicate the activation free energies ($\Delta G^{\ddagger}$) of the catalyzed and uncatalyzed reactions.

It should be kept in mind that although an enzyme can increase the rate at which a reaction occurs, it (or any other catalyst) cannot alter the overall equilibrium constant. Since K_{eq} is equal to the ratio of the rate constants for the forward and reverse processes, the catalyst increases both of these rate constants without changing their ratio.

KINETICS OF ENZYME-CATALYZED REACTIONS

Kinetic analysis was used to characterize enzyme-catalyzed reactions even before enzymes had been isolated in pure form. As a rule, kinetic measurements are made on purified enzymes *in vitro*. But the properties so determined must be referred back to the situation *in vivo* to ensure they are physiologically relevant. This is important because the rate of an enzymatic reaction can depend strongly on the concentrations of the substrates and products, and also on temperature, pH, and the concentrations of other molecules that activate or inhibit the enzyme. Kinetic analysis of such effects is indispensable to a comprehensive picture of an enzyme.

The usual procedure for measuring the rate of an enzymatic reaction is to mix enzyme with substrate and observe the formation of product or disappearance of substrate as soon as possible after mixing, when the substrate concentration is still close to its initial value and the product concentration is small. The measurements usually are repeated over a range of substrate concentrations to map out how the initial rate depends on concentration.

Spectrophotometric techniques are used commonly in such experiments because in many cases they allow the concentration of a substrate or product in the mixture to be measured continuously as a function of time : Measurements of reactions that occur in less than a few seconds require special techniques to speed up the mixing of the enzyme and substrate. One way to achieve this is to place solutions containing the enzyme and the substrate in two separate syringes.

A pneumatic device then is used to inject the contents of both syringes rapidly into a common chamber that resides in a spectrophotometer for measuring the course of the reaction. Such an apparatus is referred to as a "stopped-flow" device becase the flow stops abruptly when the movement of the pneumatic driver is arrested. In this type of apparatus it is possible to make kinetic measurements within about 1 ms after mixing of enzyme and subtrate imply that/the velocity of an uncatalyzed reaction increases indefinitely with an increase in the concentration of the reactants.

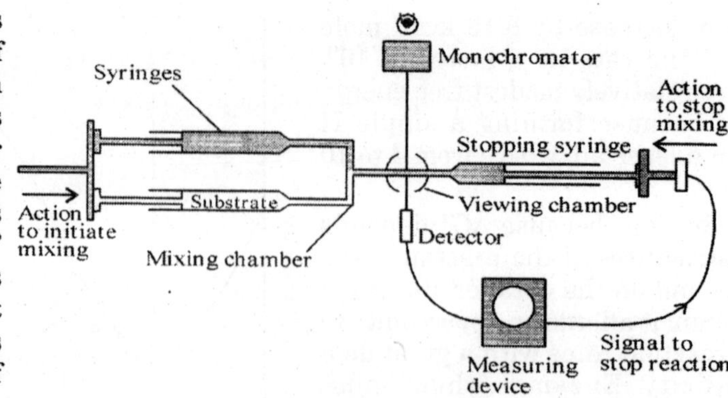

Fig. 1.5 The stopped-flow apparatus for measuring enzyme-catalyzed reactions very soon after mixing enzyme and substrate.

With enzyme-catalyzed reactions, something very different is observed. The rate usually increases linearly with substrate concentration at low concentrations, but then levels off and becomes independent of the concentration at high concentrations. The explanation for this hyperbolic dependence on substrate concentration is straightforward. For an enzyme to affect $\Delta G^{\ddagger}$, the substrate must bind to a special site on the protein, the *active site.* At very low concentrations of substrate, the active sites of most of the enzyme molecules in the solution are unoccupied. Increasing the substrate concentration brings more enzyme molecules into play, and the reaction speeds up. At high concentrations, on the other hand, most of the enzyme molecules have their active sites occupied, and the observed rate depends only on the rate at which the bound reactants are converted into products. Further increases in the substrate concentration then have little effect.

Work of Henri-Michaelis-Menten

The hyperbolic saturation curve that is commonly seen with enzymatic reactions led Leonor Michaelis and Maude Menten in 1913 to develop a general treatment for kinetic analysis of these reactions. Following earlier work by Victor Henri, Michaelis and Menten assumed that an enzyme–substrate complex (ES) is in equilibrium with free enzyme (E) and substrate (S), and that the formation of products (P) proceeds only through the ES complex:

$$E + S \underset{k_{-1}}{\overset{k}{\rightleftharpoons}} ES \xrightarrow{k_2} E + P \qquad \text{....(1.13)}$$

Their objective was to relate the reaction rate to observable quantities and interpretable molecular parameters.

Because, the slow step in the reaction described by equation (1.13) is assumed to be formation of E and P from ES, the velocity of the reaction should be

$$v = \frac{d[P]}{dt} = k_2[ES] \qquad \text{...(1.14)}$$

A maximum velocity (V_{max}) is obtained when all of the enzyme is in the form of the enzyme–substrate complex. Since the total concentration of enzyme $[E_t]$ is equal to [E] + [ES],

$$V_{max} = k_2[E_t] = k_2([E] + [ES])$$

Dividing equation (1.14) by equation (1.15) gives

$$\frac{v}{V_{max}} = \frac{[ES]}{[E] + [ES]} \quad ...(1.16)$$

The fraction on the right-hand side can be evaluated by making use of the dissociation constant of the ES complex in equation (1.13), K_d:

$$K_d = \frac{[E][S]}{[ES]} \text{ and } [ES] = \frac{[E][S]}{K_d} \quad ...(1.17)$$

Substituting this value of [ES] into equation (1.16) and rearranging, gives

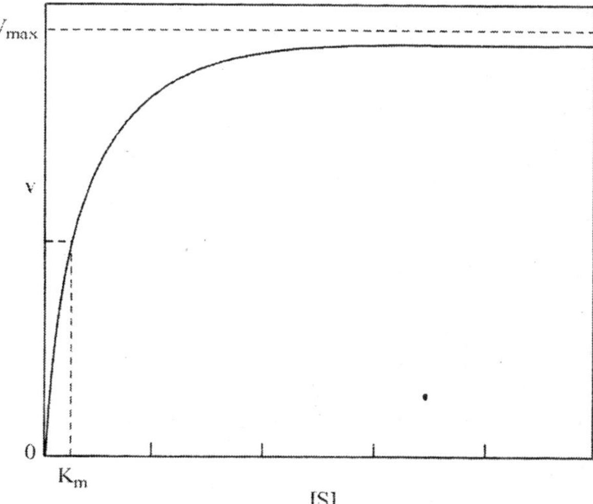

Fig. 1.6. The reaction velocity *v* as a function of the substrate concentration [S] for an enzyme-catalyzed reaction. At high substrate concentrations the reaction velocity reaches a limiting value, V_{max}. K_m is the substrate concentration at which the rate is half maximal.

$$v = \frac{V_{max}[S]}{K_d + [S]} \quad ...(1.18)$$

Equation (1.18) is the *Henri-Michaelis-Menten equation* which relates the reaction velocity to the maximum velocity, the substrate concentration, and the dissociation constant for the enzyme–substrate complex. Usually substrate is present in much higher molar concentration than enzyme, and the initial period of the reaction is examined so that the free substrate concentration [S] is approximately equal to the total substrate added to the reaction mixture.

Steady-State

Rather than discussing the implications of at this point, it is useful to develop a more general expression that avoids the assumption that the enzyme-substrate complex is in equilibrium with free enzyme and substrate. To develop this expression, we introduce the concept of the *steady state*, which was first proposed by G. E. Briggs and J. B. S. Haldane in 1925. The steady state constitutes the time interval when the rate of the reaction is approximately constant. The system usually reaches a steady state soon after enzyme and substrate are mixed, following a brief period when the concentration of the enzyme-substrate complex (ES) builds up. The ES concentration then remains almost constant for the duration of the steady state, while the concentrations of the substrate and product continue to change substantially.

Let's now write out the reaction of equation (1.13) in more detail.

$$E + S \underset{k_{-1}}{\overset{k}{\rightleftharpoons}} ES \underset{k_{-2}}{\overset{k}{\rightleftharpoons}} E + P \quad ...(1.19)$$

In this more complete scheme, the ES complex forms from E and S with rate constant k_1. ES can either dissociate again with rate constant k_{-1} or go on to P with k_2. If we confine ourselves to measuring the initial rate of the reaction in the steady state, we can continue to neglect regeneration of ES from E and P (the step involving k_{-2}) because the concentration of P will be too small for this back-reaction to occur at a significant rate. For the rate of formation of ES, v_f, we then can write

$$v_f = k_1[\text{E}][\text{S}] \qquad \text{...(1.20)}$$

Similarly, the rate of disappearance of ES, v_d, is

$$v_d = k_{-1}[\text{ES}] + k_2[\text{ES}] \qquad \text{...(1.21)}$$

If the concentration of ES is virtually constant during the steady state, the rates of formation and disappearance of ES must be nearly equal, $v_f = v_d$. We therefore can describe the situation in the steady state by combining equations (1.20) and (1.21).

$$k_1[\text{E}][\text{S}] = (k_{-1} + k_2)[\text{ES}] \qquad \text{...(1.22)}$$

or

$$\frac{[\text{E}][\text{S}]}{[\text{ES}]} = \frac{k_{-1} + k_2}{k_1} = k_m \qquad \text{...(1.23)}$$

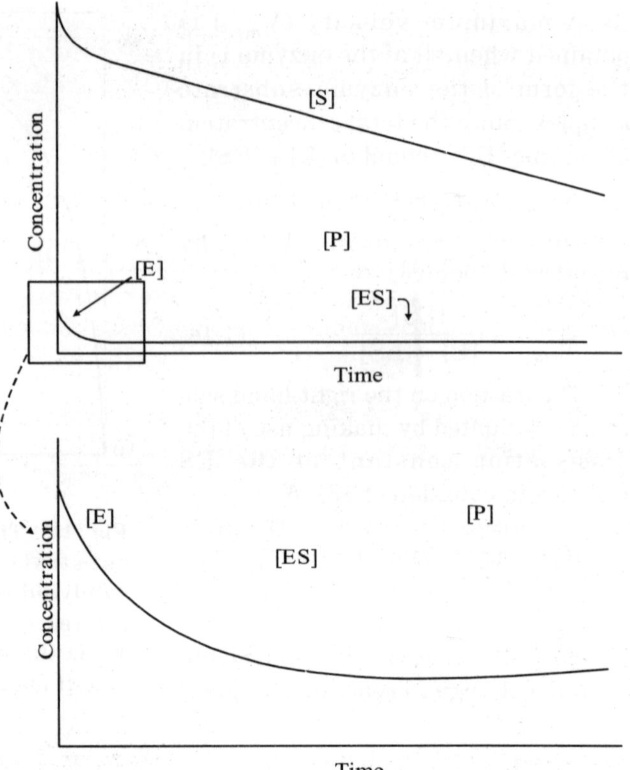

Fig. 1.7. Concentrations of free enzyme (E), substrate (S), enzyme-substrate complex (ES), and product (P) over the time course of a reaction. The shaded portion of the top graph is shown in expanded form in the bottom graph. After a brief initial period (usually less than a few seconds) the concentration of ES remains approximately constant for an extended period. The steady-state approximation is applicable during this second period. Most measurements of enzyme kinetics are made in the steady state.

The constant K_m defined in equation (1.23) is called the *Michaelis constant* and is one of the key parameters in enzyme kinetics. It is a simple matter to proceed from this point to an expression comparable to the Henri-Michaelis-Menten equation (1.18), but with K_m in place of K_s. First, rearranging equation (1.23) gives

$$[\text{ES}] = \frac{[\text{E}][\text{S}]}{k_m} \qquad \text{...(1.24)}$$

With this expression for [ES] we can follow the same procedure that led to equation (1.18), this time arriving at the *Briggs-Haldane equation* for the reaction velocity.

$$v = \frac{V_{\max}[\text{S}]}{[\text{S}] + K_m} \qquad \text{...(1.25)}$$

Equations (1.25) is identical to (1.18) except that the more complex constant K_m has replaced K_d. The term "*Michaelis-Menten equation*" is often used for either expression.

For the purpose of graphical representation of exprimental data, it is convenient to rearrange. Taking the reciprocals of both sides of equation (1.25) gives

$$\frac{1}{v} = \frac{1}{V_{max}} + \frac{K_m}{V_{max}} \frac{1}{[S]} \qquad ...(1.26)$$

This expression indicates that a plot of $1/v$ versus $1/[S]$ fits a straight line with a slope of K_m/V_{max}. Such a plot is known as a *Lineweaver-Burk*, or *double-reciprocal*, plot. The intercept of the line on the ordinate occurs at $1/v = 1/V_{max}$, and the intercept on the abscissa occurs at $1/[S] = -1/K_m$. V_{max} and K_m thus can be determined readily from the graph.

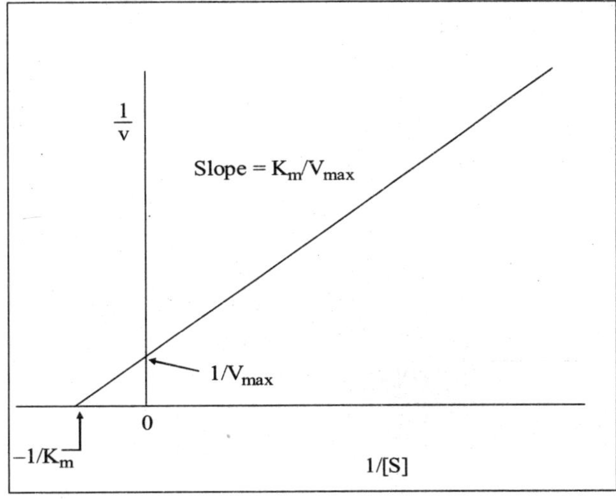

Fig. 1.8. A plot of the reciprocal of the rate ($1/v$) as a function of the reciprocal of the substrate concentration ($1/[S]$) fits a straight line. Extrapolating the line to its intercept on the ordinate (infinite substrate concentration) gives $1/V_{max}$. Extrapolating to the intercept on the abscissa gives $-1/K_m$.

Significance of the Michaelis Constant, K_m

The Michaelis constant K_m has the dimensions of a concentration (molarity), because k_{-1} and k_2, the two rate constants in the numerator of equation (1.23), are first-order rate constants with units expressed per second (s^{-1}), whereas the denominator k_1 is a second-order rate constant with units of $M^{-1}S^{-1}$. To appreciate the meaning of K_m, suppose that $[S] = K_m$. The denominator in equation (1.25) then is equal to 2[S], which makes the velocity $v = V_{max}/2$. Thus, *the K_m is the substrate concentration at which the velocity is half maximal.* K_m values for several enzyme-substrate pairs are given in table 1.1.

The values vary over a wide range but typically lie between 10^{-6} and 10^{-1} M. With enzymes that can act on several different substrates, K_m, can vary substantially from substrate to substrate. With the enzyme chymotrypsin, for example, the K_m for the substrate glycyltyrosinamide is about 50 times that for the substrate *N*-benzoyltyrosinamide. Does the K_m indicate how tightly a particular substrate binds to the active site of an enzyme? Not necessarily. According to equation (1.23).

$$K_m = \frac{k_{-1} + k_2}{k_1} \qquad ...(1.23')$$

whereas the dissociation constant of the ES complex is

$$K_d = \frac{k_{-1}}{k_1} \qquad ...(1.27)$$

Table 1.1. The Michaelis Constants for Some Enzymes

Enzyme and Substrate	$K_m (M)$
Catalase H_2O_2	1.1
Hexokinase	
Glucose	1.5×10^{-4}
Fructose	1.5×10^{-3}
Chymotrypsin	
N-Benzoyltyrosinamide	2.5×10^{-3}
N-Formyltyrosinamide	1.2×10^{-2}
N-Acetyltyrosinamide	3.2×10^{-2}
Glycyltyrosinamide	1.2×10^{-1}
Aspartate aminotransferase	
Aspartate	9.0×10^{-4}
α-Ketoglutarate	1.0×10^{-4}
Fumarase	
Fumarate	5.0×10^{-5}
Malate	2.5×10^{-5}

Comparing these two expressions, we see that K_m must always be larger than K_d. If $k_2 >> k_{-1}$, then $K_m \approx k_2/k_1$, which is much greater than K_s. On the other hand, if $k_{-1} >> k_2$, then K_m is approximately equal to K_s. In the latter circumstance, a smaller K_m means a smaller dissociation constant, which implies tighter binding to the enzyme. (This is the limiting case assumed in the Henri-Michaelis-Menten treatment, for when $k_{-1} >> k_2$, the enzyme-substrate complex is in equilibrium with the free enzyme and substrate.) But whether or not this is a valid approximation depends on the particular enzyme and the substrate. For a multistep reaction the relationship between K_m and the rate constants can be considerably more complex.

Significance of the Turnover Number, k_{cat}

The *turnover number* of an enzyme, k_{cat}, is the maximum number of molecules of substrate that could be converted to product each second per active site. Because the maximum rate is obtained at high substrate concentrations, when all the active sites are occupied with substrate, *the turnover number is a measure of how rapidly an enzyme can operate once the active site is filled.* This is given simply by

$$k_{cat} = \frac{V_{max}}{[E_t]} \qquad \text{...(1.28)}$$

Turnover numbers for some representative enzymes are listed in table 7.2. The enormous value of 4×10^7 molecules/s achieved by catalase is among the highest known; the low value for lysozyme is at the other end of the spectrum. As is the case with K_m, the relationship of K_{cat} to individual rate constants, such as k_2 and k_3, depends on the details of the reaction mechanism.

Significance of the Specificity Constant, k_{cat}/K_m

Under physiological conditions, enzymes usually do not operate at saturating substrate concentrations. More typically, the ratio of the substrate concentration to the K_m is in the range of 0.01 – 1.0. If [S] is much smaller than K_m, the denominator of the Briggs-Haldane equation [equation (1.25)] is approximately equal to K_m, so that the velocity or the reaction becomes

$$v \approx \frac{V_{max}}{[K_m]}[S] \quad (\text{when } [S] << K_m)$$

$$= \frac{k_{cat}}{K_m}[E_t][S] \qquad ...(1.29)$$

The ratio k_{cat}/K_m is referred to as the *specificity constant*. Equation (1.29) indicates that the *specificity constant provides a measure of how rapidly an enzyme can work at low [S].* Table 1.3 gives the values of the specificity constants for some particularly active enzymes. The specificity constant k_{cat}/K_m is useful for comparing the relative abilities of different compounds to serve as a substrate for the same enzyme. If the concentrations of two substrates are the same, and are small relative to the K_m values, the ratio of the rates when the two substrates are present is equal to the ratio of the specificity constants. Another use of the specificity constant is for comparing the rate of an enzyme-catalyzed reaction with the rate at which random diffusion brings the enzyme and substrate into contact.

We mentioned previously that if every collision between a protein and a small molecule results in a reaction, the maximum value of the second-order rate constant is on the order of 10^8 to 10^9 $M^{-1}s^{-1}$. Some of the values of k_{cat}/K_m in table 1.3 are in this range. The reactions these enzymes catalyze proceed at nearly the maximum possible speed, given a fixed, low concentration of substrate and given the restriction that the enzyme and substrate have to find each other by diffusion. The only practical way to go much faster is to have the substrate generated right on the enzyme or in its immediate vicinity, so that little diffusional motion is necessary.

Table 1.2. Values of k_{cat} for Some Enzymes

Enzyme	$K_{cat}(s^{-})$
Catalase	40,000,000
Carbonic anhydrase	1,000,000
Acetylcholinesterase	14,000
Penicillinase	2,000
Lactate dehydrogenase	1,000
Chymotrypsin	100
DNA polymerase I	15
Lysozyme	0.5

Enzymatic Reactions

Enzymes that catalyze reactions with two or more substrates work in a variety of ways. In some cases, the intermolecular reaction occurs when all the substrates are bound in a common enzyme—substrate complex; in others, the substrates bind and react one at a time. A frequent application of kinetic measurements is to distinguish between such alternatives. Consider a reaction in which two substrates, S_1 and S_2, are converted to products P_1 and P_2. One way for the reaction to occur is for S_1 to bind to the enzyme first, forming the binary complex ES_1. Binding of S_2 can then form the ternary complex ES_1S_2, which gives rise to the products

$$E \xrightarrow{S_1} ES_1 \xrightarrow{S_2} ES_1S_2 \xrightarrow{P_1 + P_2} E \qquad \text{...(1.30)}$$

This process is referred to as an *ordered pathway*. An alternative is a *random-order pathway*, in which, the two substrates can bind to the enzyme in either order. Still another scheme is for S_1 to bind to the enzyme and be converted to P_1, leaving the enzyme in an altered form, E′. S_2 then binds to E′ and is converted to P_2, returning the enzyme to its original form.

$$E \xrightarrow{S_1} ES_1 \xrightarrow{P_1} E' \xrightarrow{S_2} E'S_2 \xrightarrow{P_2} E \qquad \text{...(1.31)}$$

Table 1.3. Enzymes for Which k_{cat}/K_m is Close to the Diffusion-Controlled Association Rate

Enzyme	*Substrate*	$k_{cat}(s^{-1})$	$K_m(M)$	$K_{cat}/K_m(M^{-1}\,s^{-1})$
Acetylcholinestrase	Acetylcholine	1.4×10^4	9×10^{-5}	1.6×10^8
Carbonic	CO_2	1×10^6	0.012	8.3×10^7
anhydrase	HCO_3^-	4×10^5	0.026	1.5×10^7
Catalase	H_2O_2	4×10^7	1.1	4×10^7
Crotonase	Crotonyl-CoA	5.7×10^3	2×10^{-5}	2.8×10^8
Fumarase	Fumarate	800	5×10^{-6}	1.6×10^8
	Malate	900	2.5×10^{-5}	3.6×10^7
Triosephorphate isomerase	Glyceraldehyde 3-phosphate	4.3×10^3	4.7×10^{-4}	2.0×10^8
β-Lactamase	Benzylpenicillin	2.0×10^3	2×10^{-5}	1×10^8

Source: From Alan Fersht, *Enzyme Structure and Mechanism,* 3d ed. Copyright © 1985 by W. H. Freeman & Company, New York. Reprinted with permission.

This process is called the *Ping-Pong mechanism* to emphasize the bouncing of the enzyme between two states, E and E′. Ping-Pong pathways are commonly observed with enzymes that contain bound coenzymes. Interconversion of the enzyme between the two forms usually involves modification of the coenzyme. Kinetic equations for these and other mechanisms can be worked out just as we have done for reactions involving only one substrate.

Techniques for doing this are described in the references at the end of the chapter; here we simply illustrate a few of the results. For the Ping-Pong mechanism [equation (1.31)], the double-reciprocal form of the final expression is

$$\frac{1}{v} = \frac{1}{V_{\max}}\left(1 + \frac{K_{m2}}{[S_2]}\right) + \frac{K_{m1}}{V_{\max}}\frac{1}{[S_1]} \qquad ...(1.32)$$

where K_{m1} is the Michaelis constant for S_1, and K_{m2} is that for S_2. This expression is similar to that for a one-substrate reaction (equation 1.26), except that the first term on the right is multiplied by the factor $(1 + K_{m2}/[S_2])$. If we measure the rate as a function of $[S_2]$, keeping $[S_2]$ constant, a plot of $1/v$ versus $1/[S_1]$ is linear, but the intercept on the ordinate (the apparent V_{max}) depends on $[S_2]$. Increasing $[S_2]$ increases the apparent V_{max}. From a series of such plots, measured at different values of $[S_2]$, we can find the true V_{max} in addition to K_{m1} and K_{m2}.

An ordered pathway results in the expression

$$\frac{1}{v} = \frac{1}{V_{\max}}\left(1 + \frac{K_{m2}}{[S_2]} + \frac{K_{m1}}{[S_1]} + \frac{K_{m2}}{[S_2]}\frac{K_{s1}}{[S_1]}\right) \qquad ...(1.33)$$

where K_{s1} is the dissociation constant for ES_1. A plot of $1/v$ versus $1/[S_1]$ at constant $[S_2]$ is still linear, but now both the slope and the intercept depend on $[S_2]$. Again, all of the kinetic parameters can be obtained from a series of plots measured at different S_2 concentrations. However, additional measurements must be made to determine which substrate binds to the enzyme first. In some cases, the substrate that binds first can be shown to form a stable enzyme-substrate complex in the absence of the other substrate.

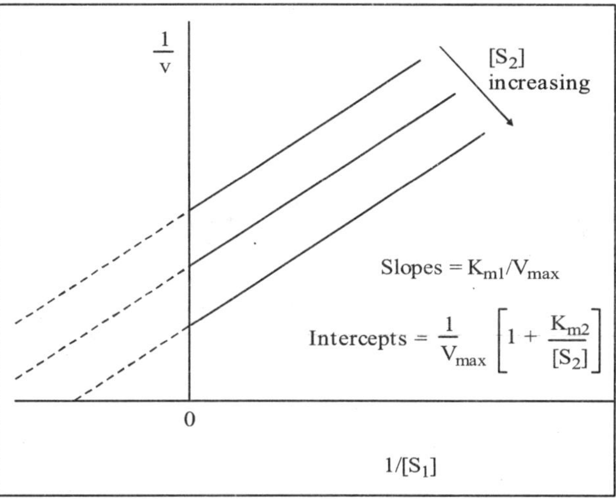

Fig. 1.9. Double-reciprocal plots ($1/v$ versus $1/[S_1]$) for the Ping-Pong mechanism. Measurements made at different values of $[S_2]$ give a set of parallel straight lines. The intercepts on the ordinate depend on $[S_2]$. K_{m1}, K_{m2}, and V_{max} can be obtained by replotting the intercepts as a function of $1/[S_2]$.

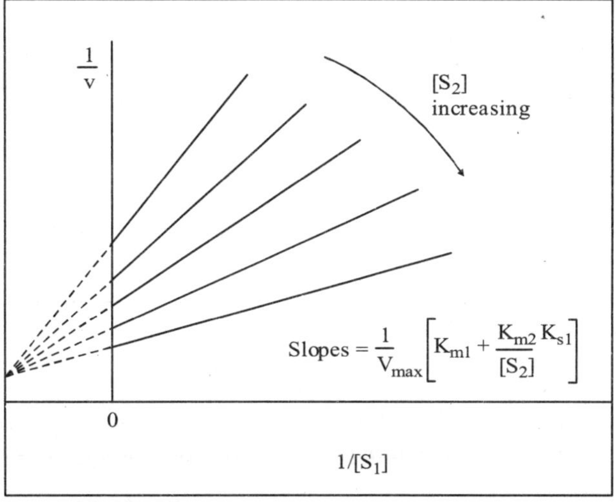

Fig. 1.10. Double-reciprocal plots for an ordered pathway. Measurements made at different fixed values of $[S_2]$ give a set of lines that intersect to the left of the ordinate. The two values of K_m, V_{max} and K_{s1} can be obtained by replotting the slopes and intercepts of these lines as functions of $1/S_2$]. A random pathway gives similar results, but can be distinguished by making such measurements for the reverse reaction ($P_1 + P_2 \rightarrow S_1 + S_2$) in addition to the forward reaction.

Factors Affecting

For most enzymes, the turnover number increases with temperature until a temperature is reached at which the enzyme is no longer stable. Above this point there is a precipitous, and usually irreversible, drop in activity. At lower temperatures, the temperature dependence of k_{cat} can be related to the activation energy of the slowest (rate-limiting) step in the catalytic pathway. With many enzymes a 10°C rise in temperature increases k_{cat} by about a factor of 2, which translates into an activation energy of about 12 kcal/ mole. Enzymes, like other proteins, are stable over only a limited range of pH.

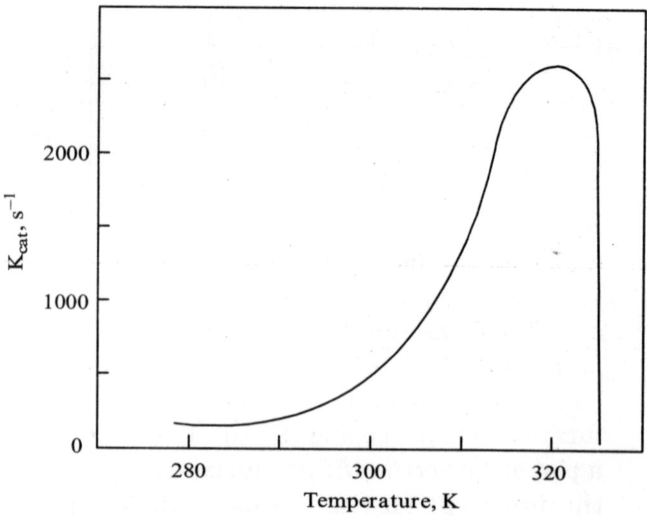

Fig. 1.11. The activity of a typical enzyme as a function of temperature. The turnover number (k_{cat}) increases with temperature until a point is reached at which the enzyme is no longer stable. The temperature at which k_{cat} is greatest should not be interpreted as the "optimum temperature" for the enzyme. Because denaturation of the enzyme occurs continuously during the measurement, the position of the maximum depends on how quickly the experimenter is able to assay the enzyme activity.

Outside this range, changes in the charges on ionizable amino acid residues result in modifications of the tertiary structure of the protein and eventually lead to denaturation. But within the range at which an enzyme is stable, both k_{cat} and K_m often depend on pH. The effects of pH can reflect the pK_a of ionizing groups on either the enzyme or the substrate.

A substrate that has an amine group, for example, may bind to the enzyme best when this group, is protonated. In many cases, however, the pH dependence reflects ionizable residues that constitute the active site on the enzyme or are essential for maintaining the structure of the active site, and the optimum pH is a characteristic more of the enzyme than of the substrate. Thus, the maximum activity of chymotrypsin always occurs around pH 8, the activity of pepsin peaks around pH 2, and acetylcholinesterase works best at pH 7 or higher. The activity of papain, on the other hand, is essentially independent of pH between 4 and 8.

ENZYME INHIBITION

Most enzymes are sensitive to inhibition by specific agents that interfere with the binding of a substrate at the active site or with conversion of the enzyme-substrate complex into products. Two of the major applications of kinetic measurements are in distinguishing between different types of inhibition and in providing quantitative information on the effectiveness of various inhibitors. Such information is essential for an understanding of how cells regulate their enzymatic activities. Comparisons of a series of inhibitors also can help to map the structure of an enzyme's active site and are a key step in the design of therapeutic drugs.

Competitive Inhibitors Bind at the Active Site

In many cases, an inhibitor resembles the substrate structurally and binds reversibly at the same site on the enzyme. This activity is called *competitive inhibition* because the inhibitor and the substrate compete for binding. The inhibitor is prevented from binding if the active site is already occupied by the substrate. As an example consider the proteolytic enzyme trypsin, which cleaves polypeptide chains at peptide linkages adjacent to basic amino acid residues.

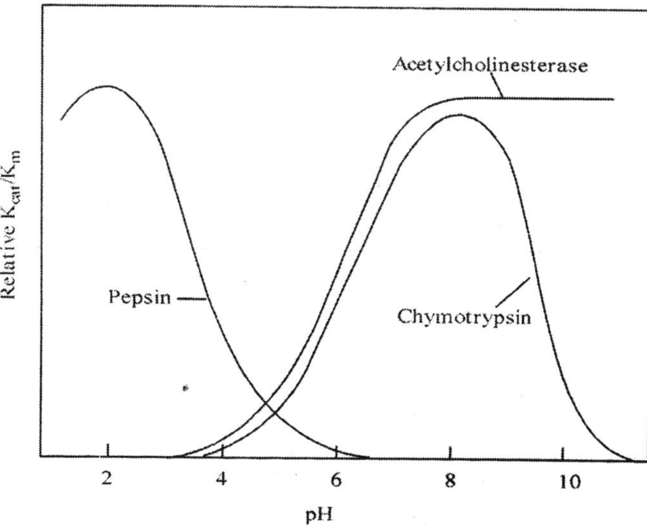

Fig. 1.12. Enzyme activity (k_{cat}/K_m) as a function of pH for three different enzymes. The optimum pH usually is a characteristic of the enzyme and not the particular substrate. Often the pH sensitivity is an indication of an ionizable group at the active site.

Trypsin is inhibited competitively by benzamidine. ic substrate-binding site on the enzyme consists of a pocket where a positively charged lysine or arginine side chain fits snugly and interacts with a negatively charged carboxylate group. When protonated, benzamidine is positively charged and has a flat, delocalized electronic structure resembling that of an arginine side chain. The binding pocket accepts and binds a benzamidine ion reversibly in place of its regular substrate.

An expression describing enzyme kinetics in the presence of a competitive inhibitor can be derived straightforwardly. Consider the reaction that we treated previously.

$$E + S \underset{k_{-1}}{\overset{k_1}{\rightleftharpoons}} ES \underset{k_{-2}}{\overset{k_2}{\rightleftharpoons}} E + P \qquad ...(1.19')$$

We now have the additional feature that the enzyme also reacts reversibly with the inhibitor (I) to give an inactive complex (EI).

$$E + I \rightleftharpoons EI \qquad ...(1.34)$$

The derivation proceeds just as the derivation that led to equations (1.25) and (1.26), except that the total enzyme concentration $[E_t]$ is now [E] + [ES] + [EI], instead of just [E] + [ES]. As a result, in place of we end up with

$$\frac{1}{v} = \frac{1}{V_{max}} + \frac{K_m}{V_{max}}\frac{1}{[S]}\left(1 + \frac{[I]}{K_i}\right) \qquad ...(1.35)$$

where K_i is the dissociation constant of the enzyme-inhibitor complex.

$$K_i = \frac{[E][I]}{[EI]} \qquad ...(1.36)$$

According to equation (1.35), a plot of $1/v$ versus $1/[S]$ is linear and passes through the same intercept on the ordinate as the plot obtained in the absence of the inhibitor ($1/V_{max}$). This is because the effect of the inhibitor disappears at infinite substrate concentration for a competitive inhibitor. The slope of the double-reciprocal plot, however, depends on the product

$$\frac{K_m}{V_{max}}\left(1+\frac{[I]}{K_i}\right)$$

instead of simply K_m/V_{maz}. As a result the apparent K_m is altered to K'_m, where

$$K'_m = K_m\left(1+\frac{[I]}{K_i}\right)$$

By measuring the slope of the plot as a function of [I] we can determine K_i.

Noncompetitive Inhibitors

Some inhibitors bind at sites other than the enzyme's active site and do not compete directly with binding of the substrate. Instead, they act by interfering with the reaction of the enzyme-substrate complex. An inhibitor that binds to an enzyme whether or not the active site is occupied by the substrate is termed a *noncompetitive inhibitor.* A noncompetitive inhibitor decreases the maximum velocity of an enzymatic reaction without affecting the K_m. The inhibitor removes a certain fraction of the enzyme from operation, no matter the concentration of the substrate.

Plots of $1/v$ versus $1/[S]$ in the presence of different concentrations of a noncompetitive inhibitor intersect at the same point on the abscissa ($-1/K_m$) but pass through the ordinate at different points still another possbility is that the inhibitor binds only to the enzyme-substrate complex and not to the free enzyme. This reaction is called *uncompetitive inhibition.* Uncompetitive inhibition is rare in reactions that involve

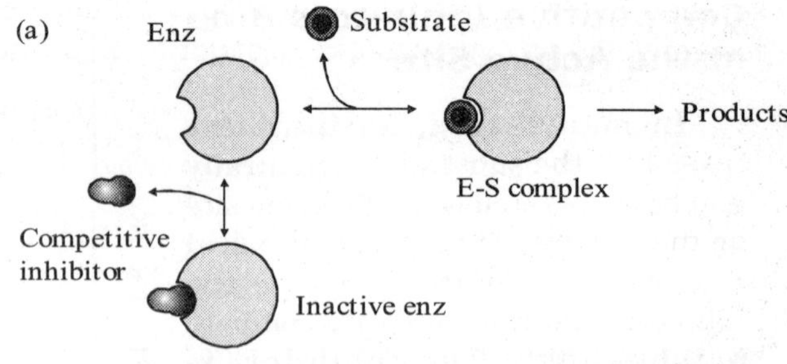

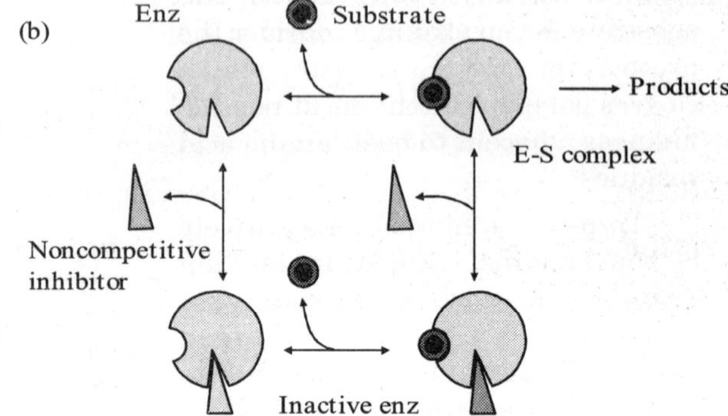

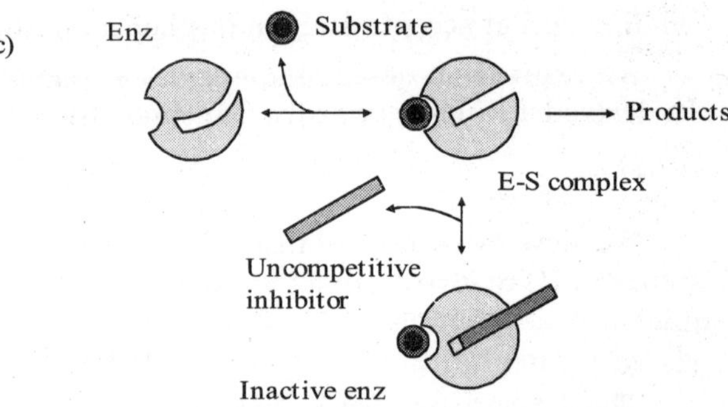

Fig. 1.13. Types of enzyme inhibition, (a) A competitive inhibitor competes with the substrate for binding at the same site on the enzyme, (b) A noncompetitive inhibitor binds to a different site but blocks the conversion of the substrate to products, (c) An uncompetitive inhibitor binds only to the enzyme-substrate complex. (E = enzyme; S = substrate.)

a single substrate but more common in reactions with multiple substrates. Plots of $1/v$ versus 1/[S] at different concentrations of an uncompetitive inhibitor give a series of parallel lines.

Table 1.4. Some Inhibitors of Enzymes that Form Covalent Linkages with Functional Groups on the Enzyme

Inhibitor	*Enzyme Group that Combines with Inhibitor*
Cyanide	Fe, Cu, Zn, other transition metals
p-Mercuribenzoate	Sulfhydryl
Diisopropylfluorophosphate	Serine hydroxyl
Iodoacetate	Sulfhydryl, imidazole, carboxyl, thioether

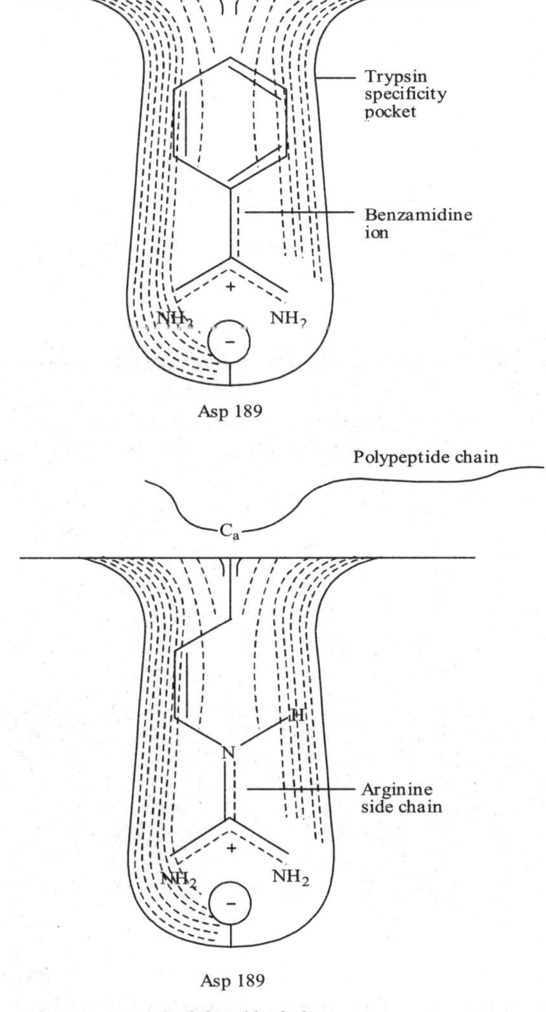

Fig. 1.14. The specificity pocket of trypsin can accommodate an arginine side chain of a polypeptide substrate or a benzamidine ion, which acts as a competitive inhibitor. Asp = aspartic acid.

Irreversible Inhibitors

Competitive, noncompetitive, and uncompetitive inhibition are all reversible. If the inhibited enzyme is dialyzed to remove the inhibitor, the enzymatic activity recovers. There are, however, inhibitors that react essentially irreversibly, usually by forming a covalent bond to an amino acid side chain or a bound coenzyme. Examples of such inhibitors are given in table 1.4. Like reversible inhibitors, irreversible inhibitors can orange either V_{max} or K_m or both.

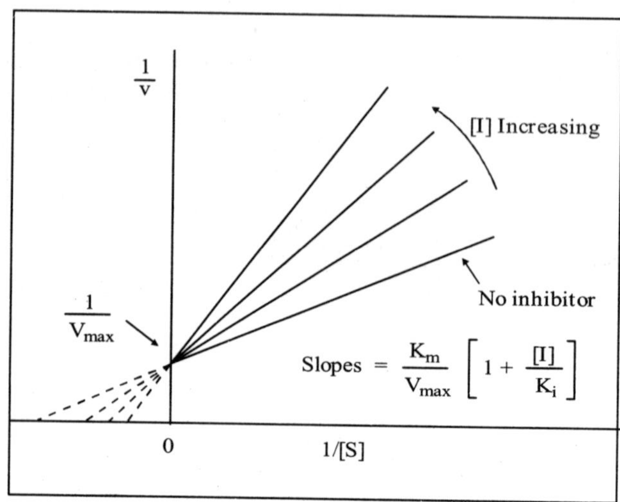

Fig. 1.15. Competitive inhibition. A scries of double-reciprocal plots (l/v versus 1/[S]) measured at different concentrations of the inhibitor (I) all intersect at the same point (1/V_{max}) on the ordinate. The slopes of the plots and the intercepts on the abscissa arc simple, linear functions of [I]K_i, where K_i is the dissociation constant of the inhibitor-enzyme complex.

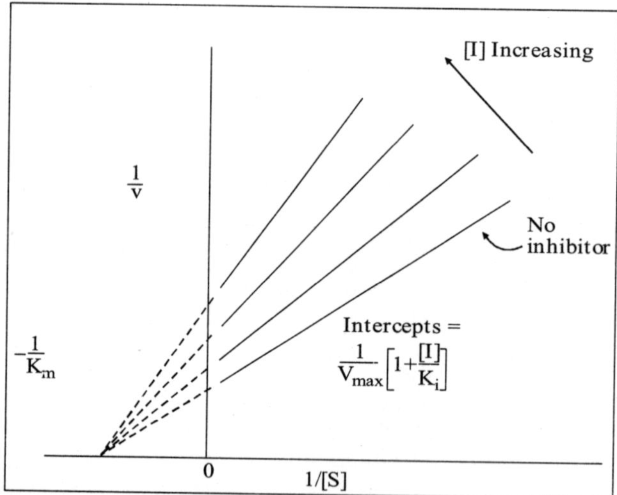

Fig. 1.16. Noncompetitive inhibition. The double-reciprocal plots pass through different points on the ordinate, but intersect at the same point (–1/K_m) on the abscissa. The slopes and the intercepts on the ordinate are linear functions of [I]/K_i.

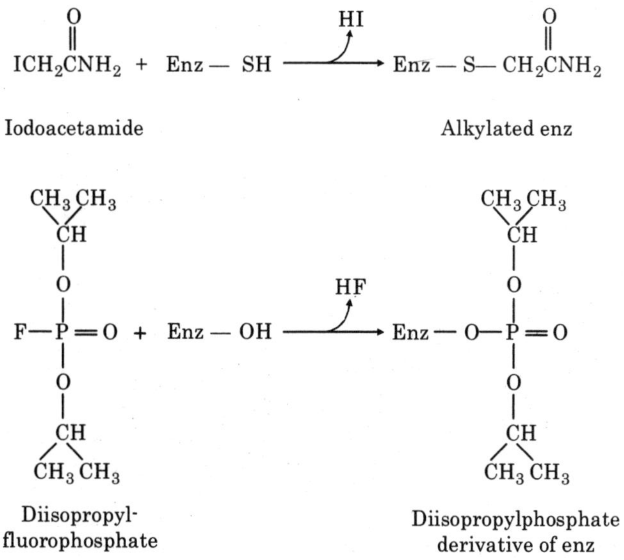

$$ICH_2\overset{O}{\overset{\|}{C}}NH_2 + Enz—SH \xrightarrow{\;HI\;} Enz—S—CH_2\overset{O}{\overset{\|}{C}}NH_2$$

Iodoacetamide Alkylated enz

$$F—P(=O)(OCH(CH_3)_2)_2 + Enz—OH \xrightarrow{\;HF\;} Enz—O—P(=O)(OCH(CH_3)_2)_2$$

Diisopropyl-fluorophosphate Diisopropylphosphate derivative of enz

Fig. 1.17. Iodoacetamide is an irreversible inhibitor of many enzymes thai contain a cysteine residue in the active site. Diisopropylfluoro-phosphate is an irreversible inhibitor of trypsin, chymotrypsin, and several related enzymes. It reacts with a serine residue at the active site.

Irreversible inhibitors often provide clues to the nature of the active site. Enzymes that are inhibited by iodoacetamide, for example, frequently have a cysteine in the active site, and the cysteinyl sulfhydryl group often plays an essential role in the catalytic mechanism. An example is glyceraldehyde 3-phosphate dehydrogenase, in which the catalytic mechanism begins with a reaction of the cysteine with the aldehyde substrate. As we discuss in trypsin and many related proteolytic enzymes are inhibited irreversibly by disopropylfluorophosphate, which reacts with a critical serine residue in the active site.

Bio-catalysts

Most of the machinery of living cells is made of enzymes. Thousands of them have been extracted from cells and have been purified and crystallized. Many others are recognized only by their catalytic action and have not yet been isolated in pure form. Most enzymes are soluble globular proteins but an increasing number of RNA molecules are also being recognized as enzymes. Many structural proteins of the cell also act as catalysts. For example, the muscle proteins actin and myosin together catalyze the hydrolysis of ATP and link the hydrolysis to movement.

Catalysis is one of the most fundamental characteristics of life. How do we recognize that a protein or RNA molecule is an enzyme? The answer is that *enzymes are recognized primarily by their ability to catalyze a chemical reaction.* For this reason, an everyday operation for many biochemists is the measurement of the catalytic activity of enzymes. Only by measuring the rates of the catalyzed reactions carefully and quantitatively has it been possible to isolate and purify these remarkable molecules. Since the beginning of biochemical investigation enzymes have held a special fascination for chemists and biologists.

How can these easily destroyed substances catalyze reactions with such speed and without formation of significant quantities of side products? Some enzymes increase the velocity of a single chemical reaction of a specific compound by a factor of as much as 10^{10}. How can a protein do this? In this chapter we'll consider both ways of measuring enzymatic activity and basic mechanisms of catalysis.

INFORMATION FROM KINETICS

The quantitative study of catalysis by enzymes, *i.e.*, the study of enzyme kinetics, is a highly developed branch of biochemistry. It is one of our most important means of learning about the mechanisms of catalysis at the active sites of enzymes. By determining *rate constants* k under a variety of conditions we can learn just how fast an enzyme can act, how tightly it binds its substrates to form the enzyme-substrate (ES) complexes essential to catalysis, how specific it is with respect to substrate structure, and how it is affected by compounds that inhibit or activate the catalysis.

Measuring the Speed of an Enzymatic Reaction

A major goal in kinetic studies is to establish a *rate equation* which describes the velocity of a reaction in terms of *kinetic constants* and other experimentally measurable parameters.

To measure the velocity of any chemical reaction precisely we must start the reaction at a definite time by rapidly mixing together the two or more reactants. Then, while keeping the mixture at an accurately constant temperature and pH, we must measure the concentration of a reactant or product after a fixed time interval, or at various times. No end of ingenuity has gone into devising ways of doing this for particular enzymes. Whatever the procedure, the information we must obtain is the *rate* at which some concentration changes with time.

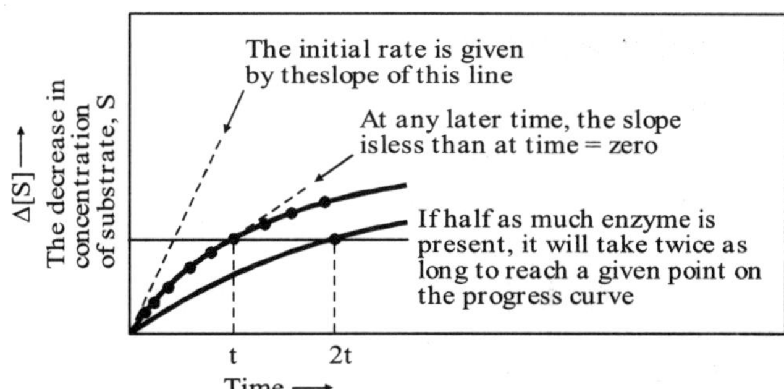

Fig. 2.1. The progress curve for an enzymatic reaction in which the substrate S is converted into products.

We can then construct a *progress curve* showing the decrease in concentration [S] of the reactant *(substrate)* or the increase in the concentration of *product* [P] with time.

The velocity v of an enzymatic reaction is defined as the rate at which a substrate disappears or at which a product is formed, the two being identical:

$$v = -d[S]/dt = d[P]/dt \quad ...(2.1)$$

Under the steady-state conditions that usually apply *the rate of increase of product will be the same as the rate of decrease of substrate.* The units of velocity are moles per liter per second (M s^{-1}) or more traditionally in enzymology moles per liter per minute. We are interested in the *instantaneous velocity,* which at any time is given by the slope of the progress cvirve. We usually want to measure the velocity immediately after the reaction is started to avoid the decrease in rate that comes from the depletion of substrate or accumulation of products.

In some case this is difficult to do with accuracy. In some chemical reactions, which involve first-order processes, a logarithmic plot of the progress curve (log[S] vs t) gives a straight line so that the initial slope need not be determined. However, most *enzyme assays* give progress curves that remain nearly linear for only short periods of time. In these cases we need very sensitive methods for detecting products. These may involve colorimetric or fluorimetric measurements or the use of radioactive substrates. One of the most sensitive approaches is to arrange the assay so that a product of the reaction serves as a catalyst for another enzymatic process, thus amplifying the amount of final product to be measured.

Another approach is to measure velocities of reactions in a very small volume, *e.g.*, 200 nanoliters, and to continuously separate and measure products using capillary electrophoresis and fluorescence detection. If the progress curve is not a straight line at the beginning of the reaction (0 time) and if the amount of compound that reacts in a fixed interval of time is taken as the rate, an erroneous answer will be obtained. Sometimes an *integrated rate expression* that describes the time course of product formation can be used. Even when the progress curve is nonlinear it is possible to estimate relative rates by noting that in most cases if one unit of enzyme yields a certain amount of product in time t_1 the same amount of product will be formed by n units of enzyme in time t_1/n.

First-order Reactions

In many chemical reactions the rate of decrease of the concentration of a given reactant [A] is found experimentally to be directly proportional to the concentration of that reactant at any given time :

$$v = -d[A]/dt = k[A] \quad ...(2.2)$$

Such a reaction is described as *first order* and the proportionality constant k is known as the *rate constant.* Such first-order kinetics is observed for *unimolecular processes* in which a molecule of A is converted into product P in a given time interval with a probability that does not depend on interaction with another molecule. An example is radioactive decay. Enzyme substrate complexes often react by unimolecular processes. In other cases, a reaction is *pseudo-first order;* comppund A actually reacts with a second molecule such as water, which is present in such excess that its concentration does not change during the experiment. Consequently, the velocity is apparently proportional only to [A].

A first-order rate constant k has units of s^{-1}. When [A] = 1, $v = k$. Thus, k is a measure of the speed in mol l^{-1} s^{-1} of the reaction of a substance at unit activity. As a first-order reaction proceeds, [A] decreases and at time t is given by any one of the following three equivalent expressions (Eq. 2.3). These are obtained by integration of eq. 2.2 which the reaction was started.

$$\begin{aligned} [A] &= A_0 e^{-kt} \\ \ln [A_0]/[A]) &= kt \\ \log [A_0] - \log[A] &= kt/2.303 \end{aligned} \quad ...(2.3)$$

Equation 2.3 is the equation of exponential decay. A charcteristic of exponential decay is that [A] is halved in a time that is independent of concentration. The *half-life* is $t_{½}$:

$$t_{½} = \ln 2/k = 0.693/k \quad ...(2.4)$$

The *relaxation time* τ for A is defined by Eq. 2.5 and represents the time required for the concentration [A] to fall to 1/*e* (or ~ 0.37) of its initial value.

$$\tau = 1/k = t_{½}/\ln 2$$

Turnover Numbers and Units of Activity

When an enzyme is catalyzing product formation at the maximum possible rate, V_{max}, we can usually assume that the active site of every molecule of enzyme contains a substrate in the form of an intermediate enzyme-substrate complex ES and that this complex is being converted to products according to Eq. 2.6 :

$$\begin{aligned} d[P]/dt &= V_{max} = k[ES] = k[E]_t \\ \text{or} \quad V_{max} &= k_{cat}[E]_t \end{aligned} \quad ...(2.6)$$

Here E_t is the total enzyme, namely, the free enzyme E plus enzyme-substrate complex ES. The equation holds only at *substrate saturation,* that is, when the substrate concentration is high enough that essentially all of the enzyme has been converted into the intermediate ES. The process is first order in enzyme but is zero order in substrate. The rate constant k is a measure of the speed at which the enzyme operates. When the concentration $[E]_t$ is given in moles per liter of *active sites* (actual molar concentration multiplied by the number of active sites per mole) the constant k is known as the *turnover number, the molecular activity,* or $\boldsymbol{k_{cat}}$. The symbol k_{cat} is also used in place of k in Eq. 2.6 for complex rate expressions in which k_{cat} cannot represent a single rate constant but is an algebraic expression that contains a number of different constants.

Turnover numbers can be measured only when the concentration of the enzyme is known. Partly for this reason the activity of an enzyme is usually given as *specific activity,* the *units of activity per milligram of protein.* One *international unit* is the amount of enzyme that produces 1 μ mol of product per minute under standard (usually optimal) conditions. The International Union of Biochemistry has recommended a larger unit, the *katal* (kat), the amount of enzyme that converts one mol s^{-1} of substrate to product.

$$1 \text{ kat} = 6 \times 10^7 \text{ international units}$$

$$1 \text{ international unit} = 16.67 \text{ nkat (nanokatals)}$$

If the enzyme is pure and is saturated with substrate under the standard assay conditions, the following relationships hold.

$$\begin{aligned} \text{Turnover No.} &= \text{katals/mol of active sites} \\ &= [\text{nkat/mg}] \times M_r \times 10^{-6}/n \\ &= [\text{international units/mg}] \times M_r \times 10^{-3}/60n \end{aligned}$$

Here M_r is the relative molecular mass of the enzyme and n is the number of active sites per molecule. Since the activity of an enzyme is dependent on both temperature and pH, these variables must be specified.

Turnover numbers of enzymes vary from <1 to ~10^6 s^{-1}. Trypsin, chymotrypsin, and many intracellular enzymes have turnover numbers of ~ 10^2 s^{-1}. The fastest enzymes, which include *catalase, carbonic anhydrase,* and *Δ^5-3-oxosteroid isomerase,* have maximum turnover numbers of 2×10^5 s^{-1} or more. Compare these reaction rates with those of a typical organic synthesis in the laboratory. A reaction mixture must often be heated for hours ($k < 10^{-3}$ s^{-1}). Many enzymes accelerate rates by factors of greater than 10^6 over those observed in the absence of an enzyme at a comparable temperature and pH. Enzymes often bring two or more substrates together, binding them at a specific location in their active sites. Because of this, rapid reactions can be catalyzed even when the reactants are present in low concentrations.

Second-order Reactions

For a chemical reaction to occur between two molecules, A and B (Eq 2.7), they must meet and collide.

$$A + B \xrightarrow{k_2} P \qquad ...(2.7)$$

The velocity of such a *second-order* process is characterized by a *bimolecular rate constant* k_2 and is proportional to the product of the concentrations of A and B:

$$v = k_2[A][B] \qquad ...(2.8)$$

The units of k_2 are M^{-1} s^{-1}. If [B] is present at unit activity, the rate is k_2[A], a quantity with units of s^{-1}. We can see that the bimolecular, or second-order, rate constant for reaction of A with B may be compared with first-order constants when the second reactant B is present at unit activity. In many real situations, reactant B is present in large excess and in a virtually constant concentration. The reaction is pseudo-first order and the experimentally observed rate constant k_2[E] is an *apparent first-order rate constant.* The bimolecular rate constant k_2 can be obtained by dividing the apparent constant by [B].

Reversible Chemical Reactions

In any reversible process, we must consider rate constants for both the forward and the reverse reactions. At equilibrium a reaction proceeds in the forward direction at exactly the same velocity as in the reverse reaction so that no change occurs. For this reason there is always a relationship between the equilibrium constant and the rate constants. For Eq 2.9, k_1 is the bimolecular rate constant for the forward reaction and k_2 the unimolecular rate constant

$$A + B \underset{k_2}{\overset{k_1}{\rightleftarrows}} P$$

$$\text{Equilibrium constant} = K = \frac{[P]}{[A][B]} = \frac{k_1}{k_2} \qquad ...(2.9)$$

for the reverse reaction. The equilibrium constant K can easily be shown, from to equal k_1/k_2 for the reaction. The student should be aware that in kinetic equations rate constants are usually numbered consecutively via subscripts and that the subscripts do not imply anything about the molecularity.

The system which is used here employs odd-numbered constants for steps in the forward direction and even-numbered constants for steps in the reverse direction. However, *many authors number the steps in the forward direction consecutively and those in the reverse direction with corresponding negative subscripts.* What relationships exist between experimentally observable rates and k_1 and k_2 for a reversible reaction? Consider first the simplest case :

$$A \underset{k_2}{\overset{k_1}{\rightleftarrows}} B \qquad ....(2.10)$$

If pure A is placed in a solution, its concentration will decrease until it reaches an equilibrium with the B which has been formed. It is easy to show that in this case [A] does not decay exponentially but $[A] - [A]_{equil}$ does. If $\log([A] - [A]_{equil})$ is plotted against time a first-order rate constant k, characteristic of *the rate of approach to equilibrium,* will be obtained. Its relationship to k_1 and k_2 is given by Eq. 2.11.

$$k = k_1 + k_2 \qquad ...(2.11)$$

The relaxation time T for approach to equilibrium can be expressed as follows :

$$\tau^{-1} = k = \tau_1^{-1} + \tau_2^{-1} = k_1 + k_2 \qquad ...(2.12)$$

and for the more complex case of Eq. 2.9 as

$$k = \tau^{-1} = k_1([A]_e + [B]_e) + k_2 \qquad ...(2.13)$$

where $[A]_e$ and $[B]_e$ are the equilibrium concentrations of A and B.

Formation and Reaction of Enzyme-Substrate Complexes

An abundance of evidence indicates that the first step in enzymatic catalysis is the combining of the enzyme and substrate reversibly to form a complex. Formation of the complex is normally reversible. ES can either break up to form enzyme and substrate again or it can undergo conversion to a product or products, often by a unimolecular process. Three rate constants are needed to describe this system for a reaction that is irreversible overall. A complete description of the kinetic behavior is fairly involved.

$$E + S \underset{k_2}{\overset{k_1}{\rightleftharpoons}} ES \xrightarrow{k_3} E + P \quad \text{...(2.14)}$$

For example, the kinetics may be different within cells, where molar concentrations of enzymes often exceed those of substrate, than in the laboratory. In most laboratory experiments the enzyme is present at an extremely low concentration (*e.g.*, 10^{-8} M) while the substrate is present in large excess. Under these circumstances the *steady-state approximation* can be used. For this approximation *the rate of formation of ES from free enzyme and substrate is assumed to be exactly balanced by the rate of conversion of ES on to P.* That is, for a relatively short time during the duration of the experimental measurement of velocity, the concentration of ES remains essentially constant.

To be more precise, the steady-state criterion is met if the absolute rate of change of a concentration of a transient intermediate is very small compared to that of the reactants and products. The *Michaelis-Menten equation* (Eq. 2.15) describes the initial reaction rate of a single substrate with an enzyme under steady-state conditions.

$$v = \frac{V_{max}}{1 + K_m/[S]} = \frac{V_{max}[S]}{K_m + [S]}$$

where

$$K_m = (k_2 + k_3)/k_1 \quad \text{...(2.15)}$$

This can be rearranged as follows :

$$\frac{V_{max}}{v} = 1 + K_m/[S] \quad \text{...(2.16)}$$

Equation 2.15 provides a relationship between the velocity observed at a particular substrate concentration and the maximum velocity that would be achieved at infinite substrate concentration. The quantities V_{max} and K_m are often referred to as the *kinetic parameters* of an enzyme and their determination is an important part of the characterization of an enzyme. Equation 2.15 can be derived by setting the rate of formation of the ES complex (k_1[E][S]) in the steady state equal to its rate of breakdown, ([$k_2 + k_3$][ES]). Rearranging and substituting K_m, as defined in Eq 2.15, we obtain Eq. 2.17.

$$[E][S] = \frac{(k_2 + k_3)}{k_1}[ES] = K_m[ES] \quad \text{...(2.17)}$$

Using this equation, together with a mass balance relationship ($[E] = [E]_t - [ES]$), we can solve for $[ES]/[E]_t$, the fraction of enzyme combined as enzyme-substrate complex.

$$\frac{[ES]}{[E]_t} = \frac{[S]}{K_m + [S]} \quad \text{...(2.18)}$$

The maximum velocity, $V_{max} = k_3[E]_t$, is attained only when all of the enzyme is converted into ES. Under other conditions $v = k_3[ES]$ and Eq. 2.19 holds.

$$[ES]/[E]_t = v/V_{max} \quad ...(2.19)$$

Substituting gives the Michaelis-Menten.

In many cases, the rate at which ES is converted back to free E and S is much greater than the rate of conversion of ES to products ($k_2 \approx k_3$). In such cases K_m equals k_2/k_1, the dissociation constant for breakdown of ES to free enzyme and substrate (sometimes called K_s). Thus, K_m *sometimes has a close inverse relationship to the strength of binding of substrate to enzyme.* In such cases, $1/K_m$ is a measure of the affinity of the substrate for its binding site on the enzyme and for a series of different substrates acted on by the same enzyme. The more tightly bound substrates have the lower values of K_m. But beware! The condition that k_3 is negligible compared to k_2 may be met with some (poorer) substrates but may not always be met by others.

From we see that K_m is always greater than or equal to (K_s) but it may be less than K_s for more complex mechanisms. A plot of velocity against substrate concentration. The position of V_{max} on the ordinate is marked, but it should be clear that the experimental velocity (v) can never attain V_{max} unless [S] is very high relative to K_m. The value of v approaches V_{max} asymptotically. Since K_m is defined as the value of [S] at which $v = V_{max}/2$, its value can be estimated from Fig. 2.2. However, K_m cannot be determined reliably because of the difficulty of establishing the value of V_{max} from a plot of this type. Notice that the curve of Fig. 2.2 is identical in form to the saturation curve for reversible binding shown in Fig. 2.1.

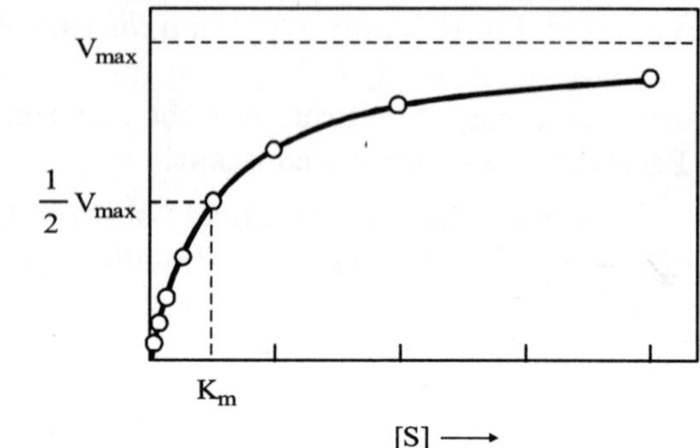

Fig. 2.2. Plot of observed velocity v vs substrate concentration [S] for an enzyme-catalyzed reaction.

Linear Forms for Rate Equations

To obtain K_m and V_{max} from experimental rate data, can be transformed by algebraic rearrangement into one of several linear forms. The popular *double-reciprocal or Lineweaver-Burk* plot of $1/v$ against 1/[S] is described. The values of K_m/V_{max} and $1/V_{max}$ can be evaluated from the slope and intercept, respectively, of this straight line plot.

$$\frac{1}{v} = \frac{1}{V_{max}} + \frac{K_m}{V_{max}}\frac{1}{[S]} \quad ...(2.20)$$

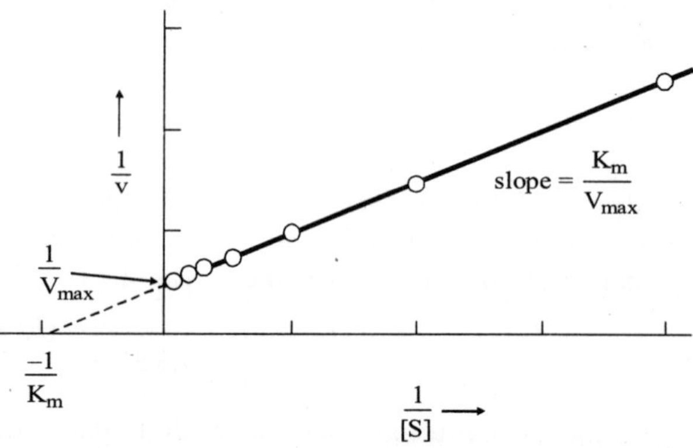

Fig. 2.3. Double-reciprocal or Lineweaver-Burk plot of $1/v$ vs 1/[S]. The intercept on the vertical axis gives $1/V_{max}$ and the slope gives K_m/V_{max}. The intercept on the horizontal axis equals $-1/K_m$.

Another linear plot, the *Eadie-Hofstee plot,* is that of v/ [S] vs v (Fig. 2.4). It is related to the Scatchard plot (Fig. 2.3) and is fitted by Eq. 2.21.

$$\frac{v}{[S]} = \frac{V_{max}}{K_m} - v\frac{1}{K_m} \quad ...(2.21)$$

While the point for [S] = 0 and v = 0 cannot be plotted, the ratio v/[S] approaches V_{max}/K_m as v approaches zero. Notice the distribution of the points. Substrate concentrations were chosen such that the increase in velocity from point to point is more or less constant, a desirable experimental situation. The points on the Eadie-Hofstee plot are also nearly evenly distributed, but those of the Lineweaver-Burk plot are compressed at one end. (However, if the substrate concentrations for successive points are selected in the ratios 1, 1/3, 1/5, 1/7, and 1/9, the spacing will be uniform on the Lineweaver-Burk plot.

A second advantage of the Eadie-Hofstee plot is that the entire range of possible substrate concentration from near zero to infinity can be fitted onto a single plot. As pointed out at the beginning of Section A, depletion of substrate with time lowers [S] from its initial value to some extent. This can have an especially adverse effect on points obtained at low [S], *e.g.*, points on the right side of Fig 2.3 or on the left side of will be more precise if the average substrate concentration over the time period of the assay, rather than that at zero time, is used.

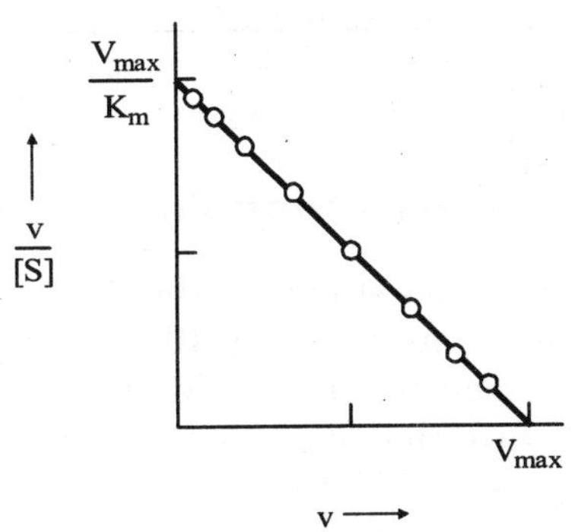

Fig. 2.4. The Eadie-Hofstee plot of v/[S] against v. The slope is $-1/K_m$; the intercept on the vertical axis is V_{max}/K_m and that on the horizontal axis is V_{max}.

Nonlinear Equations and Integrated Rate Equations

In discussions of the *control* of metabolism through regulation of enzymatic activity, it is often better to plot v against log [S]. This plot also has the virture that the entire range of attainable substrate concentrations can be plotted on one piece of paper if the point for [S] = 0 (at minus infinity) is omitted. The same scale can be used for all enzymes. The plot is S-shaped, both for simple cases that are represented and for enzymes that bind substrate cooperatively. Thus, the classification or "hyperbolic" vs "sigmoidal" is lost. However, the degree of cooperativity can be directly measured from the midpoint slope of the curve. A disadvantage is that it is awkward to measure V_{max} from a plot of this type, and it may be preferable to obtain it from a linear plot.

Alternatively, computer-assisted methods can be used to obtain both K_m and V_{max} and to fit a curve to the experimental pointsjtsin Fig. 2.5. An attractive alternative to the use of initial velocities and linear plots is to measure the kinetic parameters V_{max} and K_m using points all along the progress curve. Various procedures for doing this have been devised. For example, the integrated.

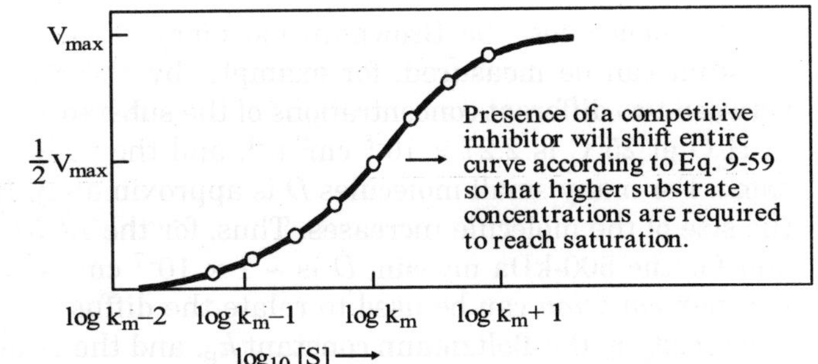

Fig. 2.5. Plot of v agains log [S] for an enzyme-catalyzed reaction.

$$V_{max}t = [P] + K_m \ln\frac{[S]_0}{[S]_0 - [P]} \qquad ...(2.22)$$

$$V_{max}t = [S]_0 - [S] + K_m \ln([S]_0/[S]) \qquad ...(2.23)$$

Here $[S]_0$ is the initial substrate concentration at $t = 0$ and [S] and [P] are the substrate and product concentrations at any later time. In one type of linear plot, based on Eq. 2.23, $([S]_0 - [S]) / t$ is plotted against $1/t \ln([S]_0/[S])$, the slope of the plot being $-K_m$ and the intercepts being V_{max} and V_{max}/K_m. Computer programs are available for analysis of enzymatic progress curves, even for complex mechanisms.

Kinetics with High Enzyme Concentrations

Laboratory studies of the kinetics ot purified enzymes are usually conducted with enzyme concentrations of 10^{-7} to 10^{-10} M, but within cells enzyme concentrations are probably often in the range 10^{-6} to 10^{-5} M, which may be higher than the concentrations of the substrates upon which the enzymes act. Be cautious in drawing conclusions about kinetics under such circumstances! Methods have been devised for handling kinetic data when the concentration of enzyme is greater than K_m, a condition that can lead to intolerably high errors if the usual equations are applied.

Diffusion and the Rate of Encounter of an Enzyme with Substrate

What determines the value of k_1. This rate constant represents the process by which the substrate and enzyme find each other, become mutually oriented, and bind to form ES. If orientation and binding are rapid enough, the rate will be determined by the speed with which the molecules can come together by diffusion. Large molecules in solution are free to travel for only a tiny fraction of their diameter as a result of their frequent collision with solvent molecules. The result is visible in the *Brownian movement* of microscopic particles suspended in a fluid. If an individual particle is followed, it is seen to undergo a "random walk," moving in first one direction then another. Albert Einstein showed that if the distances transversed by such particles in a given time Δt are measured, the mean square of these Δx values Δ^2 can be related by to the diffusion constant D (which is usually given in units of cm^2s^{-1}).

$$\Delta^2 = 2D\,\Delta T \qquad ...(2.24)$$

For molecules, the Brownian movement cannot be observed directly but the diffusion constant can be measured, for example, by observing the rate of spreading of a boundary between two different concentrations of the substance. The diffusion constant for $^1H^2HO$ (HDO) in H_2O at 25°C is 2.27×10^{-5} cm^2 s^{-1}, and the values for the ions K^+ and Cl^- are about the same. For many small molecules D is approximately 10^{-5} cm^2 s^{-1}, and the value decreases as the size of the molecule increases. Thus, for the 13.7-kDa ribonuclease, D = 1.1×10^{-6} cm^2 s^{-1}, and for the 500-kDa myosin, D is ~ 1×10^{-7} cm^2 s^{-1}. For the spherical particle the *Stokes-Einstein equation* can be used to relate the diffusion coefficient to the radius, the coefficient of viscosity η, the Boltzmann constant k_B, and the temperature (Kelvin) T.

$$D = \frac{k_B T}{6\pi\eta r} \qquad ...(2.25)$$

To estimate the rate constant for a reaction that is controlled strictly by the frequency of collisions of particles, we must ask how many times per second one of a number n of particles will be hit by another of the particles as a result of Brownian movement. The problem was analyzed in 1917 by Smoluchowski, who considered the rate at which a particle B diffuses toward a second particle A and disappears when the two collide. Using Fick's law of diffusion, he concluded that the number of *encounters* per milliliter per second.

$$\text{Number of encounters/ml s}^{-1} = 4\pi(D_A + D_B)(r_A + r_B)\, n_A n_B \qquad ...(2.26)$$

Here D_A and D_B are the two diffusion coefficients, r_A and r_B are the radii (in Å) of the two particles, and n_A and n_B are the numbers of particles per milliliter. The number of encounters per liter per second is $(4\pi/1000)\,(D_A + D_B)\,(r_A + r_B)\,N^2[A][B]$, where N is Avogadro's number. Dividing this frequency by N gives the rate of collision v in M s^{-1}, a velocity that can be equated with k_v [A] [B], where k_v is a second-order rate constant which determines the *diffusion controlled limit,* that is, the maximum possible rate of a reaction.

$$v = \frac{4\pi}{1000}(D_A + D_B) + (r_A + r_B)[A][B]N = k_D[A][B] \qquad ...(2.27)$$

This equation can be rearranged to Eq. 2.28.

$$k_D = \frac{4\pi}{1000}(D_A + D_B) + (r_A + r_B)M^{-1}s^{-1} \qquad ...(2.28)$$

While this equation is thought to overestimate the diffusion-limited rate constant slightly, it is a good approximation. If the diffusing particles are approximately spherical, diffusion constants D_A and D_B can be calculated from Eq. 2.25, and Eq. 2.28 becomes Eq. 2.29.

$$k_D \approx \frac{2RT}{3000\eta}\left[2 + \frac{r_A}{r_B} + \frac{r_B}{r_A}\right] \qquad ...(2.29)$$

The value of k_v does not vary greatly as the ratio of radii r_A/r_B is changed, and in most cases it may be assumed that $r_A/r_B \approx 1$, in which case the equation simplifies.

$$k_D \approx \frac{8RT}{3000\eta}M^{-1}s^{-1} \qquad ...(9.30)$$

For water at 25°C, the coefficient of viscosity η is ~0.01 poise (1 poise = 10^{-5} newton cm^{-2}) which leads, according to Eq. 2.30, to k_D ~0.7×10^{10} M^{-1} s^{-1}. It was shown by Debye that this rate constant must be multiplied by a correction factor when charged particles rather than uncharged spheres diffuse together. This factor may be of the order of 5–10 for a substrate and enzyme carrying two or three charges and may act to either increase or decrease reaction rates.

Since the viscosity of cytoplasm is 0.02 – 0.03 poise, will be reduced correspondingly. A simple alternative derivation of Eq. 2.30 was developed by Dexter French, who also provided the author with most of this discussion of encounter theory. Consider a small element of volume ΔV swept out by a particle as it moves through the solution for a distance equal to its own radius. This element of volume will equal πr^3:

$$\Delta V = \pi r^2 \times r = \pi r^3 \quad ...(2.31)$$

It will be swept out in a time Δt which can be calculated from Eq. 2.24 as $r^2 / 2D$. Substituting the value of D given by Eq. 2.25, we obtain for Δt :

$$\Delta t = \frac{3\pi\eta r^3}{k_B T} \quad ...(2.32)$$

Division of Eq. 2.31 by Eq. 2.32 gives the volume swept out per second by one particle :

$$\Delta V / \Delta t = k_B T / 3\eta \text{ cm}^3 \text{ s}^{-1} \quad ...(2.33)$$

Since the collision radius for two particles of equal size is two times the particle radius, the effective volume swept out will be four times that given by Eq. 2.33. Since both particles are diffusing, the effective diffusion constant will be twice that used in obtaining Eq. 2.28. Thus, the effective volume swept out by the particle in a second will be eight times that given by Eq. 2.33. The volume swept out by one mole of particles is equal to k_D (recall that the second-order rate constant has dimensions of liter mol^{-1} s^{-1}).

Thus, when converted to a moles per liter basis and multiplied by 8, Eq. 2.33 should (and does) become identical with the Smoluchowski equation (Eq. 2.30). The volume given by Eq. 2.33 is about 1.4×10^{-11} cm^3, which could be represented approximately by a cube 2.4 μm on a side. If we compare this volume with that of a cell or of an organelle, we see that in one second an enzyme molecule will sweep out a large fraction of the volume of a small cell, mitochondrion, chloroplast, etc.

The "Cage Effect" and Rotation of Molecules

It is of interest to compare the bimolecular rate constant for encounters calculated by the Smoluchowski theory with the corresponding bimolecular rate constant for molecular collisions given by the kinetic theory of gases.

$$k(\text{Collision}) = \frac{N(r_A + r_B)^2}{1000}\left[8\pi k_B T\left(\frac{1}{m_A} + \frac{1}{m_B}\right)\right]^{1/2} \quad ...(2.34)$$

Here m_A and m_B are the masses of the two particles. This rate constant is also relatively independent of molecular size and for spheres varies from (4 to 11) × 10^{11} M^{-1} s^{-1}, over an order of magnitude greater than the rate constant for encounters. In a solution, molecules still collide at about the same rate as in a gas so that *100 to 200 collisions occur between two particles for each encounter.* However, during the time of the single encounter, the particles are together in a solvent "cage." While in this solvent cage, both substrate and enzyme molecules undergo random rotational motions.

Successive collisions bring them together in different orientations, one of which is likely to lead to a sufficiently close match of complementary surfaces (of substrate and binding site) so that formation of a "productive" ES complex takes place. Molecular rotation in a solution is described quantitatively by diffusion laws (analogous to Fick's laws) for which a *rotary diffusion constant* θ is defined. Consider a group of molecules all oriented the same way initially, then undergoing rotary diffusion until their orientations become random. If we measure the orientation of each molecule by an angle we see that initially the value of cos α is 1 but that when the angles become random the mean value of cos α averaged over all molecules is zero. The rotary relaxation time τ is the time required for the mean value of cos α to fall to $1/e$ (which occurs when α = 68.5°). For a sphere θ is given by Eq. 2.35.

Initial orientation → Later orientation

$$\theta = \frac{1}{2}\tau = \frac{k_B T}{8\pi\eta r^3} \qquad ...(2.35)$$

Ellipsoidal or rod-shaped molecules have two different rotary diffusion constants while, if the dimensions of the molecules are different along all three axes, three constants must be specified.

From Eq. 2.35 we can calculate (if η = 0.01 poise) the following values for a small spherical molecule (substrate) of ~1 nm length and for a spherical enzyme of 5 nm diameter :

$$r = 0.5 \text{ nm} \quad \theta \approx 1.3 \times 10^9 \text{ s}^{-1}$$

$$r = 2.5 \text{ nm} \quad \theta \approx 1.0 \times 10^7 \text{ s}^{-1}$$

We see that smaller molecules rotate much faster than large ones and that rotational relaxation times for small proteins are of the same order of magnitude as k_D for diffusion-limited encounter. However, for very large molecules, especially long rods, the rotary relaxation time about a short axis may be a large fraction of a second.

The Rate of Substrate Binding

At very low substrate concentrations the Michaelis-Menten equation (Eq. 2.15) simplifies as follows :

$$v = (V_{max}/K_m)[S] \qquad ...(2.36)$$

Since $V_{max} = k_{cat}$ $[E]_t$ (Eq. 2.6) and at low [S] most of $[E]_t$ is free E, we obtain the following equation:

$$v = (k_{cat}/k_m)[E][S] \qquad ...(2.37)$$

From this we see that k_{cat}/K_m is *the apparent second-order rate constant for the reaction of free enzyme with substrate.* As such it cannot exceed the diffusion controlled limit k_D of Eqs. 2.28 to 2.30 which falls in the range of $10^9 - 10^{11}$ M^{-1} s^{-1}. Experimentally observed values of k_{cat}/K_m are always less than this limit, indicating that a certain time is required for a substrate molecule to become oriented and seated in the active site. However, for several real enzymes values of k_{cat}/K_m of 10^7 to 3×10^8 M^{-1} s^{-1} have been observed. For triose phosphate isomerase Albery and Knowles obtained a value of 4×10^8 M^{-1} s^{-1}, so close to the diffusion controlled limit that these authors regard this enzyme as a *nearly perfect catalyst,* one that could not have evolved further because it is already catalyzing the reaction with substrate at almost the maximum velocity that is possible.

The Displacement of Bound Ligands by Substrate

Jenkins pointed out that in many instances a substrate must displace another ligand from the active site to form the ES complex. For example, a binding site for an ionic substrate often already contains an ion, either of the product P or some other ion X which Jenkins calls a *substrate surrogate*. For this situation Eq. 2.14 must be replaced by the following set of equations:

$$\mathrm{Ex + S} \underset{k_2}{\overset{k_1}{\rightleftharpoons}} \mathrm{X + ES} \xrightarrow{k_3} \mathrm{EP} \quad \begin{cases} \xrightarrow{k_5 \text{ (product release)}} \mathrm{P} \text{ (with S)} \\ \xrightarrow{k_4} \mathrm{P} \text{ (with X)} \end{cases} \qquad ...(2.38)$$

Here k_2 and also k_4 and k_5, are second-order rate constants. The release of product, as determined by k_4 and k_5, may be rate-limiting. At zero time the reverse reactions may be ignored, and steady-state analysis shows that the Michaelis-Menten equation will be replaced by Eq. 2.39. Here, D is a constant and A is also constant if X is present at a fixed concentration.

$$\frac{V_{max}}{v} = 1 + \frac{k_2[\mathrm{X}] + k_3}{k_1[\mathrm{S}]} + \frac{k_3(1/k_5 - 1/k_1)}{k_4[\mathrm{X}]/k_5 + [\mathrm{S}]}$$

$$= 1 + K_m/[\mathrm{S}] + \mathrm{D}/(\mathrm{A} + [\mathrm{S}]) \qquad ...(2.39)$$

When k_5, the rate constant for displacement of product by substrate, is very small, this equation simplifies to Eq. 2.40.

$$\frac{V_{max}}{v} = 1 + k_3/k_4[\mathrm{X}] + (k_2[\mathrm{X}] + k_3)/k_1[\mathrm{S}]$$

$$= 1 + k_3/k_4[\mathrm{X}] + K'_m/[\mathrm{S}] \qquad ...(1.40)$$

At high concentrations of X the second term becomes negligible and the equation becomes identical to the Michaelis-Menten equation except that K_m', the apparent Michaelis constant, now increases as [X] increases. This is exactly what is observed for competitive inhibitors where

$$K'_m = \frac{k_2[\mathrm{X}] + k_3}{k_1} \qquad ...(2.41)$$

Thus substrate surrogates at high concentrations are often competitive inhibitors. However, at low [X] the second term in Eq. 2.41 may be important and X *then serves as an activator.* Under many circumstances, Eq. 2.39 does not simplify further and a nonlinear relationship between 1/ v and 1/ [S] is observed, with the shape of the curves being influenced by the values of constants A and D.

Reversible Enzymatic Reactions

For some enzyme-catalyzed reactions the equilibrium lies far to one side. However, many other reactions are freely reversible. Since a catalyst promotes reactions in both directions, we must consider the action of an enzyme on the reverse reaction. Let us designate the maximum velocity in the forward direction as V_f and that in the reverse direction as V_r. There will be a Michaelis constant for reaction of enzyme with product K_{mp}, while K_{ms} will refer to the reaction with substrate. As in any other chemical reaction, there is a relationship between the rate constants for forward and reverse enzyme-catalyzed reactions and the equilibrium constant.

This relationship, first derived by the British kineticist J.B.S. Haldane and proposed in his book *Enzymes* in 1930, is known as the *Haldane relationship.* It is obtained by setting $v_f = v_r$ for the condition that product and substrate concentrations are those at equilibrium. For a single substrate-single product system it is given by Eq. 2.42.

$$K_{eq} = V_f K_{mp} / V_r K_{ms} \qquad ...(2.42)$$

Because the Haldane relationship imposes constraints on the values of the velocity constants and Michaelis constants, it is of some value in understanding regulation of metabolism. Consider the case that the maximum forward velocity V_f is high and that K_{ms} has a moderately low value (fairly strong binding of substrate). If the reaction is freely reversible ($K_{eq} \sim 1$) and the velocity of the reverse reaction V_r is about the same as that of the forward reaction, it will necessarily be true (Eq. 2.42) that the product P will also be fairly tightly bound. If $V_r << V_f$, the value of K_{mp} will have to be much *lower* than that of K_{mA}. In such a situation P will remain tightly bound to the enzyme and since V_r is low, it will tend to clog the enzyme. Such *product inhibition* may sometimes slow down a whole pathway of metabolism,

In such a case, the only way that an enzymatic sequence can keep going in the forward direction is for product P to be removed rapidly by a subsequent reaction with a second enzyme. The first enzyme may be thought of as possessing a kind of "one-way valve" that turns off the flow in a metabolic pathway when the concentration of its product rises.

Reactions of Two or More Substrates

Enzymes frequently catalyze the reaction of two, three, or even more different molecules to give one, two, three, or more products. Sometimes all of the substrate molecules must be bound to an active site at the same time and are presumably lined up on the enzyme molecule in such a way that they can react in proper sequence. In other cases, the enzyme may transform molecule A to a product, and then cause the product to react with molecule B.

The number of variations is enormous. The order in which two molecules A and B find to an enzyme to form a complex EAB may be completely *random* or it may be *oligatorily ordered.* Both situations occur with real enzymes. Cleland introduced a widely used method of depicting the possibilities. For example, Eq. 2.43 shows the reaction of A and B in an *ordered sequence* to form the complex EAB which is then isomerized to EPQ, the complex formed by binding the two products P and Q to the enzyme. The rate constant to the left of each vertical arrow of above each horizontal arrow refers to the reaction in the forward direction is indicated by

$$\mathrm{E} \xrightarrow[]{\ k_1 \downarrow A \uparrow k_2\ } \mathrm{EA} \xrightarrow[]{\ k_3 \downarrow B \uparrow k_4\ } \mathrm{EAB} \underset{k_6}{\overset{k_5}{\rightleftharpoons}} \mathrm{EPQ} \xrightarrow[]{\ k_7 \uparrow P \downarrow k_8\ } \mathrm{EQ} \xrightarrow[]{\ k_9 \uparrow Q \downarrow k_{10}\ } \mathrm{E} \qquad ...(2.43)$$

the arrow while the other constants (to the right or below the arrows) refer to the reverse reactions. The velocity in the forward direction for an enzyme with ordered binding is given by Eq. 2.44*a*)

$$v_f = \frac{V_f[A][B]}{K_{eqA}K_{mB} + K_{mB}[A] + K_{mA}[B] + [A][B]} \qquad ...(2.44\ a)$$

which may also be written in the reciprocal form (Eq. 2.44*b*) :

$$\frac{1}{v_f} = \frac{1}{V_f}\left(1+\frac{K_{mA}}{[A]}+\frac{K_{mB}}{[B]}+\frac{K_{eqA}K_{mB}}{[A][B]}\right) \quad ...(2.44\ b)$$

An alternative form of this equations proposed by Dalziel, is sometimes used.

$$\frac{[E]_t}{v_f} = \phi_0+\frac{\phi_1}{[A]}+\frac{\phi_2}{[B]}+\frac{\phi_{12}}{[A][B]} \quad ...(2.45)$$

The kinetic parameters of Eq. 2.44 are V_f, the maximum velocity in the forward direction, the two Michaelis constants, K_{mB} and K_{mA}, and the equilibrium constant K_{eqA}, for reversible dissociation of the complex EA and which is equal to k_2/k_1. The relationship between the parameters of Eq. 2.44 (K_m's, V's, and K_{eqA}'s) and the rate constants $k_1 - k_{10}$ is not obvious. However, remember that the parameters are experimental quantities determined by measurements on the enzyme.

Some-times, but not always, it is possible to deduce some of the values of individual rate constants from the experimental parameters. An equation similar to 2.44*a* can be written for the velocity v_r of the reaction of P and Q. Also an equation can be written in similar form for $v_f - v_r$, *i.e.*, the instantaneous velocity of the reaction in any mixture of all four components A, B, P, and Q. The kinetic parameters of Eq. 2.44 *b* are often obtained from experimental data through the use of reciprocal plots. However, Eq. 2.44 *b* is linear only if the concentration of one or the other of the substrates A and B is kept constant. For this reason a series of experiments is usually performed in which [A] is varied while [B] is held constant. Then [A] is held constant and [B] is varied. Each of these experiments leads to a family of lines whose slopes and intercepts are measured.

The slopes and intercepts ot this family of curves are then plotted against the reciprocal of the second concentration, i.e., the one that was held fixed. From a set of these *secondary plots*, V_f and one of the Michaelis constants can be determined. Using two sets of secondary plots, all of the constants of Eq. 2.44 may be established. Alternatively, a computer can be used to examine all of the data at once and to obtain the best values of the parameters.

The latter approach is desirable because estimates of the standard deviations of the parameters can be obtained. However, the user must take care to ensure that the experimental errors are correctly estimated and are not simply estimates of how well the computer has fitted the points on the assumption that they contain no error. The meaning of the kinetic parameters may be slightly difficult to grasp. V_f is the velocity that would be obtained if both [A] and [B] were high enough to saturate the enzyme. Each K_m corresponds to that for a simple system in which the concentration of the second substrate is at a high, saturating value.

For the bimolecular reaction that we have considered, there are two Haldane relationships:

$$K_{eq} = \frac{V_f K_{mP} K_{dQ}}{V_r K_{dA} K_{mB}} = \left(\frac{V_f}{V_r}\right)^2 \frac{K_{dP} K_{mQ}}{K_{mA} K_{dB}} \quad ...(2.46)$$

Of these, only the first is ordinarily used.

"Ping-Pong" Mechanisms

A common type of mechanism that is especially prevalent for enzymes with tightly bound cofactors has been dubbed *ping-pong* because the enzyme alternates between two forms E and E′ (Eq. 2.47). Substrate A reacts via complex EA to form EC, a modified enzyme that often

$$E \xrightarrow[k_2]{k_1\ (A)} EA \underset{k_4}{\overset{k_3}{\rightleftharpoons}} E'P \xrightarrow[k_6]{k_5\ (P)} E' \xrightarrow[k_8]{k_7\ (B)} E'B \underset{k_{10}}{\overset{k_9}{\rightleftharpoons}} EQ \xrightarrow[k_{12}]{k_{11}\ (Q)} E \qquad \text{...(2.47)}$$

Ping-pong mechanism for reaction of A and B to form P and Q

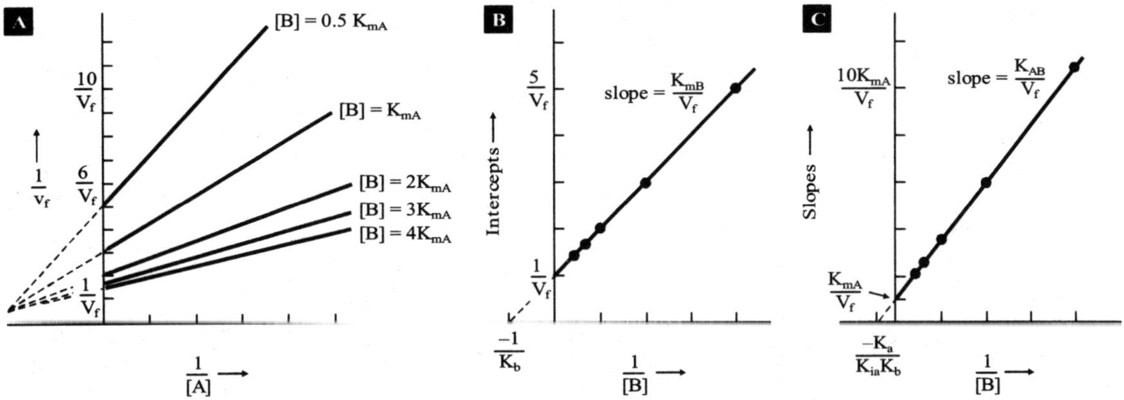

Fig. 2.6 Reciprocal plots used to analyze kinetics of two-substrate enzymes. (A) Plot of 1 / v_f against 1 / [A] for a series of different concentrations of the second substrate B. (B) A secondary plot in which the intercepts from graph A are plotted against 1/ [B]. (C) Secondary plot in which the slopes from graph A have been plotted against 1/ [B]. The figures have been drawn for the case that $K_{mA} = 10^{-3}$ M, $K_{me} = 2\ K_{mA}$, and $K_{AB} = K_{eqA}$, K_{MB} (Eq. 2.46) = $K_{mA}/200$ and [A] and [B] are in units of moles per liter. Eadie-Hofstee plots of v_f/ [A] vs v_f at constant [B] can also be used as the primary plots. The student can easily convert Eq. 2.44 to the proper form analogous to Eq. 2.21.

contains an altered coenzyme. At the same time A is changed to product P still bound to the enzyme. P dissociates leaving E′ which is then able to react with the second substrate B and to go through the second half of the cycle during which E′ is converted back to E. An example is provided by the aminotransferases in which the coenzyme pyridoxal phosphate is interconverted with pyridoxamine phosphate. The rate equations of the ping-pong mechanism resemble that for the ordered bimolecular reaction but each has one less term.

$$v_f = \frac{V_f[\text{A}][\text{B}]}{K_{mB}[\text{A}] + K_{mA}[\text{B}] + [\text{A}][\text{B}]} \qquad \text{...(2.48 } a\text{)}$$

or

$$\frac{1}{v_f} = \frac{1}{V_f}\left(1 + \frac{K_{mA}}{[\text{A}]} + \frac{K_{mB}}{[\text{B}]}\right) \qquad \text{...(2.48 } b\text{)}$$

A diagnostic feature of the ping-pong mechanism is that the families of lines (in double reciprocal plots) obtained when one substrate is held constant while the other is varied no longer intersect as they do in Fig. 2.6A but are parallel. They must be truly parallel and the experimentalist must be aware that nearly parallel lines may sometimes be observed for sequential reactions. Thus, if K_{eqA} of Eq. 2.44 b is small enough the of that equation will be small compared to the other terms and the equation will be approximated "By Eq. 2.48*b*.

The reaction will appear to be ping-pong even though it is sequential and the reaction proceeds through the ternary complex EAB. One less kinetic parameter can be obtained from an analysis of the data for a ping-pong mechanism than can be obtained for ordered reactions. Nevertheless, in Eq. 2.47, twelve rate constants are indicated. At least this many steps must be considered to describe the behavior of the enzyme. Not all of these constants can be determined from a study of steady-state kinetics, but they may be obtained in other ways.

Isomechanisms

A catalyst functions over and over again without being altered. However, during a single turnover it is likely to undergo a temporary change. For example, if an enzyme assists in removing a proton from a substrate, some functional group of the enzyme will be protonated during part of the catalytic cycle. This proton must dissociate before another catalytic cycle begins, and in some instances this dissociation may be a rate–limiting step. Northrup and coworkers developed the use of product inhibition to identify such isomechanisms.

Dead-End Complexes

In the steady state of action of an enzyme with ping-pong kinetics, part of the enzyme is in form E and part in form E′. Ideally, E would have affinity only for A and Q, while E′ would have affinity only for B and P. However, in many real situations E also binds B and P weakly; similarly, E′ binds A and Q. This is because the products and reactants usually have structural features in common. The propensity of enzymes with ping-pong kinetics to form *dead-end complexes* (also called abortive complexes) may sometimes have a regulatory function. In Eq. 2.49 the reactions of Eq. 2.47 have been rearranged to depict a situation in which product P normally undergoes a sequence of further reactions. However, if P accumulates to a high enough concentration it can react reversibly to form a dead-end complex EP. This is an effective form of product inhibition which can be relieved only by a lowering of the concentration of P through its further metabolism.

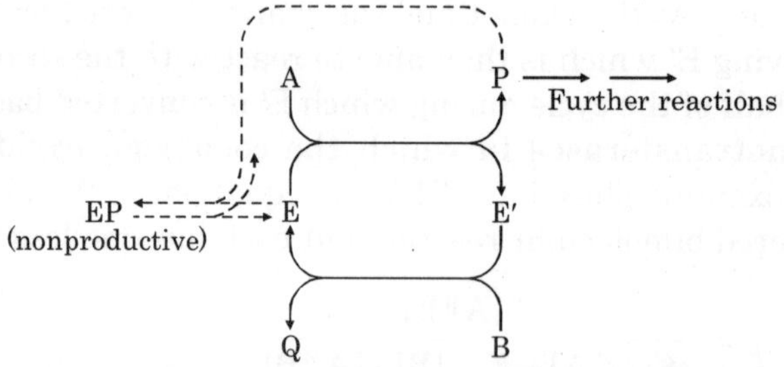

A P → → Further reactions
EP ⇠⇢ E E′ ...(2.49)
(nonproductive)
Q B

Handling Rate Equations for Complex Mechanisms

While steady-state rate equations can be derived easily for the simple cases discussed in the preceding sections, enzymes are often considerably more complex and the derivation of the correct rate equations can be extremely tedious. The *topological theory of graphs,* widely used in analysis of electrical networks, has been applied to both steady-state and nonsteady-state enzyme kinetics. The method employs diagrams of the type shown in Eq. 2.50. Here

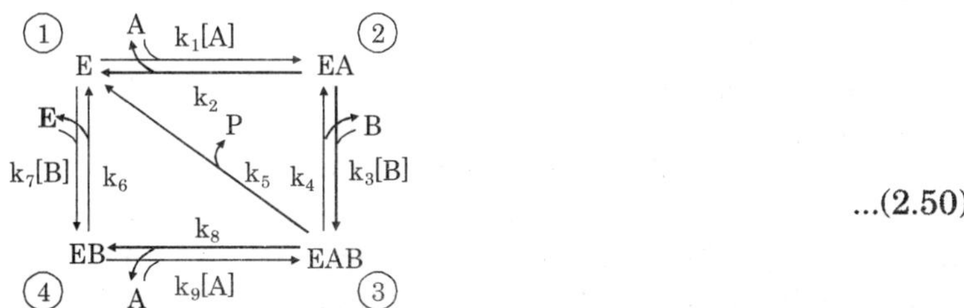

the reaction of an enzyme with two substrates A and B with a *random* order of binding is depicted. (In contrast, Eq. 2.43 shows the case of ordered binding of two substrates). When complex EAB is formed, it can decompose to free enzyme and to the single product P. Each one of the *nodes,* which are numbered 1-4 in the diagram, corresponds to a single form of the enzyme. The appropriate first-order rate constant or apparent first-order constant is placed by each arrow.

The methods provide easy rules for deriving from such a scheme the steady-state rate equation. The importance of the simplified schematic methods is apparent when one considers that the steady-state rate equation for Eq. 2.50 would have 6 terms in the numerator and 12 terms in the denominator. In the more complex case in which EAB breaks down to two products P and Q with a random order of release, the rate equation contains 672 terms in the denominator. In such cases it is worthwhile to enlist the help of a computer in deriving the equation.

The Rapid Equilibrium Assumption

Rate equations for enzymes are often simplified if a single step, *e.g.*, that of reaction of complex EAB to product in Eq. 2.50, is *rate limiting*. If it is assumed that all reaction steps preceding or following the rate-limiting step are at equilibrium, the equation for random binding with a two-substrate and two-product reaction simplifies to one whose form is similar to that obtained for ordered binding. In the absence of products P and Q Eq. 2.44 will correctly represent the steady-state rate equation corresponding to Eq. 2.50. However, this simplification may not be valid for a very rapidly acting enzyme.

Isotope Exchange at Equilibrium

Consider the reaction of substrates A and B to form P and Q. If both reactants and both products are present with the enzyme and in the ratio found at equilibrium no net reaction will take place. However, the reactants and products will be continually interconverted under the action of the enzyme. Now if a small amount of highly labeled reactant (A* or B*) is added, the rate at which isotope is transferred from the labeled reactant into one or the other of the products can be measured. In general, a label in one of the substrates will appear in only one of the products. The rate of exchange of isotopically labeled glucose

$$A + B \rightleftharpoons P + Q \qquad ...(2.51)$$

(glucose*) with glucose 6-phosphate is catalyzed by the enzyme hexokinase.

The exchange rate is plotted against the concentration of glucose 6-phosphate with the ratio [glucose] / [glucose 6-phosphate] constant at 1/19, such that an equilibrium ratio for reactants and products is always maintained. As can be seen from the graph, this exchange rate increases monotonically as substrate concentrations are increased. This is also true for the rate of ATP-ADP exchange. The fact that both exchange rates increase continuously indicates random binding of substrates.

The inequality of the two maximal exchange rates suggests that release of glucose 6-phosphate may be slower than that of ADP. Fig. 2.7B shows similar plots for lactate dehydrogenase. In this case, after an initial rise (that is not regarded as significant), the pyruvate*–lactate exchange reaches a high constant value as the amount of pyruvate is increased (with a constant [pyruvate] / [lactate] ratio of 1/35). However, the NAD*–NADH exchange increases rapidly at first but then drops abruptly as the pyruvate and lactate concentrations continue to increase. This suggests an ordered mechanism in which NAD^+ and NADH represent A and Q, respectively, and pyruvate and lactate represent B and P. As the concentrations of B and P become very high, the enzyme shuttles back and forth between EA and EQ, but these two complexes rarely dissociate to give free enzyme and A or Q. Hence, the A*–Q exchange rate drops. In other cases a label may be transferred from A into P or from B into Q. Information on such exchanges has provided a valuable criterion of mechanism.

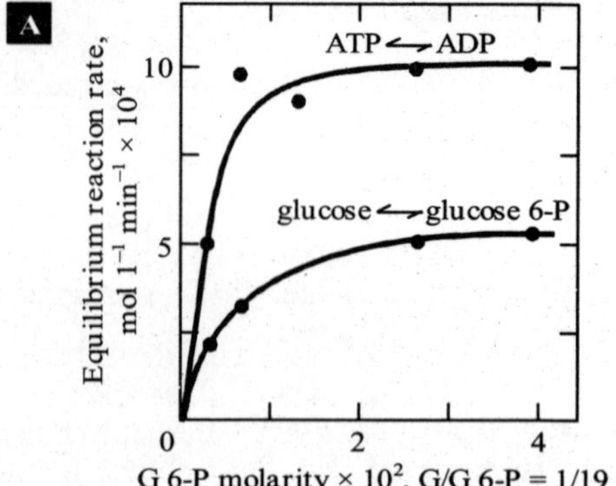

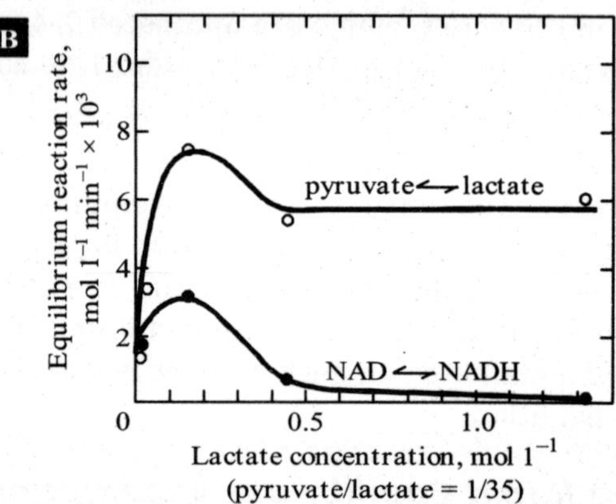

Fig. 2.7. (A) Effect of glucose and glucose 6-phosphate concentrations on reaction rate of yeast hexokinase at equilibrium. Reaction mixtures contain 1-2.2 mM ATP, and 25.6 mM ADP at pH 6.5. (B) Effect of lactate and pyruvate concentrations on equilibrium reaction rates of rabbit muscle lactate dehydrogenase. Reaction mixtures contained 1.7 mM NAD^+, and 30 - 46 µM NADH in Trisnitrate buffer, pH 7.9, 25°C.

Kinetics of Rapid Reactions

The fastest steps in an enzymatic process cannot be observed by conventional steady-state kinetic methods because the latter cannot be applied to reactions with half-times of less than about 10 s. Consequently, a variety of methods have been developed to measure rates in the range of 1 to 10^{13} s^{-1}.

Flowing Substrates Together

One of the first rapid kinetic methods to be devised consists of rapidly mixing two flowing solutions together in a special mixing device and allowing the resulting reaction mixture to move at a rate of several meters per second down a straight tube. At a flow velocity of 10 m s^{-1} a solution will move 1 cm in 10^{-3} s. Observations of the mixture are made at a suitable distance, *e.g.*, 1 cm, and with various flow rates. Using spectrophotometry or other observation techniques, the formation or disappearance of a product or reactant can be followed. The special advantage of this technique is that observation can be made slowly. However, it may require large amounts of precious reactant solutions, *e.g.*, those of purified enzymes.

In the *stopped flow* technique two solutions are mixed rapidly by the flow technique during a period of only 1–2 (or a few) milliseconds. A ram drives the solutions from syringes through a mixing chamber into an observation chamber. After the flow stops light absorption, fluorescence, conductivity, or other property, is measured. A means of rapid observation of changes during the reaction is essential. For example, light absorption may be measured by a photomultiplier with data being collected by a computer. Relaxation times as short as a few milliseconds or less can be observed in this way.

Observing Relaxation

Kinetic measurements over periods of tens of microseconds or less can be made by rapidly inducing a small displacement from the equilibrium position of a reaction (or series of reactions) and observing the rate of return (relaxation) of the system to equilibrium. Best known is the *temperature jump* method devised by Eigen and associates. Over a period of about 10^{-6} s a potential difference of ~100 kV is applied across the experimental solution. A rapid electrical discharge from a bank of condensers passes through the solution (without any sparking) raising the temperature 2–10 degrees.

All the chemical equilibria for which $\Delta H \neq 0$ are perturbed. If some property, such as the absorbance at a particular wavelength or the conductivity of the solution, is measured, very small relaxation times can be determined. While it may not be intuitively obvious, if the displacement from equilibrium is small, the rate of return to equilibrium can always be expressed as a first-order process. In the event that there is more than one chemical reaction required to reequilibrate the system, each reaction has its own characteristic relaxation time. If these relaxation times are close together, it is difficult to distinguish them; however, they often differ by an order of magnitude or more. Therefore, two or more relaxation times can often be evaluated for a given solution.

In favorable circumstances these relaxation times can be related directly to rate constants for particular steps. For example, Eigen measured the conductivity of water following a temperature jump and observed the rate of combination of H^+ and OH^- for which at 23°C equals 37 x 10^{-6} s. From this, the rate constant for combination of OH^- and H^+ was calculated as follows :

$$H^+ + OH^- \rightarrow H_2O \qquad \text{...(2.52)}$$

$$k = 1/\{\tau([OH^-] + [H^+])\} = 1.3 \times 10^{11}\ M^{-1}s^{-1} \qquad \text{...(2.53)}$$

Pressure jump and electric field jump methods have also been used, as have methods depending upon periodic changes in some property. For example, absorption of ultrasonic sound causes a periodic change in the pressure of the system.

Rapid Photometric Methods

Another useful method has been to discharge a condenser through a flash tube over a period of 10^{-12} to 10^{-4} s, causing a rapid light absorption in a sample in an adjacent parallel tube. Following the flash, changes in absorption spectrum or fluorescence of the sample can be followed. The availability of intense lasers as light sources has made it possible to follow the results of light flashes of 5–10 picosecond duration and to measure extremely short relaxation times.

Some Results

Rapid kinetic methods have revealed that enzymes often combine with substrates extremely quickly, with values of k_1 in Eq. 9–14 falling in the range of 10^6 to 10^8 M^{-1} s^{-1}. Helix-coil transitions of polypeptides have relaxation times of about 10^{-8} s, but renaturation of a denatured protein may be much slower. The first detectable structural change in the vitamin A-based chromophore of the light-operated proton pump bacteriorhodopsin occurs in ~5×10^{-8} s, while a proton is pumped through the membrane in ~ 10^{-4} s. Interconversion between chair and boat forms of cyclohexane derivatives may have $\tau \sim 10^{-5}$ s at room temperature, while rotation about a C–N bond in an amide linkage may be very slow with τ ~0.1 s. The nonenzymatic hydration of the aldehyde pyridoxal phosphate via Eq. 2.1 occurs with $\tau = 0.01 - 0.1$ s, depending upon the pH.

Cryoenzymology

An alternative to studying rapid reactions is to cool enzymes to a subzero temperature (down to –100°C) where reactions proceed more slowly. The enzyme must be dissolved in a suitable "cryosolvent," often containing 50–80% by volume of an organic solvent or solvents such as methanol, ethanol, dimethyl sulfoxide, dimethyl formamide, or ethylene glycol. Those containing methanol are especially desirable because of their low viscosities. Kinetics can be studied by various spectroscopic methods and stopped-flow, temperature jump, and other rapid-reaction techniques can be applied.

One goal of cryoertzymology is to stabilize otherwise unstable intermediates. X-ray crystallographic measurements can also be made at these low temperatures and it should be possible to observe structures of intermediate ES complexes. A problem is that the forms of the complexes stabilized may be side-products rather than true intermediates. However, a combination of low-temperature Laue X-ray diffraction and a laser-induced temperature jump may be feasible.

The Effect of pH on Enzymatic Action

Because proteins contain many acidic and basic groups it is not surprising that the activity of enzymes often varies strongly with pH. However, it is usually found that the state of protonation of only a few groups has a strong effect on activity. This is understandable because most ionized groups are on the outer surface of protein molecules and most are not close to an active site. Protonation or deprotonation of those groups will hardly ever have a major influence on events in the active site.

Often only one or two ionizable groups have highly significant effects. For many enzymes a plot of V_{max} against pH is a bell-shaped curve. The *optimum* rate is observed at some intermediate pH, which is often, but not always, in the range of pH 6–9. This type of curve can be interpreted most simply by assuming the presence in the active site of two ionizable groups and three forms of the enzyme with different degrees of protonation: E, EH, and EH_2.

$$
\begin{array}{ccccc}
 & EH_2 & & EH_2S & \leftarrow \text{Both ionizable groups protonated} \\
 & K_{1E} \updownarrow & & K_{1ES} \updownarrow & \\
S + & EH & \underset{k_2}{\overset{k_1}{\rightleftharpoons}} & EHS & \xrightarrow{k_3} EH + \text{products} \\
 & K_{2E} \updownarrow & & K_{2ES} \updownarrow & \\
 & E & & ES & \leftarrow \text{No protons on either ionizable group}
\end{array}
\qquad ...(2.54)
$$

Let us designate the acid dissociation constants (K_a) for the two groups in the enzyme as K_{1E} and K_{2E} and those of the ES complex K_{1ES} and K_{2ES}. However, these consecutive K_a values cannot necessarily be assigned to single groups. They may belong to a system of interacting groups.

Each may be the sum of more than one microscopic K_a and there may be extensive tautomerism within the active site. The rate constants k_1, k_2 and k_3 define the rates of formation and breakdown of the ES complex. If it is assumed that only from EHS reacts to give products, the bell-shaped curves of Fig. 2.8 are obtained. The frequent observation of such curves supports the model of Eq. 2.54.

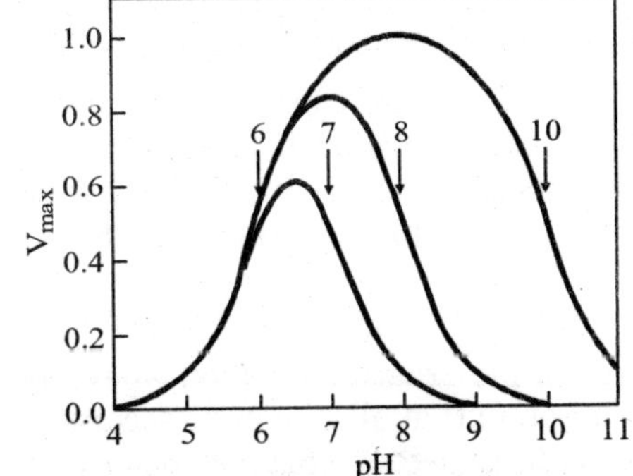

Fig. 2.8. Expected dependence of V_{max} on the pH according to Eq. 2.54 with $k_2[E_t] = 1$, $pK_{1ES} = 6$, and pK_{2ES} as given on the graph.

It also suggests that the two ionizable groups may be intimately involved in catalysis, one as an acid and the other as a conjugate base. For the simple case illustrated in Eq. 2.54, the pH dependence of the initial maximum velocity, the apparent Michaelis constant, and V_{max}/K_m are given by Eqs. 2.55 to 2.57.

$$V_{max} = \frac{k_3[E]_t}{(1+[H^+]/K_{1ES} + K_{2ES}/[H^+])} = \frac{k_3[E]_t}{F_{2ES}} \qquad ...(2.55)$$

$$K_m = \frac{(k_2+k_3)}{k_1} \cdot \frac{(1+[H^+]/K_{1E} + K_{2E}/[H^+])}{(1+[H^+]/K_{1ES} + K_{2ES}/[H^+])} = K_m F_{2E}/F_{2ES} \qquad ...(2.56)$$

$$\frac{V_{max}}{K_m} = \frac{k_1 k_3}{(k_2+k_3)} \cdot \frac{1}{(1+[H^+]/K_{1E} + K_{2E}/[H^+])} = \frac{k_3[E]_t}{K_m F_{2E}} \qquad ...(2.57)$$

The denominators (which are the Michaelis pH functions given by the first three terms of Eq. 2.6) represent the fraction of enzyme or of ES complex in the monoprotonated state. The pH dependence of enzymatic action is often more complex than that shown in Fig. 2.8 and given by the foregoing equations. However, it is easy to write Michaelis pH functions for enzymes with any number of dissociable groups in both E and ES and to write appropriate equations of the type of Eqs. 2.55 to 2.57. Bear in mind that if the *free substrate* contains groups dissociating in the pH range of interest, a Michaelis pH function for the free substrates will also appear in the numerator of Eq. 2.56. If the pH dependence of the enyzme is regulated by a conformational change in the protein, there may be a cooperative gain or loss of more than one proton and the Michaelis pH function must reflect this fact. This can sometimes be accomplished by addition of a term related to Eq. 2.45. For more information see Dixon and Webb, Cleland, or Kyte. Plots of log V_{max} (or log-specific activity) and –log K_m or log (V_{max}/K_m) versus pH yield graphs of the type shown in Fig. 2.9.

The curved segments of the graphs that extend for about 1.5 pH units on either side of each pK_a are asymptotic to straight lines of slope 1 when a single proton is involved or of a higher slope for multiple cooperative proton dissociation. The straight lines can be extrapolated and intersect at the pK_a values. (However, it is better to fit a complete curve of theoretically correct shape to the points.) Note that the curved line always passes below or above the intersection point at the value of log 2 = 0.30 except in the case of cooperative proton dissociation when it is closer. Upward turns in the curve of log K_m vs pH correspond to pK_a values in the free enzyme or substrate downward turns to pK_a values in ES. This approach to analysis of pH dependence has been adopted widely but often incorrectly.

For example, many published curves have very sharp bends in which the curved portion covers less than 3 pH units and the curve is much closer than 0.30 to the extrapolated point. This suggests cooperative proton binding and an *apparent* pK_a that is related to K of Eq. 2.21. Cooperativity is always a possibility if a conformational change in the protein is involved.

The simple treatment given above is based on the assumption that all proton dissociations are rapid compared to k_{cat}, that enzyme in only one state of protonation binds substrate, and that ES in only one state of protonation yields products. These assumptions are not always valid. It also assumes that both binding and dissociation of substrate are rapid, that is, to use Cleland's terminology the substrate is not "sticky."

For a sticky substrate that dissociates more slowly than it reacts to form products ($k_3 > k_2$; Eq. 2.54), the values of pK_{1E} will be lowered and pX_{1E} of Eq. 2.53 will be raised by log $(1 + k_3/k_2)$. In addition to the articles by Cleland, other detailed treatments of pH effects have been prepared by Brocklehurst and Dixon and Tipton and Dixon.

Fumarate hydratase (fumarase), catalyzes the reversible hydration of fumaric acid to malic acid. It was one of the first enzymes whose pH dependence was studied intensively. A bell-shaped pH dependence for both forward and reverse reactions was observed by Alberty and coworkers and, using Eqs. 2.54 and 2.55, the two apparent pK_a values were measured.

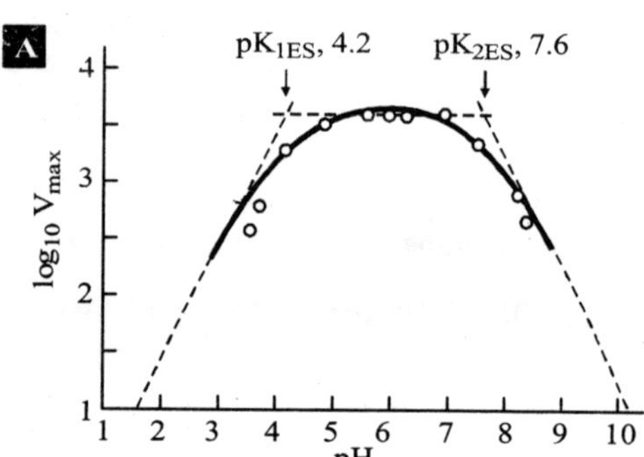

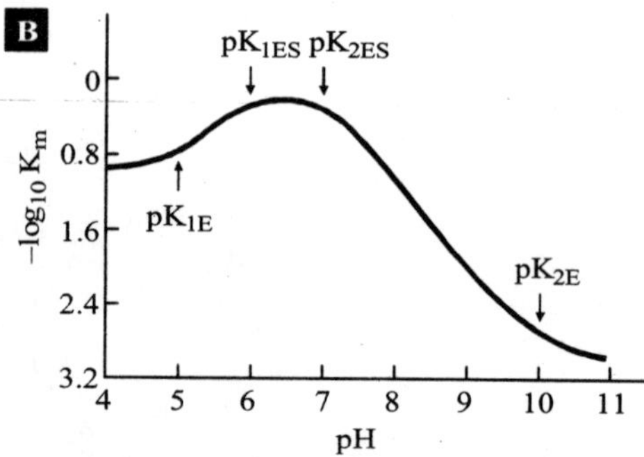

Fig. 2.9. (A) Plot of log V_{max} vs pH for a crystalline bacterial α-amylase. (B) Theoretical curve of log K_m vs pH for Eq. 2.56 with pK_{1E} = 5, pK_{2E} = 10, pK_{1ES} = 6, and pK_{2ES} = 7.

INHIBITION AND ACTIVATION OF ENZYMES

The action of most enzymes is inhibited by many substances. Inhibition is often specific, and studies of the relationship between inhibitor structure and activity have been important to the development of our concepts of active sites and of complementarity of surfaces of biomolecules. Inhibition of enzymes is also the basis of the action of a very large fraction of important drugs. Inhibition may be ***reversible or irreversible,*** the latter leading to permanent inactivation of the enzyme. Often, but not always, irreversible inhibition is preceded by reversible binding of the inhibitor at a complementary site on the enzyme surface.

Competitive Inhibitors

Inhibitors with close structural similarities to a substrate tend to bind to the substrate site. In truly competitive inhibition, substrate and inhibitor not only compete for the same site but also their binding is reversible and mutually exclusive. The affinity of the inhibitor for the enzyme is expressed quantitatively through the ***inhibition constant*** K_i which is the *dissociation constant of the enzyme inhibitor complex EI:*

$$K_i = [E]/[I]/[E] \qquad ...(2.58)$$

Using the steady state assumption for the mechanism shown in Eq. 2.14, and writing a mass balance equation that includes not only free enzyme and ES but also EI we obtain an equation relating rate to substrate concentration. It is entirely analogous to Eq. 2.15 but K_m is replaced by an apparent Michaelis constant, K'_m:

$$K_m' = K_m\left(1+\frac{[I]}{K_i}\right) \qquad ...(2.59)$$

The relationships between v and [S] and between $1/v$ and 1/ [S] are as follows :

$$\frac{v}{[S]} = \frac{V_{max}}{K_m'} - \frac{v}{K_m'} \qquad ...(2.60)$$

$$\frac{1}{v} = \frac{1}{V_{max}} + \frac{K_m}{V_{max}[S]}\left(1+\frac{[I]}{K_i}\right) \qquad ...(2.61)$$

A commonly used test for competitive inhibition is to plot $1/v$ vs 1 / [S] (Eq. 2.61), both in the absence of inhibitor and in the presence of one or more fixed concentrations of I. The result, in each case, is a family of lines of varying slope that converge on one of the axes at the value $1/V_{max}$. We see that the maximum velocity is unchanged by the presence of inhibitor. If sufficient substrate is added, the enzyme will be saturated with substrate and the inhibitor cannot bind.

The value of K_i can be calculated using Eq. 2.61 from the change in slope caused by addition of inhibitor. The effect of a fixed concentration of a competitive inhibitor on a plot of v against log [S] is to shift the curve to the right, *i.e.*, toward higher values of [S], but without any change in shape (or in the value of V_{max}). Another plot, introduced by Dixon, is that of 1 / v versus [I] at two or more fixed substrate concentrations. The student should be able to demonstrate that this plot contains a family of straight lines that intersect at a point to the left of the origin with coordinates [I] = $-K_i$ and $1/v = 1/V_{max}$. This plot may fail to distinguish certain types of inhibition discussed in the next section. Competitive inhibition is extremely common and has great significance for metabolic control and for the effects of drugs and of poisons.

Simple ions are often competitive inhibitors. Since many biochemical substances carry negative charges, anions such as Cl^-, HCO_3^-, HPO_4^{2-}, and acetate frequently act as competitive inhibitors. Two special classes of competitive inhibitors are characterized by *slow binding* and *slow, tight binding* to active sites. Among the very tight-binding inhibitors, which may also be slow to dissociate from active sites, are transition state inhibitors.

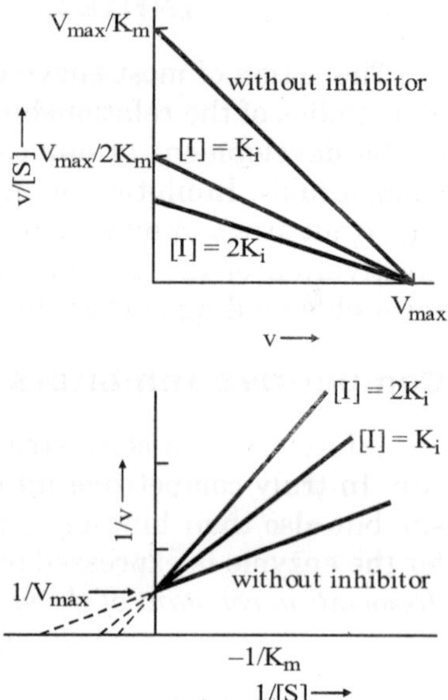

Fig. 2.10. Effect of a competitive inhibitor on the Eadie–Hofstee plot (top) and on a double reciprocal plot (bottom). The apparent K_m (Eq. 2.59) is increased by increasing but V_{max} is unchanged.

Noncompetitive Inhibition and Activation; Allosteric Sites

If an inhibitor binds not only to free enzyme but also to the enzyme substrate complex ES, inhibition is *noncompetitive*. In this case, S and I do not mutually exclude each other and both can be bound to the enzyme at the same time. Why does such an inhibitor slow an enzymatic reaction? In most instances, the structure of the inhibitor does not show a close similarity to that of substrate, which suggests that the binding of inhibitors is at an *allosteric site,* that is, at a site other than that of the substrate. The inhibition of the enzyme may result from a distortion of the three-dimensional structure of the enzyme which is caused by the binding of the inhibitor. This distortion may be transmitted to the active site even though the inhibitor binds far from that site. In some cases two distinctly different confarmers of the protein may exist, one binding substrate well and the other binding inhibitor well.

In other instances the bound inhibitor may interfere with the catalytic action by partially overlapping the active site. In either case the ES complex reacts to give product in a normal way, but the ESI complex reacts more slowly or not at all. Binding of a substance to an allosteric site sometimes has the effect of *increasing* the activity of an enzyme rather than inhibiting it. This may occur because the *activator* stabilizes the conformation that binds substrate best. The quantitative treatment of such activation is similar to that of inhibition; allosteric inhibitors and activators are often considered together and are referred to as *modifiers* or *effectors*. A general scheme is given by Eq. 2.62.

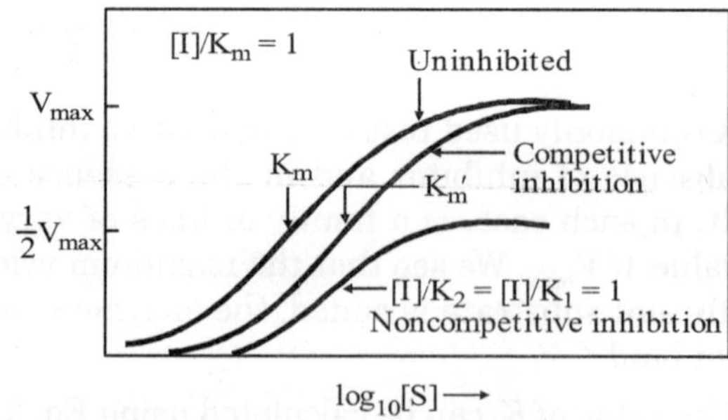

Fig. 2.11. Plots of *v* vs. log [S] for competitive and non-competitive inhibition.

$$
\begin{array}{ccc}
E + S & \underset{}{\overset{K_{ds}}{\rightleftharpoons}} & ES \xrightarrow{k_4} \text{product} \\
+ & & + \\
M \text{ (modifier)} & & M \\
\Big\downarrow\!\Big\uparrow K_1 & & \Big\downarrow\!\Big\uparrow K_2 \\
EM + S & \underset{}{\overset{K'_{ds}}{\rightleftharpoons}} & ESM \xrightarrow{k_4} \text{product}
\end{array}
\qquad \text{...(2.62)}
$$

Here K_1 and K_2 are equilibrium constants for dissociation of M from EM and ESM, respectively, while K_{ds} and K'_{ds} are the dissociation constant of ES and of ESM (to S and EM), respectively. Notice that, K'_{ds} is not independent of the others :

$$K'_{ds} = K_{ds}K_2/K_1 \qquad \text{...(2.63)}$$

We have already considered competitive inhibition which is obtained when $K_2 = 0$ (and therefore $K'_{ds} = 0$). For this case, M is always an inhibitor and no activation is possible. Notice that the inhibition will appear competitive even if M binds at an allosteric site as in Fig. 2.13 or if the inhibited form does not react at all with substrate. Noncompetitive inhibition will be observed if ESM is formed but does not react, *i.e.*, if $k_4 = 0$. Then the rate equation in reciprocal form will be given by Eq. 2.64.

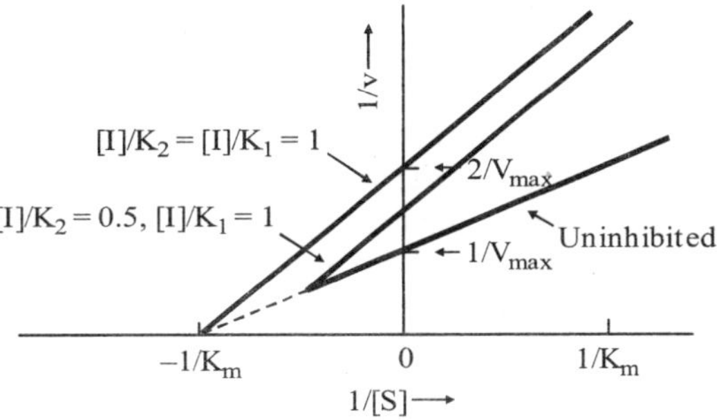

Fig. 2.12. Double reciprocal plots for two cases of non-competitive inhibition.

$$\frac{1}{v} = \frac{1}{V_{max}}\left(1+\frac{[I]}{K_2}\right)+\frac{K_m}{V_{max}[S]}\left(1+\frac{[I]}{K_1}\right) \qquad \text{...(2.64)}$$

We see that $1 / V_{max}$ is multiplied by a term containing [I] and K_2. It is a characteristic of noncompetitive inhibition that the maximum velocity is decreased from that observed in the absence of the inhibitor. Also, no matter how high the substrate concentration, inhibition cannot be reversed completely. Fig. 2.12 shows a plot of $1/v$ against $1/[S]$ at a series of fixed values of [I]. For the case that $K_1 = K_2$ (*classical noncompetitive inhibition*), a family of reciprocal plots that intersect on the horizontal axis at a value of $-1/K_m$ is obtained.

On the other hand, if K_1 and K_2 differ (the general case of noncompetitive inhibition), the family of curves intersect at some other point to the left of the vertical axis and, depending upon the relative values of K_1 and K_2, either above or below the horizontal axis. The example illustrated is for $K_2 = 0.5\ K_1$; that is, for the binding of M to ES being twiceas strong as that to E. Fig. 2.11 shows inhibition data for both the noncompetitive and the competitive cases plotted vs log[S]. The shift of the midpoint to the right in each case reflects the tendency of the inhibitor to exclude the substrate from binding, while the lowered value of the maximum velocity in the case of noncompetitive inhibition results from the failure of the substrate to completely displace the inhibitor from the enzyme even as [S] becomes very high. If an inhibitor binds only to ES and not to *E, i.e.*, $K_1 = 0$, a family of parallel double-reciprocal plots of $1/v$ vs 1/[S] will be obtained. This case is referred to as *uncompetitive inhibition.* Multisubstrate enzymes with either ordered sequential or ping-pong mechanisms also often give parallel line plots with inhibitors. Uncompetitive inhibitors of liver alcohol dehydrogenase could be used to treat cases of poisoning by methanol or ethylene glycol. The aim is to prevent rapid oxidation to the toxic acids HCOOH and $HOCH_2COOH$, which lower blood pH, while the alcohols are excreted.

Uncompetitive inhibitors have an advantage over competitive inhibitors as therapeutic agents in that the inhibition is not overcome when the substrate concentration is saturating. If rate constant k_4 of Eq. 2.62 is not zero we may observe either inhibition or activation. If $k_3 = k_4$ the effect of a modifier will be to alter the apparent K_m by either increasing it (inhibition) or decreasing it (activation). The maximum velocity will be unchanged. Monod *et al.* referred to enzymes showing such behavior as *K systems*. On the other hand, if $K_1 = K_2$ and k_3, differs from k_4 we have a purely *V system*. In the general case a modifier affects both the apparent K_m and V_{max}. Although in the foregoing discussion it has been assumed that activators bind at allosteric sites, ions and other small molecules (substrate surrogates) that act as competitive inhibitors at high concentrations may be activators at low concentrations. Activation of enzymes by specific metallic ions is often observed.

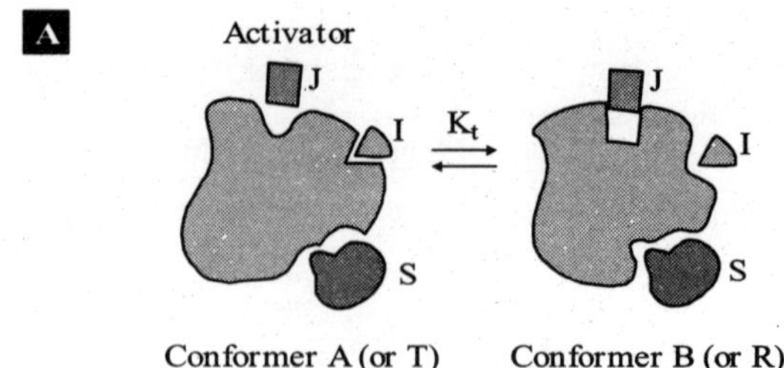

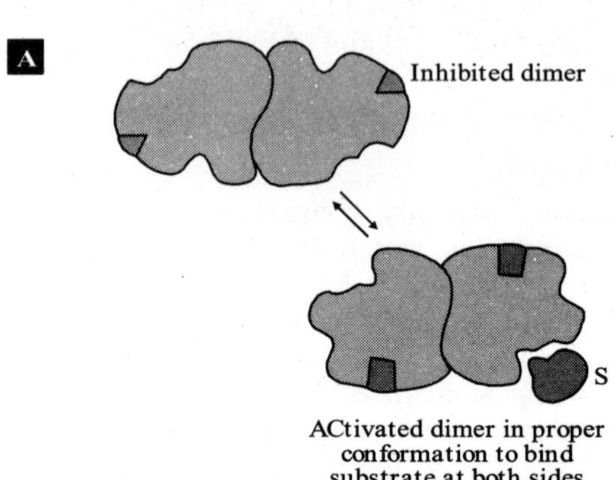

Fig. 2.13. (A) An enzyme with binding sites for allosteric inhibitor I and activator J. Conformer A binds inhibitor I strongly but has little affinity for activator J or for substrate S. Conformer B binds S and catalyzes its reaction. It also binds activator J whose presence tends to lock the enzyme in the "on" conformation B. Conformers A and B are designated T and R in the MWC model of Monod, Wyman, and Changeux. (B) Inhibited and activated dimeric enzymes.

In many instances the metallic ion is properly regarded as a *second substrate* which must bind along with the first substrate before reaction can occur. Alternatively, the complex of the organic substrate with the metal ion can be considered the "true substrate." Thus, many enzymes act upon the magnesium complex of ATP. The enzymes can either be regarded as three-substrate enzymes requiring Mg^{2+} + ATP^{4-} and another substrate or as two-substrate enzymes acting upon $ATPMg^{2-}$ and a second substrate.

Inhibitors in the Study of Mechanisms

A substrate analog will frequently inhibit only one of the two forms of a multisubstrate enzyme with a ping-pong mechanism. Reciprocal plots made for various inhibitor concentrations consist of a family of parallel lines reminiscent of uncompetitive inhibition. Observation of such parallel line plots can support a ping-pong mechanism for an enzyme but cannot prove it because in some cases parallel lines are observed for inhibition of enzymes acting by an ordered sequential mechanism. The following question arises naturally for any ordered bimolecular reaction (Eq. 2.43). Of the two substrates required by the enzymes, which one binds to the enzyme first? If the concentration of one substrate is kept constant while varying concentrations of an inhibitory analog of that substrate are added and $1/v$ is plotted against the reciprocal of the concentration of the other substrate, parallel lines are obtained if, and only if, the substrate of fixed concentration is B, the substrate is adding second in the binding sequence, *and* if I is its analog.

The substrate binding first (A) is the one whose concentration was varied in the experiment. Product inhibition can also provide information about mechanisms. For example, if $1/v$ is plotted against $1/[A]$ in the presence and absence of the product Q, the product will be found to compete with A and to give a typical family of lines for competitive inhibition. On the other hand, a plot of $1/v$ vs $1/[B]$ in the presence and absence of Q will indicate noncompetitive inhibition if the binding of substrates is ordered. In other words, only the A–Q pair of substrates are competitive. Product inhibition is also observed with enzymes having ping-pong kinetics as a result of formation of nonproductive complexes.

Allosteric Effectors in the Regulation of Enzyme Activity

The binding of a substance at an allosteric site with the induction of a conformational change forms the basis for many aspects of regulation. The term *allostery* (allosterism) usually refers to the effects of allosteric modifiers, which may be either inhibitors or activators, on oligomeric enzymes. However, as we have already seen, monomeric enzymes may also be subject to allosteric regulation by modifiers. Consider a monomer that contains binding sites for substrate, inhibitor, and activator and which exists in conformations A and B. Let us assume that conformer B binds both substrate and activator well but that it binds inhibitor poorly or not at all.

On the other hand, A binds inhibitor well but binds substrate and activator poorly. This simple combination of two conformers with different binding properties provides a means by which enzymes can be turned "on" or "off" in response to changing condition. If an inhibitory substance builds up to a high concentration within a cell, it binds to conformation A; if the inhibitor concentration is high enough, virtually all of the enzyme will be locked in the inactive conformation A. The enzyme will be turned off or at least reduced to a low activity. On the other hand, in the presence of a high concentration of activator the enzyme will be turned on because it is locked in the B conformation. The relative concentrations of inhibitor, activator, and substrate within a cell at any given time will determine what fraction of the enzyme is in active conformation B. It is this interplay of inhibitory and activating effects that provides the basis for much of the regulation of cell chemistry. The effects of inhibitors or activators on the kinetics of the monomeric enzyme of Fig. 2.13 can be described by Eq 2.62 to 2.64.

Separate terms for both inhibition and activation can be included. The equilibrium between the two conformers can also be indicated explicitly according to Eq. 2.30. However, for monomeric enzymes it is usually not profitable to try to separate the two constants K_t and K_{BX} which describe the conformational change and binding of substrate or activator, respectively in Eq. 2.30. Most intracellular enzymes are oligomeric, and the binding of allosteric effectors leads to additional interesting effects. Binding constants or dissociation constants must be defined for both inhibitor and activator to both conformers A and B. Since all species must be taken into account in the mass balance, the equations are complex. However, the Monod-Wyman-Changeux (MWC) model gives a relatively simple picture. The saturation curve for an oligomeric enzyme following this model may be derived from Eq. 2.39 and is given by the following expression :

$$Y = \frac{Lc\alpha(1+c\alpha)^{n-1} + \alpha(1+\alpha)^{n-1}}{L(1+c\alpha)^n + (1+\alpha)^n} \quad \text{...(2.65)}$$

Here, L is the *allosteric constant* which is given (for a dimer) by Eq. 2.36. The constant *c* is the ratio of dissociation constants K_{BS} and K_{AS} for the two conformers :

$$c = K_{BS}/K_{AS} \quad \text{...(2.66)}$$

Notice that in this chapter *dissociation* constants (K_d) of ES complexes are being used, whereas the equations are all written in terms of association constant (K_r). The parameter α is defined as follows, where K_{BS} equals $1/K_{BX}$

$$\alpha = [S]/K_{BS} \quad \text{...(2.67)}$$

To take account of effects of inhibitor and activator the *ratios* of dissociation constants of I from BI and AI and of activator J from BJ and AJ are defined as in Eq. 2.68. Likewise, "normalized concentrations" of I and J are defined as β and γ, respectively.

$$d = \frac{K_{BI}}{K_{AI}} > 1 \qquad e = \frac{K_{BJ}}{K_{AJ}} < 1 \quad \text{...(2.68)}$$

$$\beta = \frac{[I]}{K_{AI}} \qquad \gamma = [J]K_{BJ} \quad \text{...(2.69)}$$

According to the MWC model, in the presence of inhibitor and activator at normalized concentrations P and y an enzyme will still follow Eq. 2.65, but the allosteric constant L will be replaced by an apparent allosteric constant L′ (Eq. 2.70). Fig. 2.14 shows plots of Y vs. log α for two different values of L′ for a tetramer with a specific value assumed for *c*. In both cases, Y approaches 1 as log a increases, but since we are dealing with noncompetitive inhibition

$$L' = L\left[\frac{(1+\beta d)(1+\gamma e)}{(1+\beta)(1+\gamma)}\right]^n \quad \text{...(2.70)}$$

at high values of L′, much of the enzyme will be in the T(A) conformation at saturation. Noncompetitive inhibition cannot be completely reversed by very high substrate concentrations. Monod *et al.* defined for an allosteric enzyme a *function of state* $\bar{R}$ which is the fraction of total enzyme in the $\bar{R}$ (B) conformation :

$$\bar{R} = \frac{(1+\alpha)^n}{L(1+c\alpha)^n + (1+\alpha)^n} = \text{function of state} \quad \text{...(2.71)}$$

In a *K* system it is the value of $\bar{R}$ that determines the velocity with which an enzyme reacts. Fig. 2.14 also shows $\bar{R}$ as a function of log [α]. Note that when L′ is low $\bar{R}$ does not approach zero even when [S] → 0. In other words, the enzyme is never completely turned off, just as when F′ is high the enzyme is never completely turned on. Fig. 2.14 may be compared with Fig. 2.11 which shows similar curves for noncompetitive inhibition of a monomeric enzyme.

The significant difference between the two figures is that saturation of the oligomeric enzyme occurs over a narrower concentration range than does that of the monomer, *i.e.*, saturation of the oligomeric enzyme, especially in the presence of inhibitor, is *cooperative*. Note that cooperative binding of substrate requires that the free enzyme be largely in conformation T (A), as it is in the presence of an inhibitor. Allosteric interactions between two identical molecules, whether of substrate or of effector, are described by Monod *et al.* as *homotropic interactions*. Such interactions lead to cooperativity or anticooperativity in binding.

Allosteric interactions between two different molecules, *e.g.*, a substrate and an activator are designated *heterotropic*. For many enzymes the MWC model is unrealistically simple. The more general treatment of binding equilibria may be applicable. However, in addition to K systems there are V systems in which a conformational change alters the maximum velocity and sometimes both substrate affinity and maximum velocity. The fact that data can be fitted to an equation is not proof that a mathematical model is correct; other models may predict the results just as well. For example, Jenkins has shown that the presence of ions that act as substrate surrogates can produce cooperative or anticooperative binding curves for substrates. Rabin suggested the following explanation of cooperativity for a monomomeric enzyme with a single substrate-binding site.

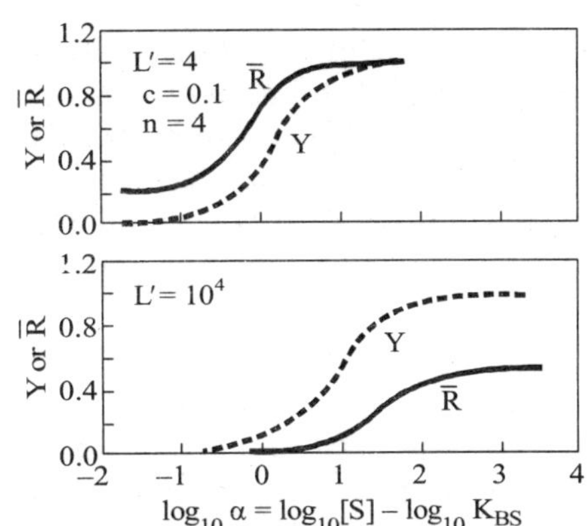

Fig. 2.14. Fractional saturation Y and "function of state" $\bar{R}$ for hypothetical tetrameric enzymes following the MWC model. Curves are calculated for two different values of the apparent allosteric constant L′ (Eq. 2.70) and for *c* = 0.1 (Eq. 2.66).

In its active conformation E the enzyme reacts with substrate rapidly and the ES complex yields product rapidly as in the upper loop of Eq. 2.72. However, a slow conformational change interconverts E and E′, a less active form with much lower affinity for substrate.

$$\begin{array}{ccccc} S & \longrightarrow & ES & \longrightarrow & E + P \\ & & \uparrow\downarrow & & \big\downarrow \text{ slow} \\ S & \longrightarrow & ES' & & E' \text{ (inactive)} \end{array} \qquad \ldots(2.72)$$

At the same time, the complex ES′, if formed, can equilibriate with ES and thereby alter the conformational state of the protein. At low substrate concentrations E′ will predominate and the enzymatic activity will be low. At high substrate concentrations most E will exist as ES and after release of product will tend to remain in the active conformation long enough to bind and act on another substrate molecule. This and other kinetic models of cooperativity are discussed by Newsholme and Start. The physiological significance of cooperative binding of substrates to enzymes is analogous to that of cooperative binding of oxygen by hemoglobin which provides for more efficient release of oxygen to tissues.

However, in the presence of excess activator an enzyme is locked in the *R* (B) conformation and no cooperativiry is seen in the binding of substrate. In this case, each binding site behaves independently. On the other hand, *there will be strong cooperativity in the binding of the activator.* The result is that control of the enzyme is sensitive to a higher power than the first of the activator concentration. Likewise, the turning off of the enzyme is more sensitive to inhibitor concentration as a result of cooperative binding of the latter. It seems likely that the evolution of oligomeric enzymes is at least partly a result of the greater efficiency of control mechanisms based on cooperative binding of effectors.

Irreversible inhibition of Enzymes

Some of our most effective drugs, such as penicillin are irreversible inhibitors of specific enzymes. Such inhibitors are also of practical importance to biochemists who wish to inhibit specific enzymes such as proteases that might otherwise destroy proteins they are studying. Transition state analogs and slow-binding inhibitors may appear to inactivate an enzyme irreversibly because they bind tightly and dissociate slowly. However, true irreversible inhibitors, such as oxidizing agents or alkylating agents, cause a more permanent chemical modification, usually at or near the active site.

Two groups of irreversible inhibitors are of special interest: affinity labeling agents and enzyme-activated inhibitors: *Affinity labeling agents* (active-site directed inactivating reagents) have two essential properties: *a high affinity for the active site* of the specific enzyme to be inhibited and *the presence of a chemically highly reactive group* which can attack a functional group in a protein. A good example is provided by derivatives of chloro-methylketones, which are reactive alkylating agents. One of these, *N*-tosylphenylalanine chloromethyl ketone (TPCK), is a potent inhibitor of chymotrypsin. In addition to the chloromethyl ketone group it contains the phenylalanine side chain, which helps direct it into the substrate binding site of chymotrypsin, and a sulfonamide linkage that mimics the normal amide linkages of the substrate. The group Y, which becomes alkylated in an S_N2-like displacement of a chloride ion is thought to be His 57 of the active site. For additional discussion see Walsh, Kyte, and Plapp.

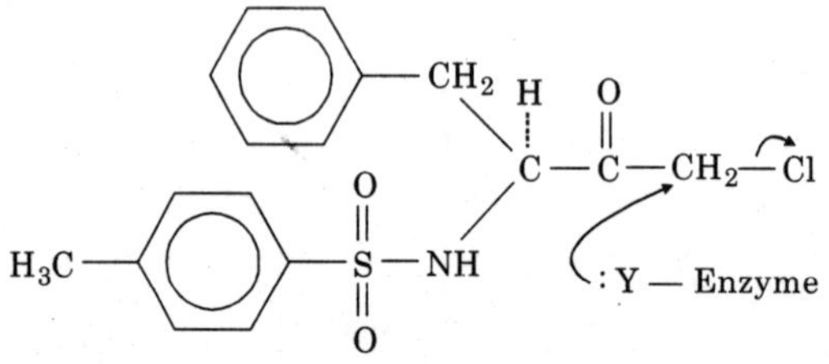

CH_2 H O
C — C — CH_2 — Cl
H_3C — S — NH
O O
: Y — Enzyme

L-N-tosylphenylalanine chloromethylketone (TPCK)

Enzyme-activated inhibitors (also called suicide substrates, k_{cat} inhibitors, or mechanism-based inhibitors) are chemically inert until they enter the active site of their target enzymes. Then, by passing through at least some of the normal stages of the catalytic action of the enzyme, they are converted to reactive intermediates that can become irreversibly bound to an enzyme. For example, a halogen atom (F, Cl, Br) together with a proton may be eliminated from an intermediate to give an unsaturated compound to which a nucleophilic side chain of the protein adds. Several of these inhibitors are discussed in later chapters. Because of their high specificity many enzyme-activated inhibitors are potential drugs. One of them, α-difluoromethylornithine, is said to bind covalently to only one protein in the body of a rat, namely, ornithine decarboxylase, the target enzyme for the drug.

THE SPECIFICITY OF ENZYMATIC ACTION

Enzymes are usually impressively specific in their action. The specificity toward substrate is sometimes almost absolute. For many years urea was believed to be the *only* substrate for the enzyme *urease* and succinate the only substrate for *succinate dehydrogenase*. Even after much searching for other substrates, only one or two closely related compounds could be found that were acted on at all. In other cases enzymes can use a class of compounds as substrates. For example, the *D-amino acid oxidase* of kidney oxidizes a variety of D-amino acids but does not touch L-amino acids. Almost as impressive as the substrate specificity of enzymes is the specificity for a given type of reaction.

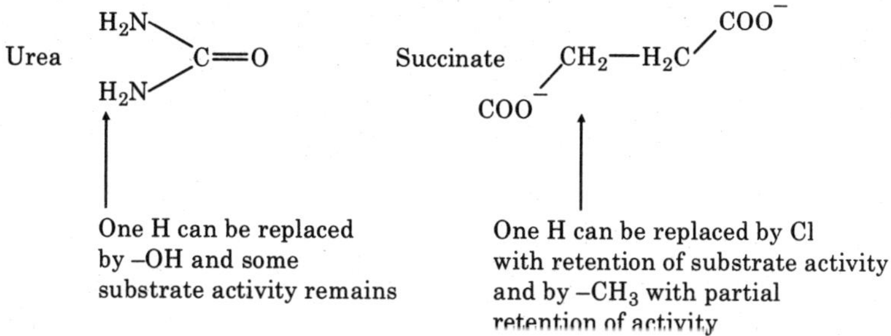

Many substrates are capable of undergoing a variety of different chemical reactions, either unimolecular or with water or some other compound present in the cell. The enzyme catalyzes only one of these reactions. Although side reactions may occur to a small extent, the most impressive thing in comparing an enzymecatalyzed reaction with an uncatalyzed organic reaction is that the latter often produces large amounts of side reaction products, but the enzymatic reaction does not.

Complementarity of Substrate and Enzyme Surfaces

Impressed by the specificity of enzymatic action, biochemists early adopted a "lock-and-key" theory which stated that for a reaction to occur the substrate must fit into an active site precisely. Modern experiments have amply verified the idea. A vast amount of kinetic data on families of substrates and related competitive inhibitors support the idea and numerous X-ray structures of enzymes with bound inhibitors or with very slow substrates have given visual evidence of the reality of the lock-and-key concept. Directed mutation of genes of many enzymes of known three-dimensional structure has provided additional proof. As anticipated, hydrophobic groups of substrates or inhibitors usually contact hydrophobic regions of the protein and the fit is tight. For example, the active site of chymotrypsin contains a "hydrophobic pocket" designed to hold a large hydrophobic side chain, thus providing the specificity observed with this enzyme.

Likewise, polar groups of substrates contact polar groups of the enzyme. The interactions are complementary, positive charges fitting against negative and with correctly formed hydrogen bonds. Trypsin, whose structure is similar to that of chymotrypsin, exerts its specificity for a positively charged side chain next to the bond cleaved by virtue of the presence of a negatively charged carboxylate at the bottom of the hydrophobic pocket. Several C = O and N–H groups of the peptide linkages in the substrate form hydrogen bonds to the edge of a F sheet in the protein, in effect making the substrate an added β strand in the sheet. Aspartate aminotransferase acts on the dicarboxylic amino acids glutamate and aspartate. A pair of arginine side chains bind the two carboxylates of the substrate while the $-NH_3^+$ of the substrate is attracted to a negatively charged group in the coenzyme pyridoxal phcsphate, which is also present in the active site.

Stereospecificity and Prochiral Centers

Most enzymes possess an infallible ability to recognize the difference between the right side and the left side of an organic substrate even when the latter has perfect bilateral symmetry. In fact, this ability is limited to *prochiral centers* of molecules and is a natural consequence of their reaction with the chiral enzyme.

Consider carbon atom number 1 of ethanol. The two attached hydrogen atoms are chemically identical and would react identically with a nonchiral reagent. Nevertheless, these atoms are no more equivalent stereochemically than are your right and your left arms. We say that the molecule has a prochiral center at C–1. A prochiral center on a tetrahedrally bonded carbon atom always contains two identical atoms or groups but *a* total of three different kinds of atoms or groups. If the priority of this H is elevated, *e.g.*, by substitution of ^2H for ^1H the configuration would be *R* (the priority of this ^2H would become *c*).

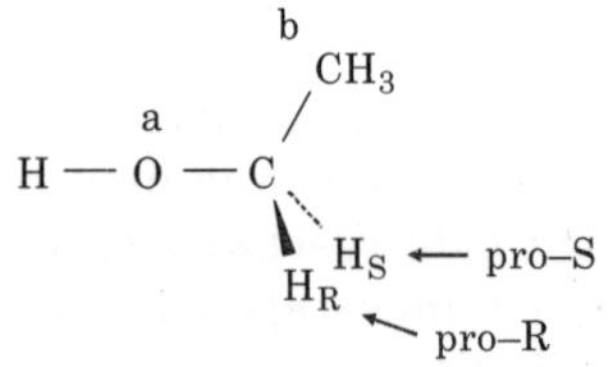

```
              b
               CH3
      a       /
H — O — C
          \  \
        HR  HS ← pro–S
          ↖ pro–R
```

The two hydrogen atoms on C–1 of ethanol are called *enantitopic* because replacement of one or the other by a fourth different kind of atom or group would produce a pair of enantiomers. This fact suggests a way of naming the positions occupied by two enantiotopic atoms or groups. We first assign priorities to all of the atoms or groups attached to the central carbon atom according to the *RS* system. Now, we ask whether the configuration will be R or S when the priority of one of the two identical groups is raised, *e.g.*, by substitution of one of the hydrogen atoms by deuterium. If the configuration becomes R, that group occupies a *pro-R* position; if the configuration becomes S, it occupies a *pro-S* position.

Referring to the preceding diagram, it is easy to see (by viewing down the bond to the group of lowest priority and applying the usual rule for determining configuration) that if the *pro-R* hydrogen (H_R) is replaced by deuterium (^2H), the configuration will be R. Conversely, replacement of H_s by deuterium will lead to the S configuration. When ethanol is oxidized by the action of *alcohol dehydrogenase,* only the *pro*-R hydrogen atom is removed. If the reaction is reversed in such a way that deuterium is introduced into ethanol from the reduced coenzyme the optically active *R*-2-deuterio ethanol is formed. The ability of an enzyme to recognize a single hydrogen of a pair of hydrogens on a CH_2 group was at first a surprise to many bio-chemists. However, it is a natural result of the complementarity of enzyme and substrate surfaces, just as the fit of a shoe is determined by the complementarity of surfaces of foot and shoe. Only a chiral catalyst can havethis ability. Another example is provided by malic acid, a chiral molecule which also contains a prochiral center.

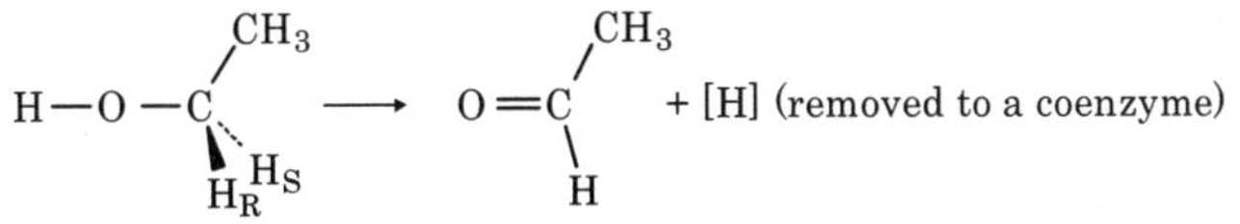

$$H{-}O{-}C(CH_3)(H_R)(H_S) \longrightarrow O{=}C(CH_3)(H) + [H] \text{ (removed to a coenzyme)}$$

In this case replacement of the *pro-R* or *pro-S* hydrogen atom by another atom or group would yield a pair of diastereoisomers rather than enantiomers. Therefore, these hydrogen atoms are *diastereotopic.* When L-malic acid is dehydrated by fumarate hydratase the hydrogen in the *pro*-R position is removed but that in the *pro-S* position is not touched. This can be demonstrated by allowing the dehydration product, fumarate, to be hydrated to malate in ^2H_2O. The malate formed contains deuterium in the *pro*-R position. If this malate is now isolated and placed with another portion of enzyme in H_2O, the deuterium is removed cleanly. The fumarate produced contains no deuterium.

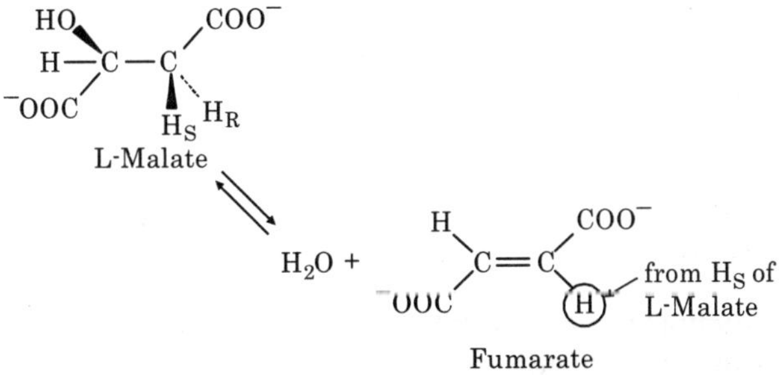

L-Malate ⇌ H_2O + Fumarate (H from H_S of L-Malate) ...(2.74)

Stereochemical Numbering

In some prochiral molecules, such as glycerol, the two ends of the main carbon chain form identical groups. Since the two ends are distinguishable to an enzyme, it is important to decide which should be labeled C–1 and which C–3. Hirschmann proposed a stereochemical numbering system in which the carbons are numbered beginning with the end of the chain that occupies the *pro*-S position.

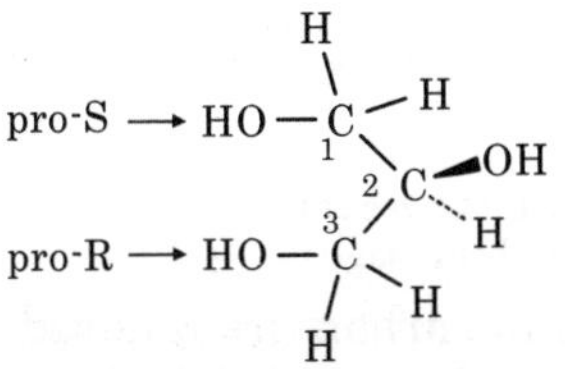

pro-S → HO—C1(H)(H); C2(OH)(H); pro-R → HO—C3(H)(H)

Stereochemical numbering for a symmetric prochiral molecule. Carbons are numbered beginning with the end of the chain that occupies the pro-S position.

In this numbering system derivatives of the parent prochiral compound are given the prefix *sn*-. Thus, glycerol phosphate, used by cells to construct phospholipids, usually bears a phosphate group on the $–CH_2OH$ in the *pro*-R position of glycerol and is therefore *sn*-3-glycerol phosphate. The *citrate ion,* a very important prochiral metabolic intermediate, has three prochiral centers at C–2, C–3, and C–4, respectively. That at C–3 distinguishes the *pro*-R and *pro*-S arms and determines the stereochemical numbering.

Citrate containing ^{14}C in the *sn*-1 position is called sn-citrate[1–^{14}C] and is the form of labeled citrate that is synthesized in living cells from oxaloacetate and [1–^{14}C] acetyl coenzyme A. The first step in the further metabolism of citrate is the elimination of the –OH group from C–3 together with the H_R proton from C–4 through the action of the enzyme *aconitate hydratase* (aconitase). In this case the proton at C–4 (in the *pro*-R arm) is selected rather than that at C–2.

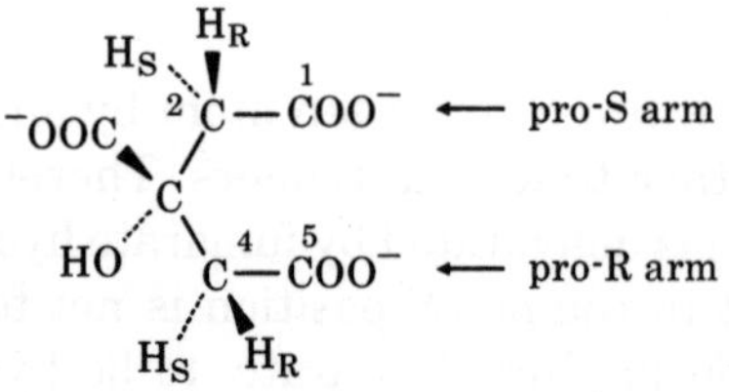

```
        HR
   HS \ |  1
      2 C—COO⁻   ←— pro-S arm
⁻OOC  /
     C                Citrate ion mi,nered
   / \  4  5          according to the
HO    C—COO⁻   ←— pro-R arm    stereochemical system
     / \
   HS   HR

      ¹COO⁻
       |
HS—²C—HR
       |
HO—³C—COO⁻     Drawn using the
       |          Fisher convention
HR—⁴C—HS
       |
      ⁵COO⁻
    Citrate
```

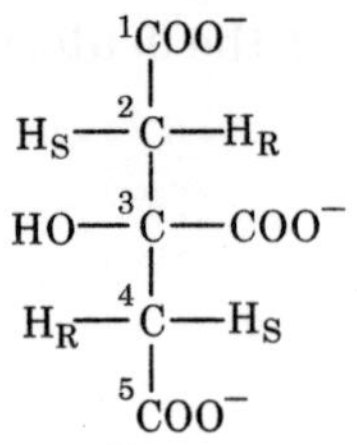

Trigonal Prochiral Centers

Planar trigonal atoms, such as those of aldehydes, ketones, and alkenes, are also prochiral if they are attached to three different kinds of atoms or groups. The two faces are enantio-topic or diastereotopic, if another chiral center is present. Hanson proposed that the faces be named. The trigonal atom is viewed from one side and the three groups surrounding the carbon atom are given priorities, *a, b,* and *c,* as in the *RS* system. If the sequence *a, b,* and *c* of priorites is clockwise, the face toward the reader is *re* (*rectus*); if counterclockwise, it is *si* (*sinister*). Priorities can often be assigned on the basis of the atomic numbers of the first atoms in the three groups (as in the example shown) but, if necessary, replica atoms must be added.

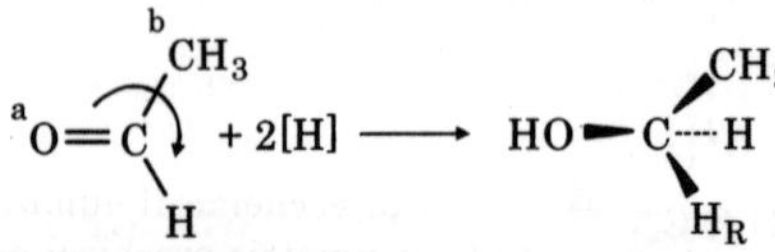

```
       b
        CH3                     CH3
 a                               /
  O=C    + 2[H]  ——→   HO—C---H
     \                         \
      H                         HR
```

re (or Re) face is toward the reader
si (or Si) face is behind the page

Replica atoms attached to an originally trigonal carbon are ignored completely, but those attached to O or N of C = O or C = N may be required to establish the priorities of the groups around the carbon atom. For C = O and C = N the faces of the O and N atoms are taken to be the same as that of the attached trigonal carbon atom. In C = C the two carbons may have their *re* faces either on the same side or on opposite sides of the group. Notice that addition of an atom of ^{2}H to the *re* face of acetaldehyde would give R-deuteroethanol (Eq. 2.75). This is the reaction catalyzed by alcohol dehydrogenase. Addition of ^{1}H to the *re* face places the entering hydrogen in the *pro*-R position.

Addition to the *si* face would place it in the *pro-S* position. Fumarate has two trigonal carbon atoms. The *si* faces of both are toward the viewer in the following structure. Referring to Eq. 2.74, we see that an HO^- ion from water adds to the *si* face of one carbon atom of fumarate to give s-malate (L-malate). At the same time, a proton combines with the *re* face of the adjacent unsaturated carbon atom to enter the *pro-R* position.

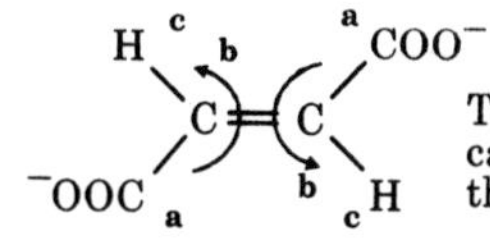

H c b a COO⁻ C=C ⁻OOC a b c H

The si faces of both trigonal carbon atoms are toward the viewer.

Fumarate

Induced Fit and Conformational Changes

Results of many X-ray studies indicate that the lock-and-key picture of enzyme action must be modified. If an enzyme is a "lock" and the substrate the "key," *the entrance of the key into the lock often induces a conformational change in the protein.* Binding of substrates may be imperfect in the initially formed ES complex but may be more nearly perfect a few nanoseconds or microseconds later as the protein readjusts its structure to accommodate the substrate. The substrate has induced a fit. For example, when an amino acid substrate binds to aspartate aminotransferase one whole domain of the enzyme moves inward, packing hydrophobic I side chains of the protein against the substrate. This strengthens the electrostatic interactions between the ion pairs that orient the substrate and align it for reaction.

Similar conformational changes have been observed for citrate synthase, glycogen phosphorylase, various kinases alcohol dehydrogenase, and a growing list of other enzymes. Accompanying changes in circular dichroism, ultraviolet spectra, and sedimentation constants are often observed. For many other enzymes the observed conformational changes are subtle. A single loop of polypeptide chain or even of a side chain moves to cover the bound substrate.

With trypsin and other serine proteases a flexible segment of the peptide chain becomes immobilized and forms more hydrogen bonds after substrate binds than before. It is probably quite unusual for a major unfolding and refolding of parts of a protein to take place. The term induced fit usually refers to substrate binding, but as substrates are interconverted within active sites successive changes in geometry and in charge distribution occur. Small conformational changes may be required at several stages of an enzymatic reaction to ensure that complementarity of substrate and enzyme is preserved.

In line with theis idea are the facts that proteins are less tightly packed at active sites than in other parts of the molecules, and that active sites often lie between domains and usually are formed by several loops of the peptide chain. Although conformational changes allow proteins to maintain good complementarity with substrates, it does not follow that substrates are therefore bound very tightly. This is easiest to understand for reversible reactions in whicha substrate is the product of the reverse reaction.

Not only binding but also the rate of release of product must be rapid. Very tight binding would retard release and cause product inhibition. Furthermore, the velocity of catalysis is the product of [S][E] × k_{cat} / K_m, where [E] is, *free enzyme*. Catalysis can be made more efficient in two ways: by increasing k_{cat} or by decreasing K_m, that is, by tighter binding. However, tighter binding will be advantageous only if K_m is not so low that the enzyme is approaching saturation with substrate. Otherwise, [E] will fall as K_m is lowered and little increase in rate will be observed. In fact, it appears that enzymes have evolved to have high values of k_{cat} and *high* values of K_m, that is, *weak binding* of substrates.

Frequently, in one confermational state of a protein the active site is open, with solvent molecules or substrate surrogates present, while in a second state of nearly equal energy it is closed around the substrate. The functional groups of the enzyme's active site can be bound either to external ions or to ionic groups of substrates and either to water or to hydrogen-bonding groups of substrates. In some cases two ionic groups in the protein may pair with each other in the open conformation and with ionic groups of the substrate in the closed conformation. Thus, the energy changes accompanying the confor-mational changes can be small but very good comple-mentarity can exist in the ES complex, an important factor in establishing specificity.

Specificity and k_{cat}

Enzyme specificity is often observed not only in binding but also in the rate at which ES is converted to products. Thus, it is the values of k_{cat} / K_m that determine specificity. Good examples are provided by chymotrypsin and related serine proteases for which substrates with the shortest chains are often bound as well as those with longer chains but react more slowly. For example, *N*-acetylphenylalanine amide binds to chymotrypsin about as well as does the longer *N*-acetylphenylalanylalanine amide but reacts only 1/47 as fast. One might anticipate that increasing the length of the substrate would make it bind more tightly because of the greater number of contacts between substrate and enzyme.

It has often been suggested that the reason that this does *not* happen is that the binding of the longer, more specific substrates distorts the enzyme and that *the binding energy is now stored in the enzyme. It is as if the enzyme contained an internal spring which would be compressed when the substrate binds.* This would keep the binding weak but the energy in the spring might then be used to increase the velocity.

Proofreading

Although enzymes may be very specific they do make mistakes. This is of particular concern for processes such as protein synthesis in which the correct amino acid is placed at each position in the sequence with an error rate that has been estimated for *E. coli* as only 1 in 10^4. It would be impossible for an enzyme designed to attach valine to its specific transfer RNA to avoid attaching the smaller alanine if discrimination between the two were based solely on the Gibbs energy differences of binding. However, it would be easier for an enzyme to exclude *larger* amino acids. This problem may be resolved by use of multistep screening.

For example, isoleucyl-tRNA synthetase does occasionally attach the smaller valine to the specific tRNA, but when it does the enzyme in a "proofreading and editing" step hydrolyzes off the incorrect amino acid. The active site for this hydrolyzing activity, whether at a different place on the enzyme surface or created by a conformational change, may be able to exclude sterically the larger isoleucyl residue while acting on the valyl-tRNA. This editing mechanism for isoleucyl-tRNA synthetase was demonstrated directly in 1998 by X-ray crystallography on complexes of the enzyme with L-isoleucine and L-valine. Both substrates fit into the ATP-requiring synthetic site but neither isoleucine nor isoleucyl-tRNA will fit into the editing site which is located in an adjacent β-barrel domain. Proofreading steps based on differing chemical properties as well as size can also be visualized.

MECHANISMS OF CATALYSIS

Kinetic studies tell how fast enzymes act but by themselves say nothing about *how* enzymes catalyze reactions. They do not give the *chemical mechanism* of catalysis, the step-by-step process by which a reaction takes place. Most of the individual steps involve the simultaneous breaking of a chemical bond and formation of a new bond. Consider a simple *displacement reaction,* that of a hydroxyl ion reacting with methyl iodide to give the products methanol and iodide ions.

$$HO^- + \underset{H\ \ H}{\overset{H}{C}}{-}I \longrightarrow HO{-}CH_3 + I^- \qquad ...(2.76)$$

The reaction can be thought of as an "attack" by the OH^- ion on the "back side" of the carbon atom of the methyl groups with a simultaneous displacement of the I^-.

The "Transition State"

A reaction such as that of Eq. 2.76 is not instantaneous, and at some time between that at which the reactants exist and that at which products have been formed the C-I bond will be stretched and partially broken and the new C–O bond will be partially formed. The structure at this point is not that of an ordinary compound and is energetically unstable with respect to both the reactant and the products. The intermediate structure of the least stability is known as the *transition state.* Although no one has actually seen a transition state structure, we might represent that for Eq. 2.76 as follows:

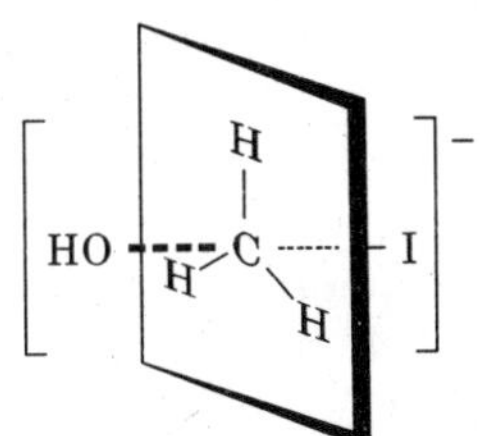

Transition state structure

The negative charge is distributed between the attacking HO^- group and the departing iodide. The bonds to the central carbon atom are no longer tetrahedrally arranged but the C–H bonds lie in a single plane and the partial bonds to the OH and iodine atom lie at right angles to that plane. It is useful in discussing a reaction mechanism to construct a *transition state diagram* in which Gibbs energy G is plotted against *reaction coordinate.* Energy *E* or enthalpy *H* may be plotted in the same way and authors frequently do not state whether *G*, *E*, or *H* is meant.

The reaction coordinate is usually not assigned an exact physical meaning but represents the progress from reactants toward products. It is directly related to the extent to which an existing bond has been stretched and broken or a new one formed. The high energy point is the transition state. A some-what more detailed idea of a transition state is obtained from a contour diagram such as that of Fig. 2.15B. Here, energy is plotted as a function of two distances, *e.g.*, the lengthening C–I bond distance and the shortening C–O distance for Eq. 2.76. The path of minimum energy across the "saddle point" representing the transition state is indicated by the dashed line. In a reaction coordinate diagram the difference in value of G between reactants and products is the overall Gibbs energy change ΔG for the reaction, while the difference in G between the transition state and reactants is $\Delta G^{\ddagger}$, the *Gibbs energy of activation.* The magnitude of $\Delta G^{\ddagger}$ represents the "energy barrier" to a reaction and largely determines the rate constant. The diagrams in Fig. 2.15 are too simple because enzymatic reactions usually occur in several intermediate steps.

There will be transition states for each step with valleys in between. The valleys correspond to intermediate species, which are sometimes very unstable. The passing from reactants to products in an enzymatic reaction can be likened to wandering through a series of mountain ranges of various heights and finally reaching the other side.

Qunatitative Transition State Theory

In the 1880s Arrhenius observed that the rate of chemical reactions varies with temperature according to Eq. 2.77 in a manner similar to the variation of an equilibrium constant with temperature. Here, k is a first-order rate constant, the quantity E_a is known as the *Arrhenius activation energy,* and the constant A is referred to as the "pre-exponential factor" or the "frequency factor".

$$k = Ae^{-E_a/RT} \quad ...(2.77)$$

The Arrhenius equation, together with studies of the effects of salts on reaction rates and observation of quantitative correlations between rates and equilibrium constants, suggested that a rate constant for a reaction might be a product of a constant term which is nearly independent of temperature and a constant $K^{\ddagger}$ which has the properties of an equilibrium constant for formation of the transition state.

Eyring made this quantitative in 1935 with his "absolute rate theory" according to which all transition states break down with a rate constant $\kappa k_B T/h$. Eyring reached this conclusion by assuming that *the rate of a chemical reaction is determined by the frequency of stretching of the bond that is being broken in the transition state.* To be more precise, it is the "normal-mode" oscillation of the transition state complex along the reaction coordinate. This frequency ν was deduced by describing the vibrational energy as $h\nu$ (from quantum mechanics) and as k_B/T (from classical mechanics) and setting them equal.

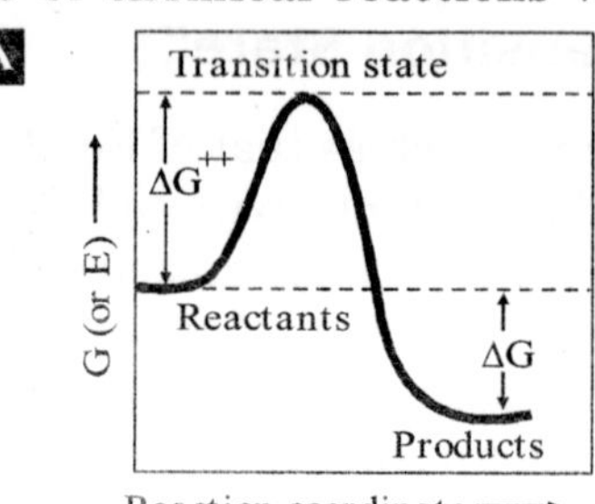

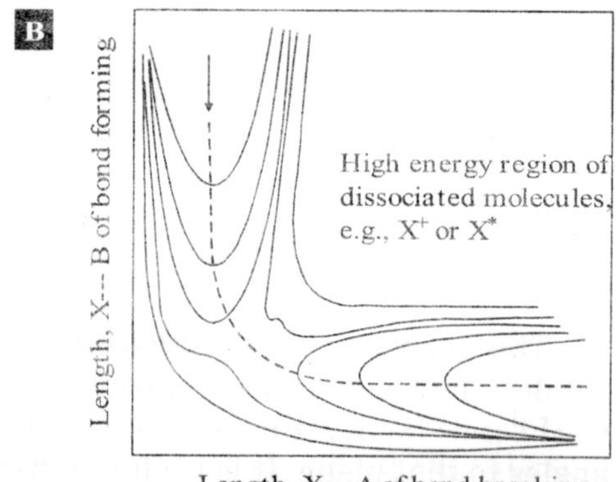

Fig. 2.15. (A) Transition state diagram illustrating Gibbs energy vs reaction coordinate for conversion of reactants to products in a chemical reaction. (B) Contour map of Gibbs energy vs interatomic bond distances for reaction B + X – A → B – X + A.

$$h\nu = k_B T \text{ and}$$
$$\nu = k_B T/h$$

k_B is the Boltzmann constant h is Planck's constant. ...(2.78)

The right side of Eq. 2.78 is usually multiplied by a *transmission coefficient* κ, which may vary from 1 to 0.1 or even much less. However, for lack of any better value, κ is usually assumed to be 1. From Eq. 2.78, at 25°C $v = 6.2 \times 10^{12}\ s^{-1}$. This is the maximum rate for a chemical reaction of molecules in the transition state. This is the rate for a single molecule and must be multiplied by the concentration of the reacting substance X in the transition state. This concentration $[X]^{\ddagger}$ is determined by the equilibrium constant $K^{\ddagger} = [X]^{\ddagger}\ [X]$. The velocity of the reaction becomes

$$\frac{-d[X]}{dt} = v = \frac{k_B T}{h}.K^{\ddagger}[X] \qquad ...(2.79)$$

The first-order rate constant is

$$k_1 = \frac{-d[X]/dt}{[X]} = K\frac{k_B T}{h}.K^{\ddagger}\ s^{-1} \qquad ...(2.80)$$

Since $\Delta G^{\ddagger} = -RT \ln K^{\ddagger}$, Eq. 2.80 can be rewritten as Eq. 2.81, in which $\Delta G^{\ddagger}$ is the Gibbs energy of activation.

$$k_1 = \frac{k_B T}{h}.e^{-\Delta G^{\ddagger}/RT} s^{-1} \qquad ...(2.81)$$

At 25°C, $\Delta G^{\ddagger}$ in kJ/ mol, the following "practical" form of Eq. 2.81 can be written.

$$k_1 = 6.2\times10^{12} e - \Delta G^{\ddagger/2.48} s^{-1}$$

$$\log k_1 = 12.79 - \Delta G^{\ddagger}/5.71$$

$$\text{and } \Delta G^{\ddagger} = 73.0 - 5.71 \log k_1$$

Equation 2.81 is approximate and a more correct statistical mechanical treatment is available. See also comment on about log k (and log k_1) being unitless. Employing Eq. 2.14, we may expand Eq. 2.81 as follows:

$$k = \left(\frac{k_B T}{h} e^{\Delta S/R}\right) e^{-\Delta H^{\ddagger}/RT} s^{-1} \qquad ...(2.82)$$

From this it appears that $\Delta H^{\ddagger}\ E_a$, the Arrhenius activation energy. A more correct treatment gives $\Delta H^{\ddagger} = E_a - RT$ for reactions in solution. However, since RT at 25°C is only 2.5 kJ mol^{-1}, the approximation that $\Delta H^{\ddagger} = E_a$ is often used. The preexponential term, in parentheses in Eq. 2.82, depends principally on $\Delta S^{\ddagger}$, the entropy change accompanying formation of the transition state.

The quantities $\Delta G^{\ddagger}$, $\Delta H^{\ddagger}$, and $\Delta S^{\ddagger}$ are sometimes measured for enzymatic reactions but useful interpretations are difficult. Equations 2.81 and 2.82 also show that a small decrease in transition state energy will give a large increase in rate; stabilization of the transition state by 5.7 kJ/mol (1.4 kcal/mol) will increase the rate 10-fold. If $\Delta G^{\ddagger} = 400$ kJ/mol, as for a strong covalent single bond, $K_1 = 5 \times 10^{-58}\ s^{-1}$ and the half-life $t_{½} = .693/k_1 = 1.3 \times 10^{57}$ s (4×10^{49} years, greater than the ~10^{10} years estimated age of the universe). If $\Delta G^{\ddagger} = 100$ kJ/ mol, log $k_1 \approx -4.7\ s^{-1}$, $k_1 = 1.9 \times 10^{-5}$, and $t_{½} = 10$ h. This is about the rate of a typical nonenzymatic "model" reaction at 25°C. If $\Delta G^{\ddagger} = 50$ kJ/mol, log $k_1 = 4$, as for *a fast enzyme.* We can conclude that *enzymes make use of relatively small energy differences in catalyzing reactions.*

The energies of numerous van der Waals interactions of hydrogen bonds and electrostatic attraction or repulsion of charges are sufficient. Nevertheless, we can see that anything that stabilizes the transition state (decreases $\Delta G^\ddagger$) will increase the rate of reaction. The role of a catalyst is to permit the formation of a transition state of lower energy (higher stability) than that for the uncatalyzed reaction. Stabilization of the transition state of a reaction by an enzyme suggests that the enzyme has a higher affinity for the transition state than it does for substrate or products, an idea that appears to have been expressed first by Haldane and popularized by Pauling.

I think that enzymes are molecules that are complementary in structure to the activated complexes of the reactions that they catalyze. The attraction of the enzyme molecule for the activated complex would thus lead to a decrease in its energy, and hence to a decrease in energy of activation of the reaction and to an increase in the rate of the reaction.

Transition State Inhibitors

Suppose that a chemical reaction of a compound S takes place with rate constant k_N through transition state T. Let the equilibrium constant for formation of T be $K_N^\ddagger$. Assume that an enzyme E can combine either with S with dissociation constant K_{ds} or with the compound in its transition state structure T with dissociation constant K_{dT}.

$$\begin{array}{ccccc} S & \overset{K_N^\ddagger}{\rightleftharpoons} & T & \longrightarrow & \text{product} \\ + & & + & & \\ E & & E & & \\ \Big\updownarrow K_{dS} & & \Big\updownarrow K_{dT} & & \\ ES & \overset{K_E^\ddagger}{\rightleftharpoons} & ES & \longrightarrow & \text{product} \end{array} \qquad ...(2.83)$$

If equilibrium is assumed for all four sets of double arrows it is easy to show that Eq. 2.84 holds.

$$K_e^\ddagger/K_N^\ddagger \;=\; K_{ds}/K_{dr} \qquad ...(2.84)$$

According to transition state theory, if the transmission coefficient $\kappa = 1$, T and ET will be transformed to products at the same rate. Thus, if the mechanisms of the nonenzymatic and enzymatic reactions are assumed the same, the ratio of maximum velocities for first-order transformation of ES and S will be given by Eq. 2.85. For some enzymes the ratio

$$k_E/k_N \;=\; K_E^\ddagger/K_N^\ddagger = K_{ds}/K_{dT} \qquad ...(2.85)$$

k_E/k_N may be 10^8 or more. Thus, if $K_{ds} \approx 10^{-3}$ the constant K_{dT} would be $\approx 10^{-11}$. The enzyme would be expected to bind the transition state structure (T) 10^8 times more tightly than it binds S. The foregoing reasoning suggests that if structural analogs of T could be found for a particular reaction, they too might be very tightly bound—more so than ordinary substrate analogs. Wolfenden listed a series of compounds which may be *transition state inhibitors* of this type and many others have been described.

Nevertheless, very tight, reversible binding is no *proof* that an inhibitor is actually binding to the transition state structure of the enzyme. For example, an inhibitor may bind to a conformational form of the enzyme that lies off to a side rather than directly on the reaction pathway for catalysis.

Describing The Transition State

How can we describe the structure of the transition state when we have no direct knowledge of its structure? We can try to predict the atomic coordinates of all of the atoms of the reacting substrates and of the protein. The transition state structure of the substrate must be between that of the last ES complex and the first EP complex formed. It should be similar to that of a transition state inhibitor. We can hope to obtain suitable X-ray or NMR structures and deduce an appropriate enzyme structure by molecular modeling.

Theoretical calculations using quantum chemical methods are of value in this effort. Lengths of the breaking and forming bonds in a transition state are often estimated using *kinetic isotope effects* (KIEs) on the velocity of the reaction. This is somewhat indirect. Nevertheless, by measuring these effects for a large number of substrates and several isotopes a picture of the transition state structure that is good enough to use as a model for design of an enzyme inhibitor may be obtained. Other physical measurements such as Raman difference spectroscopy of isotopically labeled inhibitors can also be of value.

Getting to the Transition State

Since some bonds must lengthen and some shorten and bond angles must also change, it may not be simple for an enzyme to catalyze necessary steps prior to the rate-limiting breakdown of the transition state. Hydrogen bonds within the protein may need to be broken and reformed with new bonding partners. Nonpolar side chains may need to shift to give a packing arrangement that will allow the movement of nuclei that must occur in forming the transition state structure. The enzyme must not only stabilize the transition state but also hold the substrate and escort it into the transition state configuration. The attacking hydroxyl ion in Eq. 2.76 carries a negative charge which becomes distributed more or less equally between the hydroxyl and the iodine atom in the transition state.

In this state the CH_3 group carries a partial positive charge as well. To provide good complementarity between an enzyme and this transition state, a change in the initial charge distribution within the enzyme will also be required. If a positively charged group initially binds the OH^- ion it may have to lose part of its charge and a group next to the iodide atom may have to gain positive charge. In the nonenzymatic reaction of Eq. 2.76 the ionic atmosphere provided by positive counterions in solution can continuously readjust to keep the negative charge effectively balanced at every step along the reaction coordinate and through the transition state. Within enzymes this adjustment may occur via redistribution of electrical charges within the polarizable network of internal hydrogen bonds. The enzyme structure must allow this. Because of the complexity of an enzymatic transition state it may be hard to compare it with the transition state of the corresponding nonezymatic reaction. Effects of pressure and of temperature also raise doubts about the simple picture of Eq. 2.83. It also follows that tight binding of substrates in the ground state does not necessarily interfere with transition state stabilization.

Microscopic Reversibility

The important statistical mechanical principle of microscopic reversibility asserts that *the mechanism of any chemical reaction considered in the reverse direction must be exactly the inverse of the mechanism of the forward reaction.* A consequence of this principle is that if the mechanism of a reaction is known, that of the reverse reaction is also known. Furthermore, it follows that *the forward and reverse reactions catalyzed by an enzyme must occur at the same active site* on the enzyme and the transition state must be the same in both directions.

The principle of microscopic reversibility is often useful when thelikelihood of a given mechanism is being considered. If a mechanism is proposed for a reversible reaction in one direction the principle of microscopic reversibility will give an unambiguous mechanism for the reverse reaction. Sometimes this reverse mechanism will be chemically untenable and, recognizing this, the enzymologist can search for a better one.

Acid and Base Catalysis

Many reactions that are promoted by enzymes can also be catalyzed by acids or bases or by both. An example is *mutarotation,* the reversible interconversion of the α- and β-anomeric forms of sugars. This reaction is catalyzed by a specific *mutarotase* and also by inorganic acids and bases. The frequently observed bell-shaped curve for the dependence of rate of catalysis on pH also suggests participation of both protonated and unprotonated acid-base groups present in enzymes.

Acidic and Basic Groups in Enzymes.

In this discussion the symbols HB or H^+B will be used for acids and B^- or B for their conjugate bases. Remember that *strong acids* have *low pK_a values* and that the *conjugate bases formed from them are weak bases.* Likewise, *very weak acids liave high pK_a values and their conjugate bases are strong.* The following are among the pK_a values that may be important in considering an enzyme mechanism. However, depending upon the environment of the side chain in a protein, these pK_a values may fall substantially outside of the indicated ranges.

–1.74	H_3O^+ in 55.5 M water
~3.6	–COOH, terminal in peptide
4.3 – 4.7	–COOH in a glutamic or aspartic acid side chain
6.4 – 7.0	Imidazole (histidine) and phosphate ($-OPO_3H^-$)
7.5 – 8.0	– NH_3^+, terminal in peptide
8.5 – 10	– SH of cysteine side chains
9.5 – 11	Phenolic –OH of tyrosine
~ 10.5	$-NH_3^+$ of lysine side chain
~ 13.6	$-CH_2OH$ of serine side chain
15.7	free HO^- in 55.5 M H_2O, 25°C.

Acid-base Catalysis of Mutarotation

The mutarotation of glucose proceeds through the free aldehyde form as an intermediate. A hydrogen atom is removed as H^+ from the oxygen at carbon 1 in step *a* and a proton (probably a different one) is transferred to the oxygen of the ring with cleavage of the O – C bond to the anomeric carbon atom. A similar process in reverse is required for step *b.* Transfers of hydrogen ions between atoms of oxygen, nitrogen, and sulfur atoms are a common feature of biochemical reactions. The bonds between hydrogen and O, N, and S atoms tend to be polarized strongly, leaving a partial positive charge on the hydrogen atoms. Consequently, the groups are weakly acidic and protons can be transferred from them relatively easily. It is reasonable to suppose that acid and base catalysis is related to these proton transfers.

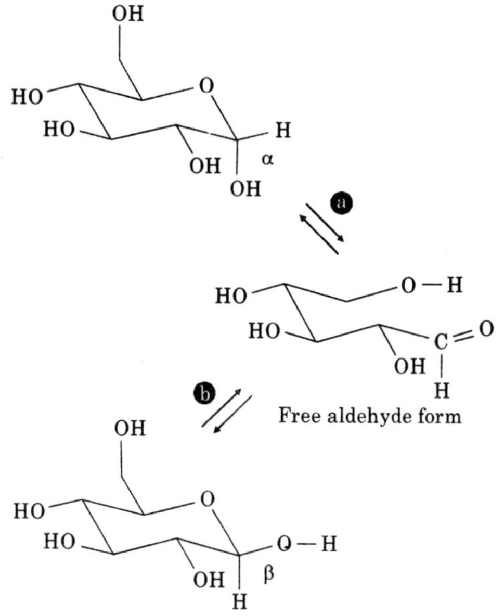

Free aldehyde form

...(2.86)

General Base and General Acid Catalysis

Base-catalyzed mutarotation might be formulated as follows: A hydroxyl ion or some other base attacks the proton on the anomeric –OH group of the sugar, removing it to form an anion and the conjugate acid BH^+ (Eq. 2.87 step *a*). The anion is isomerized to a second anion with the ring opened (step *b*). Addition of a proton (transfer of a proton from H_3O^+) produces the free aldehyde form of the sugar (step *c*). The catalytic base B: might be HO^- or a weaker base such as ammonia or even water. For reactions the rate is proportional only to the concentration of OH and the presence of other weaker bases has no effect. Such catalysis is referred to as *specific hydroxyl ion catalysis.*

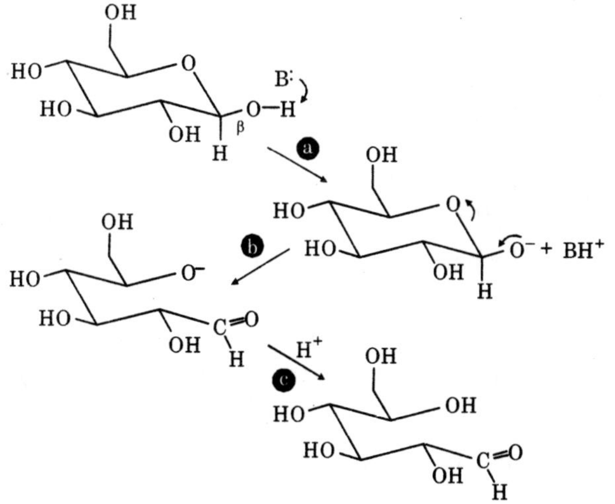

...(2.87)

More commonly, the rate is found to depend both on $[OH^-]$ and on the concentration of other weaker bases. In such cases the apparent first-order rate constant (k_{obs}) for the process is represented by a sum of terms (Eq 2.88). The term k_{H_2O} is the rate in pure water and represents catalysis occurring by the action of water alone as either an acid or a base. The

$$k_{obs} = k_{H_2O} + K_{OH^-}[OH^-] + k_B[B] \quad ...(2.88)$$

last two terms represent the contributions to the catalysis by OH^- and by the other base, respectively. The term $k_B[B]$ represents *general base catalysis.* Catalysis of mutarotation by acids occurs if an acid donates a proton to the oxygen in the sugar ring as shown in Eq. 2.98.

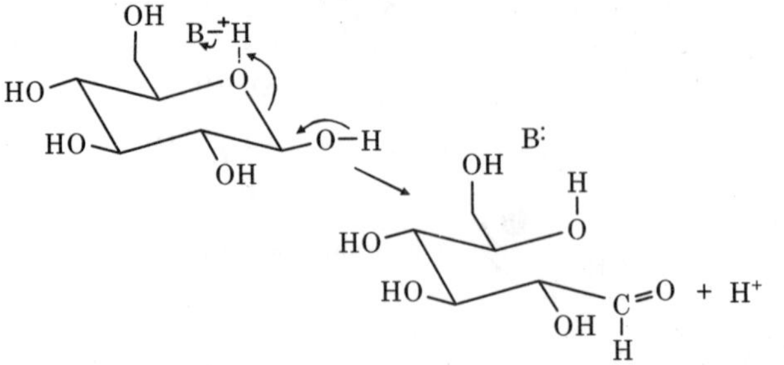

...(2.89)

Again, either *specific acid catalysis* (by H_3O^+) or *general acid catalysis* is possibter. Enzymes are not able to concentrate protons or hydroxyl ions to the point that *specific* base or acid catalysis would be effective.

However, either general can be accomplished by in their normal states of protonation at the pH of the cell. Thus, if one of the enzymes must be dissociated to its conjugate base and a second must be protonated for reaction to take place it is reasonable to suppose that these two groups participate in acid-base catalysis. Of the acidic and basic groups present in proteins, the imidazole group of histidine would appear to be the most ideal both for general base catalysis and, because both forms may exist in nearly equal amounts at pH 7, for general acid catalysis. However, carboxyl, thiol, lysyl, tyrosyl, and N-terminal amino groups are all thought to function in various enzymes. The mutarotase of *E. coli* has a turnover number of 10^4 s^{-1}. The plot of –log K_m vs pH indicates two pK_a values in the free enzyme at 5.5 and 7.6, while the plot of log V_{max} yields a single pK_a of 4.75 for the ES complex.

The latter might represent a carboxyl group on an imidazole group in its conjugate base form. Why doesn't the group having pK_a 7.6 in the free enzyme also show a pK_a in the ES complex? Either the group has no catalytic function so that EH_2S of Eq. 2.54 reacts to form products just as fast as does HES or the pK_a is so strongly shifted by substrate binding that it is not detected in the log V_{max} plot.

The Bronsted Relationships

The effectiveness of a specific base as a general base catalyst can usually be related to its basicity (pK_a) via the *Brönsted equation.* Here, k_B is defined by Eq. 2.88 and G_B is a constant

$$\log k_B = \log G_B + \beta(pK_a) \quad ...(2.90)$$

for a particular reaction. A similar equation relates the constant k_A for general acid catalysis

$$\log k_{HB} = \log G_A - \alpha(pK_a) \quad ...(2.91)$$

to pK_a. These equations are linear Gibbs energy relationships similar to the ones discussed.

For the Bronsted equations to hold, the Gibbs energy of activation for the reaction must be directly related to the basicity or acidity of activation for the reactionmust be directly related to the basicity or acidity of the catalyst. The exponents β and α of Eqs. 2.90 and 2.91 measure the *sensitivity* of a reaction toward the basicity or acidity of the catalyst.

It is easy to show that as β and α approach 1.0 general base or general acid catalysis is lost and that the rate becomes exactly that of specific hydroxyl ion or specific hydrogen ion catalysis. As β or α approach zero basic or acidic catalysis is indetectable. Thus, general base or general acid catalysis is most significant when β or α is in the neighborhood of 0.5. Under these circumstances it is easy to see how a moderately weak basic group, such as the imidazole group of histidine, can be an unusually effective catalyst at pH 7. To determine α or β experimentally a plot of log k_B or log k_{HB} vs pK_a (a *Brönsted plot*) is made and the slope is measured.

Statistical corrections should be applied for dicarboxylic acids and for ammonium ions from which one of three protons may be lost from the nitrogen atom. General base or general acid catalysis implies an important feature of any mechanism for which it is observed, namely, that removal of a proton or addition of a proton is involved in the rate-determining step of a reaction.

Concerted Acid-base or Tautomeric Catalysis

A third possible type of catalysis requires that a base and an acid act *synchronously* to effect the breaking and formation of bonds in a single step. Thus, tetramethyl-glucose mutarotates very slowly in benzene containing either pyridine (a base) or phenol (an acid). However, when both pyridine and phenol are present, mutarotation is rapid. This suggested to Swain and Brown a *concerted mechanism* in which both an acid and a base participate. During the reaction shown, the acid BH^+ is converted to its conjugate base B and the base B′ to its conjugate acid H^+B'.

It might seem that these agents, having been altered by the reaction, are not serving as true catalysts. However, a simple proton exchange will restore the original forms and complete the catalytic cycle. In aqueous solutions, water itself might act as the acid or the base or even both in concerted catalysis. The original experimental evidence tor concerted acid-base catalysis of the mutarotation in benzene is now considered unsound and concerted acid-base catalysis has been difficult to prove for nonenzymatic reactions in aqueous solution.

However, measurements of kinetic isotope effects seem to support Swain and Brown's interpretation. Concerted acid-base catalysis by acetic acid and acetate ions may have been observed for the enolization of acetone and it may be employed by enzymes. Swain and Brown showed that a more effective catalyst for the mutarotation of sugars than a mixture of an acid and a base can be designed by incorporating the acidic and basic groups into *the same molecule*. Thus, with 0.1 M tetramethylglucose in benzene solution, 0.001 M α-hydroxypyridine is 7000 times as effective a catalyst as a mixture of 0.001 M pyridine + 0.001 M phenol. Swain and Brown suggested the following completely concerted reaction mechanism for the *polyfunctional catalyst α-hydroxypyridine* in which the hydrogen-bonded complex formed is analogous to an enzyme-substrate complex. The product of the catalyst is 2-pyridone, a tautomer of

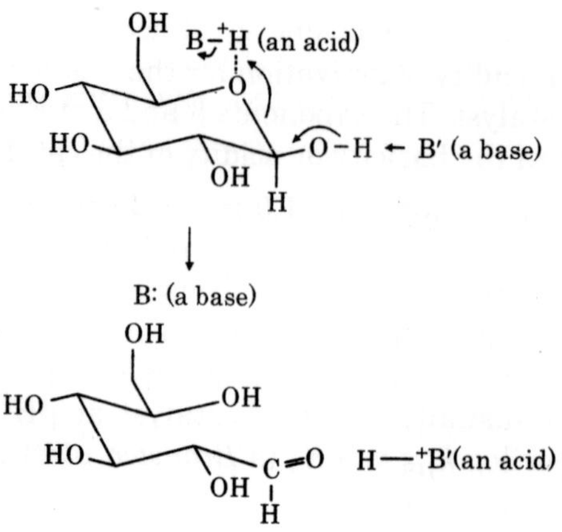

...(2.92)

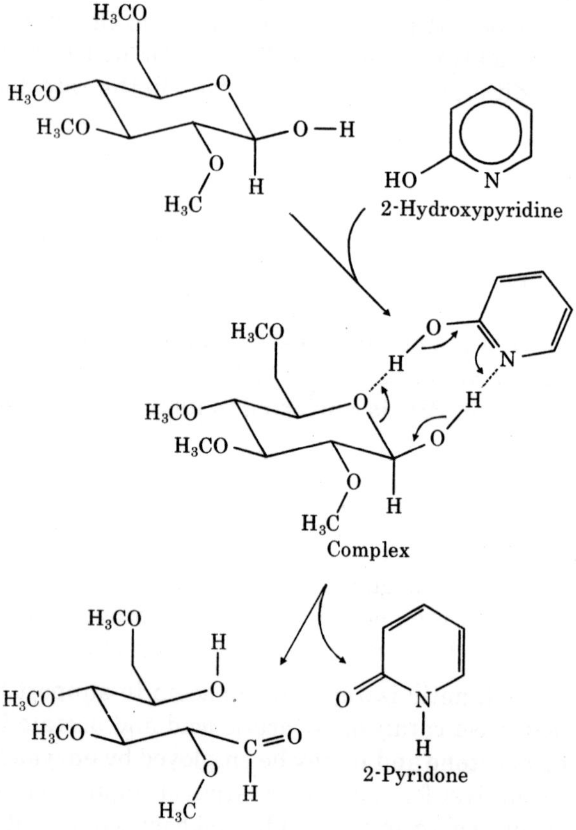

...(2.93)

2-hydroxypyridine, with which it is in a rapid reversible equilibrium. Rony called catalysis of the type illustrated in Eq. 2.93 tautomeric catalysis and suggested that its efficiency lies not simply in the close proximity of acidic and basic groups in the same molecule but also in the ability of the catalyst to repeatedly cycle between the two tautomeric states. For an enzyme the tautomerization of the free catalyst could sometimes be rate determining.

Transfer of Protons within Active Sites

An extensive network of hydrogen bonds runs throughout most proteins and may be especially complex within active sites. The network often runs through bound substrate molecules and immobilized water molecules in the active site cavity. This network arises in part because of the frequent occurrence of acidic and basic catalytic groups in active sites and by the fact that many substrates contain polar groups.

The structure of an enzyme often seems to be more rigid in a complex with substrates or inhibitors than in the free state. Does this network of linked hydrogen bonds play a role in catalysis? If so, what? Wang suggested that rapid transfer of protons along rigidly and accurately held hydrogen bonds in the ES complex may be an essential feature of enzymatic catalysis. This conclusion is supported by many more recent observations. A remaining question is whether this proton transfer can take place in such a way that the transition state barrier is lowered.

Ultrafast Proton Transfer

The diffusion-controlled limit for second-order rate constants is ~10^{10} M^{-1} s^{-1}. In 1956, Eigen, who had developed new methods for studying very fast reactions, discovered that protons and hydroxide ions react much more rapidly when present in a lattice of ice than when in solution. He observed second-order rate constants of 10^{13} to 10^{14} M^{-1} s^{-1}. These represent rates almost as great as those of molecular vibration. For example, the frequency of vibration of the OH bond in water is about 10^{14} s^{-1}. The latter can be deduced directly from the frequency of infrared light absorbed in exciting this vibration.

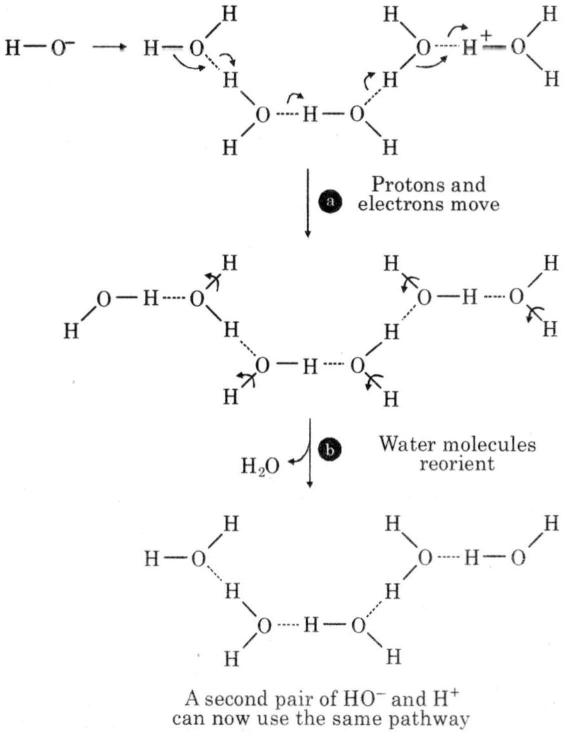

(a) Protons and electrons move

(b) Water molecules reorient

A second pair of HO^- and H^+ can now use the same pathway

A second pair of HO^- and H^+ can now use the same pathway ...(2.94)

Frequency v equals wave number (3710 cm^{-1} for –OH stretching) times c, the velocity of light (3×10^{10} cm s^{-1}). This ultrafast transfer of protons can be explained as follows: The OH^- ion and the proton, which is combined with a water molecule to form H_3O^+, are both hydrogen-bonded to adjacent water molecules. In ice a chain of hydrogen-bonded water molecules links the hydroxide and the hydrogen ions. By synchronous movement of electron pairs from the OH^- ion and from each of the water molecules in the chain the neutralization can take place during the time of one molecular vibration. The positions of the oxygen atoms remain unchanged at the end of the reaction but the protons that were engaged in hydrogen bond formation have moved toward the left and are now attached to different oxygen atoms.

The central chain of water molecules can be restored to its original state if each molecule rotates around one of its single bonds, swinging a hydrogen from right back to left (Eq. 2.94, step b). This must be a slower process which may occur one molecule at a time. The rotation of the leftmost water molecule would leave an empty space between two O-atoms, a "fault" in the H-bonded chain of the ice structure. This fault can be corrected by rotation of the second molecule; however, this would create a fault at the right of the second water molecule.

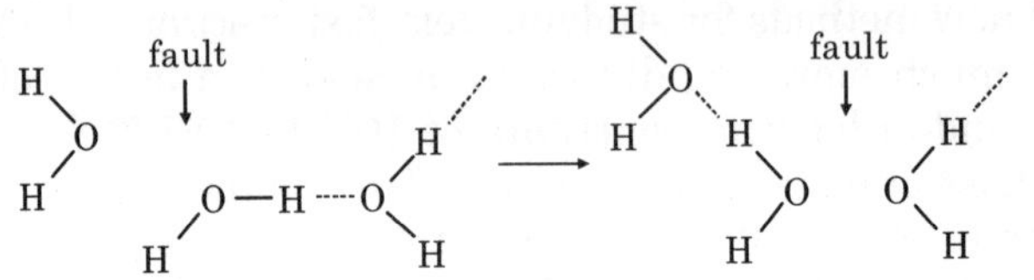

fault ... fault ...(2.95)

This would induce rotation of the third molecules etc., causing the fault to migrate from left to right across the entire chain of water molecules and leaving the chain ready to function again in transferring another proton. Because of the disorder present in ice, an array of water molecules such wouldn't revert to its exact original form in step b. However, active sites of enzymes are highly structured and proton transfers may occur with precision. For example, a synchronous shift of protons in an array of carboxylic acid, imidazolium, and phosphate groups can be envisioned readily. The net effect of the process is to transfer a proton from one end of the chain to the other with facile tautomerization reactions providing the pathway.

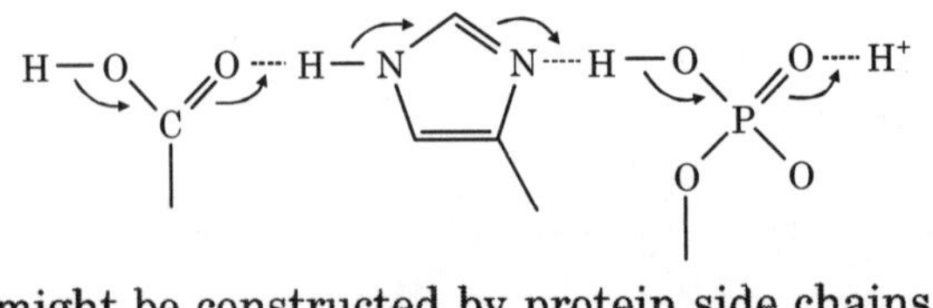

...(2.96)

Such a pathway might be constructed by protein side chains to join the two sides of an active center promoting a concerted acid-base catalyzed reaction such as that of Eq. 2.92. Other tautomerization processes are possible within proteins if the existence of less stable minor tautomers of selected amide groups in the peptide back-bone is allowed. Nagle and Morowitz suggested a process similar to that of Eq. 2.94 but involving side chains of serine residues.

Proton Transfer Rates

Consider the reversible reaction of a proton acceptor B with acid H-A. Eigen pointed out that the reaction will be fastest if the two reactants form a hydrogen-bonded complex. The hydrogen bonding shortens the distance from the proton to B and allows for very rapid transfer of the proton from A to B within the hydrogen-bonded complex (step b). The activation energy is close to zero. The complex dissociates in step c to form the products. The reactions are reversible, even though they have been indicated by unidirectional arrows. The hydrogen-bonded complexes A–H—B and A — H–B can be formed readily if A and B are oxygen or nitrogen bases such as – COO^-, – OH, –NH_2, or imidazole. In such cases, as in ice, the interconversion of the two complexes is very fast.

$$\text{A—H} \xrightarrow[\text{B}]{k_1 \; (a)} \text{A—H}\cdots\text{B} \xrightarrow{k_2 \; (b)} \text{A}\cdots\text{H—B} \xrightarrow[\text{A}]{k_3 \; (c)} \text{H—B} \quad ...(2.97)$$

The overall rate of the proton transfer will then be limited by diffusion of B and H–A together or by diffusion of A and H–B apart. It might seem that these two processes should also both be very fast. However, the rates will be determined by the concentration ratio [A–H—B]/[A—H–B]. If [A–H—B] equals [A— H–B] the rate of dissociation of A–H–B will be half what it is if nearly all of the complex is A—H–B. If A–H—B predominates by 1000 to 1, the rate will be slowed much more.

The equilibrium constant for the reaction of Eq. 2.97 will be :

$$K_{eq}^{AB} = K_a^{HA}/K_a^{HB} = k_f/k_r$$

or

$$\log K_{eq} = pK_a^{HA} - pK_a^{HB} = \log k_f - \log k_r \quad ...(2.98)$$

Here k_f and k_r refer to the rates in the forward and reverse directions. If the pK_a's of HA and HB are equal k_f and k_r will be the same, but if they are very far from equal the reaction will be slowed in one direction. If proton transfer is a step in an enzymatic reaction it may be slowed enough to become rate limiting. The difference in pK_a values between the proton donor and the proton acceptor in Eq. 2.97 can be expressed as the Gibbs energy change which at 25°C is equal to ±5.71 × ΔpK_a.

This is often referred to as the *thermodynamic barrier* ΔG° to a reaction and $\Delta G^\ddagger$ can be expressed as the sum of the thermodynamic barrier ΔG° plus an *intrinsic barrier* $\Delta G^\ddagger_{int}$. For the proton transfer of Eq. 2.97 the intrinsic barrier (for step *b*) is thought to be near (zero so that $\Delta G^\ddagger = 5.71\ \Delta pK_a$. From this we can conclude that two pK_a values can be as much as eight units apart and $\Delta G^\ddagger$ will still be less than 50 kJ / mol, low enough to permit rapid enzymatic reactions. However, for transfer of a proton from a C–H bond to a catalytic group, for example, to form an enolate ion for an aldol condensation, the intrinsic barrier is known to be about 50 kJ/mol. In this case, to allow rapid enzymatic reaction either the thermodynamic barrier must be very low, as a result of closely matching pK_a values, or the enzyme must lower the intrinsic barrier. It may do both.

Marcus Theory

Discussion of intrinsic barriers is often approached using a quantitative theory proposed by Marcus. It was first applied to electron transfer but has been used for a great variety of nonenzymatic and enzymatic reactions. As used by Gerlt and Gassman, the Marcus formalism describes the reaction coordinate as an inverted parabola whose shape is determined by the overall Gibbs energy change ΔG° and the intrinsic barrier $\Delta G^\ddagger_{int}$. The value of the Gibbs energy at any point on the curve is designated G and the reaction coordinate x is taken as 0 for the reactants and 1 for the products.

$$G = -4\,\Delta G^\ddagger_{int}\,(x-0.5)^2 + \Delta G°\,(x-0.5)$$

$$\text{Limits: } \Delta G°/4 \le \Delta G^\ddagger_{int} \le 4\,\text{AG}° \quad ...(2.99)$$

Differentiation of Eq. 2.99 yields the position of the transition state coordinate $x^{\ddagger}$ as follows:

$$x^{\ddagger} = 0.5 + \Delta G°/8\Delta G^{\ddagger}_{int} \quad ...(2.100)$$

and it follows that

$$\Delta G^{\ddagger} = \Delta G^{\ddagger}_{int}(1 + \Delta G°/4\Delta G^{\ddagger}_{int})^2 \quad ...(2.101)$$

Diffusion-Controlled Dissociation of Protons

The direct proton transfer between C–1 and C–2 during the action of sugar isomerases may seem puzzling. How can a highly mobile proton remain attached to a group in the enzyme for a millisecond or more instead of being transferred out to a solvent molecule? This can mean that the enzyme promotes the transfer of a hydride ion or of a hydrogen atom rather than a proton. If so, the observed proton exchange with solvent would be an unimportant side reaction. On the other hand, could the group in the enzyme that removes the proton be out of contact with the aqueous medium and thus able to hold onto the proton more tightly? In recent years, it has been recognized that neither of these explanations may be necessary.

An imidazole group is a likely proton-carrying gro up at many active sites and it is thought that a proton not be expected to transfer out from an imidazo group with a rate constant greater than ~ 10^3 s^{-1}.

The argument is as follows: The rate of donation of a proton from H_3O^+ to imidazole (reverse of Eq. 2.102) is known to be diffusion controlled with a rate constant of 1.5×10^{10} $M^{-1}s^{-1}$.

$$ImH^+ + H_2O \rightleftharpoons [ImH^+...OH_2] \rightleftharpoons Im + OH_3^+ \quad ...(2.102)$$

Very fast diffusion controlled
$k \sim 1.5 \times 10^{10}$ M^{-1} s^{-1}

The equilibrium constant for Eq. 2.102, calculated from the pK_a of 7.0 for imidazole, is 10^{-7} M. Since K_{eq} is also the *ratio of the overall rate constants for the forward and reverse reactions,* we see that for the forward reaction $k_f = 10^{-7} \times 1.5 \times 10^{10} = 1.5 \times 10^3$ s^{-1}. This slow rate results from the fact that in the intermediate complex (in brackets in Eq. 2.102) the proton is on the imidazole group most of the time.

For a small fraction of the time it is on the coordinated molecule H_2O but reverts to being on the imidazole many times before the imidazole and OH_3^+ separate. Because of this unfavorable equilibrium within the complex, the diffusion-controlled rate of proton transfer from a protonated imidazole to water is far less than for proton transfer in the reverse direction.

Coupled Proton Transfers

Enzymatic reactions often require the transfer of two or more protons. They may move individually, one proton at a time or they may move synchronously in a *coupled* or concerted process. Such coupled movement is generally not possible for heavier nuclei. However, studies of solvent isotope effects using a *proton inventory* technique. Have provided evidence favoring coupled proton transfers for a variety of enzymes. Movement of protons along hydrogen-bonded paths, as well as electron transfer, may take place with some participation of *quantum mechanical tunneling.* Coupling to vibrational modes of the hydrogen-bonded protons may provide *vibration-assisted tunneling.* These reactions are associ-ated with unusually large kinetic isotope effects.

Unusually Strong Hydrogen Bonds

The strength of a hydrogen bond is thought to be directly related to the length, which is ordinarily taken to be the distance between the two surrounding heavier atoms. Hydrogen bonds are sometimes classified on the basis of N–H---O distances as:

<0.25 nm	very strong
0.25-0.265 nm	strong
>0.28 nm	weak
>0.37 nm	no van der Waals contact but electrostatic interaction still occurs

N–H---O distances may be a little longer than these.

The 0.276-nm hydrogen bonds in ice are regarded as moderately strong. However, if one of the oxygen atoms in an O–H---O hydrogen bond carries a negative charge, as in the maleate monoanion, it will be shorter. Although the proton will be closer to one oxygen atom than to the other, it will be able to move between them by passing a very low transition state barrier.

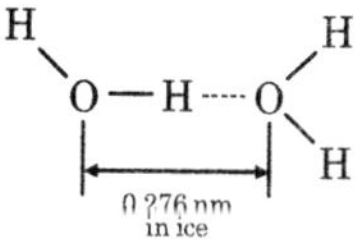

0.276 nm in ice

For this to occur the microscopic pK_a values for the two groups (when they are protonated) must be similar; for the maleate anion they are identical.

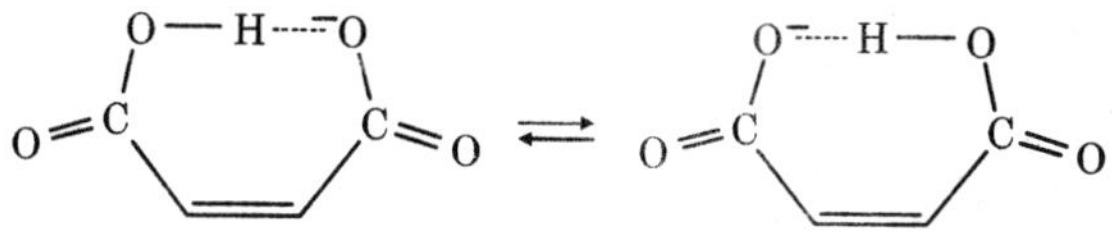

Monohydrogen maleate

The *strength of* a hydrogen bond can be measured for hydrogen-bonded complexes in the gas phase and range from 10–100 kJ/ mol and even higher or such complexes as $(FHF)^-$. It is more difficult to establish the strength in the liquid state or within the active site of an enzyme, but shorter hydrogen bonds are usually stronger than longer ones. Hydrogen bond distances in crystals of small molecules can be measured precisely X-ray or neutron diffraction but greater uncertainty is present in distances within proteins.

Lengths and probably strengths of hydrogen bonds in proteins can be measured by NMR methods. For example, solid-state NMR measurements on crystals of amino acids and other carboxylic acids have provided a plot of ^1H chemical shift vs hydrogen bond length. For very short (0.240–nm to 0.25–nm) hydrogen bonds the ^1H chemical shift may be as great as 21 ppm. See also Fig. 2.30 and associated discussion. When dissolved in an $^1H_2O - {}^2H_2O$ mixture strongly hydrogen-bonded protons within a protein become enriched in ^1H. The $^2H/^2H$ ratio of the hydrogen-bonded protons provides another measure of the hydrogen bond strength.

A short hydrogen bond is also present in such cations as the following:

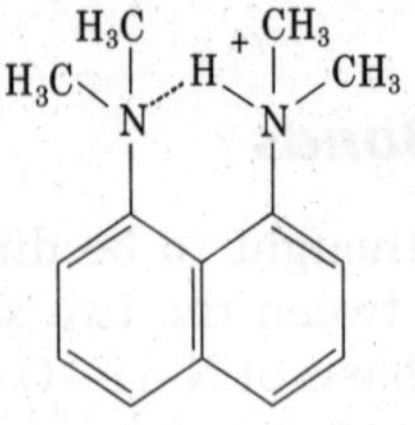

$$\text{[1,8-bis(dimethylamino)naphthalene, protonated: } (H_3C)_2N^{+}\cdots H\cdots N(CH_3)_2\text{]}$$

The pK_a for dissociation of its proton is 12.3 and the hydrogen-bonded proton is probably located in the *center* of the bond with both amino groups sharing the charge. Enols can also form unusually strong "resonance-assisted" hydrogen bonds :

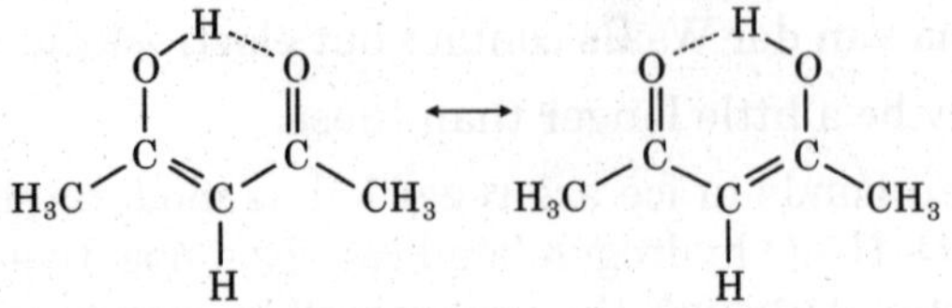

$$H_3C{-}C(OH){=}CH{-}C(=O){-}CH_3 \longleftrightarrow H_3C{-}C(=O){-}CH{=}C(OH){-}CH_3$$

The structure can be thought of as a resonance-stabilized enolate anion with a proton bound between the two oxygen atoms and equidistant from them. Do these short *"low-barrier hydrogen bonds"* have a special significance in enzymology? Proposals that they contribute to stabilization of transition states have received some support and aroused controversy. In later chapters we will examine specific enzymes in which low-barrier hydrogen bonds have been observed.

Covalent Catalysis

In addition to participating in acid–base catalysis, some amino acid side chains may enter into covalent bond formation with substrate molecules, a phenome-non that is often referred to as covalent catalysis. When basic groups participate this may be called *nucleophilic catalysis.* Covalent catalysis occurs frequently with enzymes catalyzing nucleophilic displacement reactions. They include the formation of an acyl-enzyme intermediate by chymotrypsin. These coenzymes combine with substrates to form reactive intermediate compounds whose structures allow them to be converted rapidly to the final products.

Proximity and Orientation

One of the earliest ideas about enzymes was that they simply brought reactants together and bound them side by side for a long enough time that the reactive groups might bump together and finally react. How important is this *proximity factor?* Page and Jencks estimated that rate enhancements by factors of 10^3 or more may be expected solely from the loss in the entropy of two reactants when they are bound in close proximity on an enzyme surface. In view of the large entropy decrease involved, *the enthalpy of binding must be high,* and if this explanation is correct the *binding of the substrates* to the enzyme provides much of the driving force for catalysis. Westheimer described this by stating that enzymes use the substrate-binding force as an *entropy trap.*

The losses of translational and rotational entropy, which Page and Jencks estimated as up to – 160 to – 210 J/deg/mol, overcome the unfavorable entropy of activation that is usual in bimolecular reactions. How precise must the orientation of substrates be for rapid reaction.

Compounds such as the acid shown in Eq. 2.103 form an internal ester (lactone) spontaneously with elimination of water.

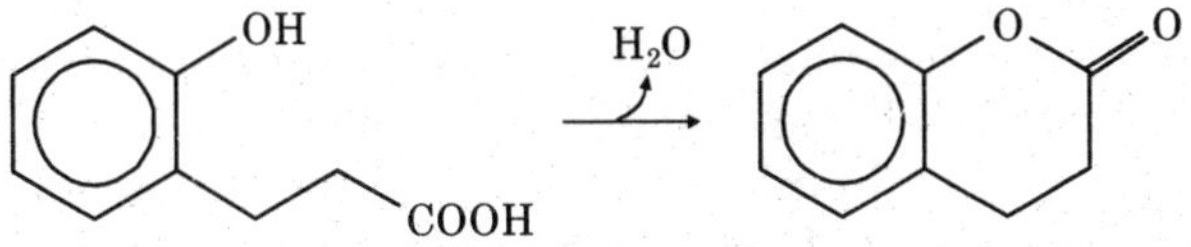

OH COOH $\xrightarrow{H_2O}$ O O ...(2.103)

However, the following compound reacts at a rate over 10^{11} times that of the acid shown in Eq. 2.102. This is presumably because its conformation is highly restricted and the –COOH is constrained to frequently collide with the –OH group. The three methyl groups interdigitate

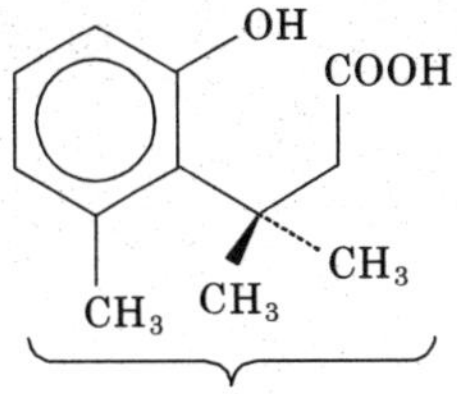

OH COOH CH_3 CH_3 CH_3

Trialkyl lock

groups interdigitate and form a *trialkyl lock*. Orientation must also play a large role in enzymatic catalysis. As previously emphasized, enzymes often orient substrates precisely by formation of multiple hydrogen bonds. Small distortions by mutation or substitution of an essential metal by a different one can have very great effects. For example, because of an altered coordination pattern isocitrate dehydrogenases with Ca^{2+} in the active site has a maximum velocity of catalysis only 2.5×10^{-3} that with the normal Mg^{2+}-containing enzyme. Bruice and associates concluded that enzymes must bring reacting groups into close proximity with orbitals of the reactants properly aligned in the ground state prior to moving to the transition state.

They suggested that enzymes preofganize the enzyme-substrate complex into a *near attack conformation* in which the positions of reacting groups, the arrangements of hydrogen bonds, and the local dielectric constant in the active site, resemble those in the transition state. In this conformation the energy barrier to the transition state may be very low. Preorganization of the complex also acts to eliminate the slow components of solvent reorganization required for reaction in aqueous solutions. The necessity for reacting groups of substrates to collide with an orientation that allows productive interaction of electronic orbitals is often called a *stereoelectronic effect.*

An example is the addition reaction of Eq. 2.74. The orbital of an unshared pair of electrons on the HO^- ion must be perpendicular to the plane of the double bond. Furthermore, if the proton becomes attached to the adjacent carbon in a synchronous or concerted manner it must enter from the opposite side. In many biochemical reactions an alcohol or amine is eliminated from a tetrahedrally bonded intermediate as in Eq. 2.104. Deslongchamps proposed a stereoelectronic theory according to which elimination of either NH—R_2 or –O—R_1 from this intermediate will depend upon the values of the torsion angles θ_1 and θ_2. The theory asserts that for rapid elimination of OR_1, unshared electron pairs on *both* the – O^- and N atoms must be antiperiplanar to the bond being broken. If rotation around the C–O bond (θ_1) occurs an orientation can be found in which an electron pair on each of the two oxygens will be antiperiplanar to the bond to NHR_2. From this orientation the latter will be eliminated.

The theory has been supported by much experimental evidence involving reaction rates and product distribution among competing reactions of small conformationally restricted organic molecules. Although more recent experiments suggest that stereoelectronic factors may be of less significance than had been assumed, even a small decrease in transition state energy can be significant in an enzyme-catalyzed reaction. Enzymes may not only orient substrates in accord with stereoelectronic principles but also be able to promote conformational alterations in intermediates that allow them to take advantage of stereoelectronic factors.

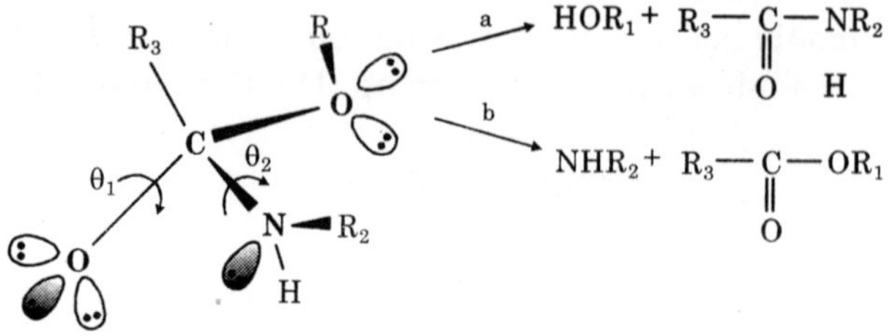

$$\xrightarrow{a} HOR_1 + R_3{-}\overset{\overset{\displaystyle O}{\|}}{C}{-}NR_2 \qquad \xrightarrow{b} NHR_2 + R_3{-}\overset{\overset{\displaystyle O}{\|}}{C}{-}OR_1 \qquad ...(2.104)$$

The Microenvironment

A substrate bound at an active site may be in an environment quite different than that in an ordinary aqueous solution. In fact, the protein often surrounds the substrate to the extent that *the protein is the solvent.* What kind of *microenvironment* does the protein provide and can it assist in the catalytic process? Charged and dipolar groups of the protein provide an electrostatic field that provides part of the binding energy and which may also assist in catalysis, a concept expressed by Quastel as early as 1926. The ends of protein helices often seem to point at active centers.

In many cases more than one N terminus its positive electrical potential or more than one C terminus (negative potential) of a helix point to an active site. These helix dipoles may be important in stabilizing transition state structures or in altering pK_a values of functional groups. Fluctuations in charge distribution within a protein and in the hydrogen-bonding pattern of the protein with solvent and substrate may also be important. For some enzymatic reactions a transition state will be favored by a medium of very low dielectric constant and a correctly constructed active site can provide just such a surrounding. Hydrophobic groups may be packed around a site where an ion pair or other ionic interaction between enzyme and substrate occurs increasing the strength of that interaction. Comformational changes may enhance such effects.

The dehydration of polar groups that must often occur upon binding of substrate may make these groups more rective. It has been suffested that the substial volume changes ($\Delta V^{\ddagger}$) that sometimes occur during formations may result largely from changes in hydration of goups on the enzyme surface and that these changes may play an important role in catalysis. In the past most enzymologists tacitly assumed that the external medium in which enzymes act must be aqueous.

However, many enzymes function well in media containing largely hydrocarbons. Enzymes in a dry, powdered form have been suspended directly in organic solvents. Under these conditions, enzymes may contain only tightly bound "structural" water, together with less than one equivalent of a monolayer of water outside the protein. The enzymes often remain active and are able to catalyze reactions with an altered substrate specificity as well as different reactions overall.

Strain and Distortion

The fact that enzymes appear to bind their substrates in such a way as to surround and immobilize them means that something other than the kinetic energy of the substrate is needed to provide energy for the ES complex to pass over the transition state barrier. What is the source of this activation energy? As with nonenzymatic reactions, it must come ultimately from the translational energy of solvent or solute molecules bombarding the complex. Can enzymes act as "energy funnels" that effectively channel kinetic energy from spots on the enzyme surface to the active site? This could either be through strictly mechanical movement or through induction by fluctuations of the electrical field. It is often suggested that enzymes assist in catalysis by distorting bond lengths or angles away from their normal values. If the distorted structure were closer to the transition state geometry than the undistorted structure, catalysis would be assisted. However, Levitt concluded that forces provided this way by a protein are small and that the protein would become distorted rather than a substrate.

On the other hand, if binding of a substrate distorts the protein, could the resultant "stored energy" be used in some way to assist in catalysis? If binding of a substrate bends a "spring" can the tension in the spring then be used to distort the enzyme to be more exactly complementary to the transition state? This concept has become popular. The amino acid side chains of proteins are usually well packed. However, neither the side chains nor the main chain are rigid and immobile.

Some regions ot the protein will contain empty spaces-packing defects. Lumry called these *mobile defects* because, as a result of fluctuations in side chain packing, they can move within a protein from one site to another for a considerable distance. Much evidence, including X-ray studies at low temperatures, supports the existence of many *Conformational substates* in proteins. Some substates may bind substrates better than others and some may allow conversion to the transition state more readily than do others. Rotation of histidine rings, peptide linkages, – OH groups, or amide side may be required and has, in fact, been observed for some enzymes.

Perhaps in one of the substates of an enzyme-substrate complex an especially favorable vibrational mode leads the complex to the transition state. In the transition state the packing of side chains may be especially tight. A mobile defect present in the active site may have moved elsewhere. The binding of substrates to the protein in the transition state will also be tighter because of the conformational alterations that have occurred. The binding energy of the substrates is now being utilized to lower the transition state barrier. The substrate is literally squeezed into the transition state configuration. Is it possible that the protein domains forming an active site act like *ferroelectric crystals,* which change their dipole moment in response to a change in electric field?

The highly polarizable hydrogen-bonded network the amide linkages, imidazole rings, guanidinium group, etc. of active sites may permit a flip-flop of the dipole moment as in domains of ferroelectric crystals. In such crystals, *e.g.*, those of the hydrogen-bonded KH_2PO_4, a 180° change of dipolemoment direction results from very small movements of heavy atoms together with larger movements of the hydrogen bonded protons. Could a similar flip-flop in a protein domain be coupled to the passage of a substrate over the transition state barrier? I have not seen any discussion of this possibility, but the structure of protein domains would seem to allow it. One could also imagine that with a small change in the hydrogen-binding arrangement of protein groups an active site could become preorganized to favor a flip-flop along a hydrogen-bonded chain in a different direction for a subsequent step in a reaction sequence.

Why Oligomeric Enzymes

As we have seen, a large fraction of all proteins exist as dimers, trimers, and higher oligomers. Oligomeric proteins raise the osmotic pressure much less than would the same number of monomeric subunits and this may be crucial to a cell. Another advantage of oligomers may be reflected in the fact that active sites of enzymes are often at interfaces between two or more subunits. This may enhance the ability of enzymes to undergo conformational rearrangements that are required during their action, just as hemoglobin changes its oxygen affinity in concert with a change in inter-subunit contacts. A curious observation is that crystals of the dimeric pig heart malate dehydrogenase bind only one molecule of the substrate NAD^+ per dimer tightly; the second NAD^+ is bound weakly. Similar *anti-cooperativity* has been reported for other crystalline dehydrogenases and various other enzymes. An intriguing idea is that anticooperativity in binding might reflect a cooperative action between subunits during catalysis. Suppose that only conformation A of a dehydrogenase bound reduced substrate and NAD^+, while conformation B bound NADH and oxidized substrate. If reduced substrate and NAD^+ were present in excess and if oxidized substrate were efficiently removed from the scene by further oxidation, the following cooperative events could occur in the mixed AB dimer.

The subunit of conformation A would bind substrates, react, and be converted to conformation B. At the same time, because of the strong AB interaction, the subunit that was originally in conformation B would be converted back to A and would be ready to initiate a new round of catalysis. Since conformation A has a low affinity for NADH, dissociation of the reduced coenzyme, which is often the slow step in dehydrogenase action, would be facilitated. Such a *reciprocating or flip-flop* mechanism, suggested first by Harada and Wolfe, is attractive because it provides a natural basis for the existence of the many known dimeric enzymes that do not exhibit evident allosteric properties.

Attempts to verify this idea have largely failed. However, recent crystallographic studies of both glyceraldehyde phosphate dehydrogenase and thymidylate synthase are consistent with the proposal. There is still a possibility that coordinated reciprocal changes in distribution of electrical charges in the two subunits may also be important.

Remarks

It appears that enzymes exert their catalytic powers by *first* bringing together substrates and binding them in proper orientations at the active site. *Second,* they often provide acidic and basic groups in the proper orientations to promote proton transfers within the substrate. *Third,* groups within the enzyme (especially nucleophilic groups) may enter into covalent interaction with the substrates to form structures that are more reactive than those originally present in the substrate. *Fourth,* the protein often closes around the substrate to immobilize it and to hold it in an environment, often lacking water, which could impede catalysis.

The enzyme is also probably able to make readjustments of its structure to provide good comple-mentarity to the substrate at every stage and especially to the transition state structure. *Fifth,* the enzyme may be able to induce strain or distortion in the substrate perhaps accompanying a conformational change within the protein. The following question is often asked: "Why are enzymes such large molecules? At least part of the answer is that an enzyme usually interacts, sometimes via special domains, with other proteins. Another part of the answer is evident when we consider that the formation of a surface complementary to that of the substrate and possessing reasonable rigidity requires a complex geometry in the peptide backbone. In addition, the enzyme must provide functional groups at the proper places to enter into catalysis. It may require a certain bulk to provide a low dielectric medium. Finally, if essential conformational changes occur during the course of the catalysis, we can only be surprised that nature has succeeded in packing so much machinery into such a small volume.

CLASSIFICATION OF ENZYMES

An official commission of the International Union of Biochemistry (IUB) has classified enzymes in the following six categories:

1. *Oxidoreductases,* Enzymes catalyzing dehydrogenation or other oxidation and reduction reactions.
2. *Transferases.* These catalyze group transfer reactions.
3. *Hydrolases.* Enzymes catalyzing transfer of groups to the HO^- ion of H_2O.
4. *Lyases.* Enzymes promoting addition to double bonds or the reverse.
5. *Isomerases.* Enzymes catalyzing rearrangement reactions.
6. *Ligases* (synthetases). Enzymes that catalyze condensation with simultaneous cleavage of ATP and related reactions.

As an example, chymotrypsin is classified EC 3.4.4.5 according to the IUB system. In this book a more mechanistically based classification is used. Because some official names are quite long, traditional trivial names for enzymes often have been retained. Remember that to be precise it is always necessary to mention the species from which an enzyme was isolated and, if possible, the strain. Also remember that almost every significant genetic difference is reflected in some change in some protein. It is possible that the enzyme you are working with is slightly different from the same enzyme prepared in a different laboratory.

A HISTORICAL NOTE ON ENZYMES

While the earliest physiologists postulated a "vital force" to explain the chemical reactions of cells, the existence of biochemical substances pro-moting reactions outside of the body was recognized at least by the early 1600s. However, the role of yeast in fermentation was still unknown and it was thought that both alcoholic fermentation and animal digestion were caused by unknown substances called *ferments.* In 1752, Reamur demonstrated the solvent power of the gastric juice of birds and by 1783 Spallanzani had extended the studies to humans and other species. In 1836 Schwann isolated the enzyme *pepsin* from gastric juice.

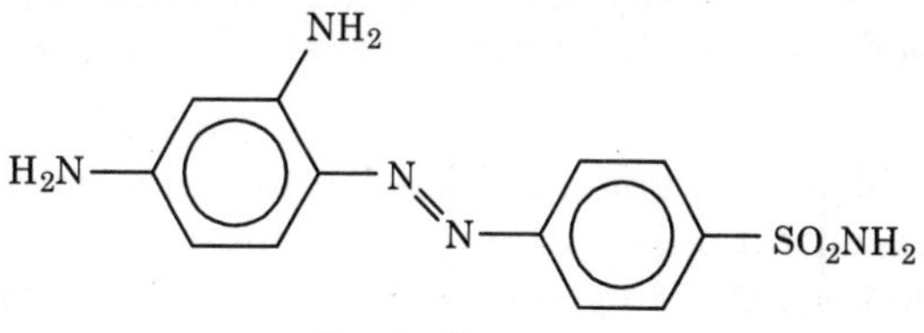

In the same year Jacob Berzelius introduced the concept of catalysis, which he developed as a result of studies of the effects of acids and bases in promoting the hydrolysis of starch and of the effects of metals on the decomposition of

Prontosil

hydrogen peroxide. Berzelius proposed the term *catalyst* from the Greek "katalysis," meaning "dissolution." Although he had been concerned primarily with inorganic catalysts, Berzelius recognized that a natural catalyst, an analase that causes the hydrolysis of starch, had already been isolated from germinating barley in 1833 and that "in living plants and animals thousands of catalytic processes take place. Some chemists and physiologists accepted Berzelius' concept of biochemical catalysis immediately, but many did not. The matter was complicated by Pasteur's discovery that yeast cells were the causative agent of alcoholic fermentation. Only after Edward Buchner's reports in 1897 that a juice formed by grinding yeast with sand and filtering could still ferment sugar was the reality of enzymes in metabolism generally accepted. The word *enzyme* (from the Greek "in yeast") was introduced earlier by F. W. Kuhrie, professor of physiology in Heidelberg and a person who had accepted Berzelius' concept.

Sulfonamides
R=H in sulfanilamide

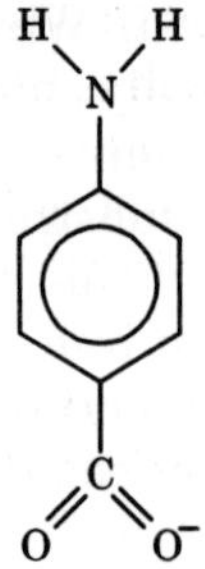

p-Aminobenzoate

GROWTH RATES OF CELLS

How do we correctly describe the rate at which a cell grows? Consider bacteria in their rapidly growing "log phase." Each cell divides after a fixed length of time, the *doubling time*. For *E. coli* this may be as short as 17 min in the early stages of growth but becomes somewhat longer as time goes on. A mean value of about 26 min. for *E. coli* at 37°C is typical. In contrast, the doubling time for mammalian cells in tissue culture is often about one day.

If a given volume of culture contains N_0 cells initially, the number N_n after n cell divisions will be:

$$N_n = 2^n N_0$$

From this equation we can calculate that a single bacterium with a generation time of 20 min can produce 2^{144} cells in 48 h of exponential growth. The *exponential growth rate constant* ***k*** is equal to *the number of doublings per unit time*. Thus, k is the reciprocal of the doubling time. It is easy to show that the number of bacteria present at time t will be given by the following equation.

$$N_t = 2^{kt} N_0$$

This can be rewritten in a form that can be used to determine k by counting the number of bacteria at zero time and at time t.

$$kt = \log_2(N_t/N_0) = \log_{10}(N_t/N_0)/0.301$$

Another way of expressing the growth is to equate the rate of increase of the number of bacteria with a growth rate constant μ multiplied by the number of bacteria present at that time.

$$dN/dt = \mu N$$

This is a general equation for an autocatalytic reaction and N could be replaced with a concentration, for example, the total content of cellular matter per liter in the medium. From the two preceding equations the following can be derived.

$$\mu = k \ln 2 = 0.69k$$

When bacteria are transferred to new medium there is usually a lag before exponential growth begins. Exponential growth eventually stops and the culture enters *stationary phase,* which is usxially followed by relatively rapid death of cells in the culture. It is often desirable to study cell growth under conditions of *continuous cultivation* in which a constant generation time is maintained but the density of cells in the medium does not increase. This can be done with a simple device known as the *chemostat.* A culture vessel containing bacteria is stirred to ensure homogeneity. Fresh culture medium continuously enters the vessel from a reservoir and part of the content of the vessel, suspended bacteria included, is continuously removed through another tube. The bacterial population in the vessel builds up to a constant level and can be maintained at the same level for relatively long periods of time.

Here are some other statistics on cell growth. Bacteria growing exponentially expand their linear dimensions by 1.5 nm and synthesize 1.6×10^7 Da of new cell material in one second. This is equivalent to ~1000 small proteins and includes 23 ribosomes and 3000 base pairs of DNA at each growing point. The much larger HeLa human tumor cell grows by 0.13 nm/s but makes 4.6×10^8 Da/s of material.

THE SULFONAMIDES AS ANTIMETABOLITES

The development of the "sulfa drugs," derivatives of *sulfanilamide,* originated with studies of the staining of protozoal parasites by Paul Erhlich. In 1932 it was shown that the red dye 2,4-diamino-azobenzene-4′-sulfonamide (Prontosil) dramatically cured systemic infections by gram-positive bacteria. Subsequent studies revealed that bacteria converted the azo dye to sulfanilamide, a compound with strong bacteriostatic activity; that is, it inhibited bacterial growth without killing the bacteria. Although sulfanilamide had been used in large quantities since 1908 as an intermediate in synthesis of dyes, its potential as an antibacterial agent had not been recognized.

In 1935, D. D. Woods found that the growth inhibition of sulfanilamide was reversed by yeast extract. From this source, in 1940 he isolated *p-aminobenzoic acid* and demonstrated that the inhibitory effect of 3×10^{-4} M sulfanilamide was overcome by 6×10^{-8} M *p*-aminobenzoate. The relationship between the two compounds was shown to be strictly competitive. If the sulfanilamide concentration was doubled, twice as much *p*-aminobenzoate was required to reverse the inhibition as before. These facts led to the formulation by Woods and by P. Fildes (in 1940) of the *antimetabolite theory.* It was proposed that *p*-aminobenzoate was needed by bacteria and that sulfanilamide competed for a site on an enzyme designed to act on *p*-aminobenzoate. We now know that the idea was correct and that the enzyme on which the competition occurs catalyzes the synthesis of dihydropteroic acid (Fig. 2.19), a precursor to folic acid.

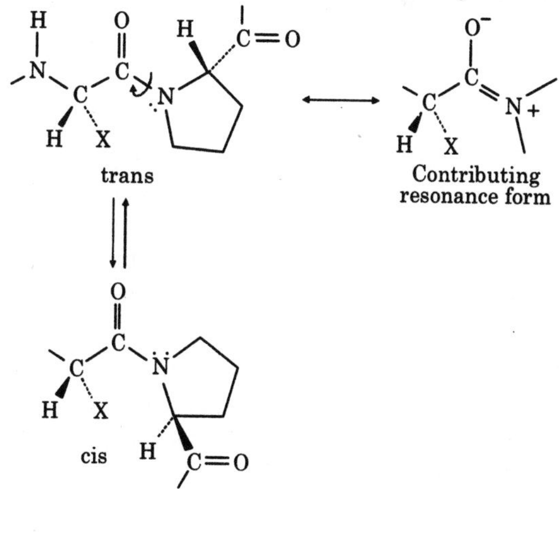

trans

Contributing resonance form

cis

While sulfanilamide itself is somewhat toxic, a variety of related drugs of outstanding value have been developed. Over 10,000 sulfonamides and related compounds have been tested for antibacterial action.

IMMUNOPHILINS AS ROTAMASES

Since its introduction into clinical use in about 1979 the immunosuppresant *cyclosporin* has been responsible for a revolution in human organ transplantation. The exact mechanism of action in suppressing T-lymphocyte-mediated autoimmune responses is still not completely clear, but cyclosporin, a cyclic lipophilic peptide from a fungus, was found to bind to specific proteins that were named *cyclophilins*. Human cyclophilin A is a 165-residue protein which associates, in the crystal form, as a decamer with five-fold rotational and dihedral symmetry. This protein is also found in almost all

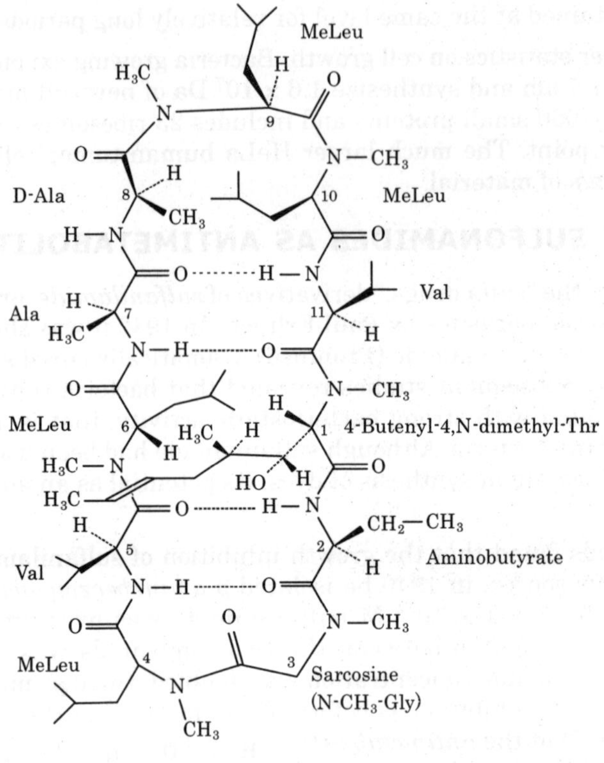

Cyclosporin A

other organisms and has a highly conserved sequence. Another immunosuppresant, a synthetic molecule known only as *FK506,* has an action similar to that of cyclosporin but binds to distinctly different proteins, the FK506-binding proteins (FKBPs), of which the 107-residue FKBP-12 is the best known. These proteins also bind another immunosuppresant, *rapamycin,* which, however, has different biological properties than does FK506. Both FK506 and rapamycin are macrocyclic compounds but with structures very different from that of cyclosporin.

It was a surprise to discover that all of the cyclophilins and FK506-binding proteins are *peptidyl prolyl cis-trans isomerases* or *rotamases.* They all catalyze the following simple and reversible reaction of a prolyl peptide linkage:

The transition state energy for the reaction is lowered by 33 kJ mol^{-1} by cyclophilin A and by 27 kJ mol^{-1} by FKBP-12 with corresponding rate increases of ~ – 6 × 10^{5}– and 5 × 10^{4} –fold, respectively.

For both enzymes the maximum velocity is independent of pH over a wide region. For cyclophilin the rate is nearly independent of the nature of residue X in the foregoing structure but FKBP prefers a hydrophobic residue.

How do these enzymes work? One possibility would be for the protein to transfer a proton to the nitrogen atom of the proline ring. This would destroy the partial double-bond character of the amide linkage and allow free rotation about a single bond. The same thing could be accomplished if a nucleophilic group from the enzyme formed a covalent adduct with the C=O of the substrate. However, this seems unlikely because there is no suitably placed nucleophile. A third possibility is that the enzyme distorts the substrate and stabilizes the transition state using only noncovalent interactions. This would account for the lack of pH dependence and observable solvent isotope effects. However, in human cyclophilin the guanidinium group of arginine 55 hydrogen bonds to the peptide C = O of the substrate's proline residue and could easily shift to place a guanidinium proton against the proline ring nitrogen effectively protonating it and permitting rotation about the peptide linkage.

Do cyclosporin and FK506 act as transition state inhibitors? If so, we might learn something about the mechanism from the three-dimensional structures of the inhibited rotamases. From these structures as well as those with simpler substrate analogs, it is seen that the substrate is desolvated and that many nonbonded (van der Waals) interactions stabilize the binding. Hydrogen bonding is also important. The presence of distinct hydrogen bonds to the peptide NH and C=O groups on the N-terminal side of the substrate X-Pro linkage and less well-defined bonding on the terminal side suggests that the C-terminal part may be rotated while the enzyme holds the N-terminal part. It can also be concluded that mechanisms are probably not identical for all rotamases.

Cyclophilins and FKBPs are large families of proteins. In E. *coli* there are two cyclophilin genes, three encoding FK506-binding proteins, and one of a third family of bacterial rotamases (parvulins) that lack sequence similarity to the other two families. Yeast contains at least five cyclophilins and even more eukaryotic FKBPs are known. There is abundant evidence that these proteins play an important role in protein folding *in vivo*.

Effectiveness of Enzymes

The Effectiveness of enzymes is due to the fact that they bring the reactants (substrates) together in a location where they are ideally oriented with respect to each other and the catalytic groups of the enzyme

Several broad themes recur frequently in enzymatic reaction mechanisms. Among the most important of these are (1) *proximity effects* (2) *general-acid and general-base catalysis,* (3) *electrostatic effects,* (4) *nucleophilic or electrophilic catalysis by enzymatic functional groups,* and (5) *structural flexibility.* All known enzymes use at least one of these themes, and most use more than one. We start by discussing these five themes in general terms and then see how they apply to three enzymes for which the mechanisms have been studied in depth.

The Proximity Effect

The idea of the *proximity* effect is that *an enzyme can accelerate a reaction between two species simply by holding the two reactants close together in an appropriate orientation.* It has long been known that intramolecular reactions involving two groups that are tied together in a single molecule are usually much faster than the corresponding intermolecular reactions between two independent molecules. The cyclization of succinic acid to form succinyl anhydride [equation (3.1)] for example, occurs much more rapidly than the formation of acetic anhydride from two molecules of acetic acid [equation (3.2)]

$$\begin{array}{l} CH_2-CO_2H \\ | \\ CH_2-CO_2H \end{array} \longrightarrow \begin{array}{l} CH_2-C(=O) \\ | \qquad\qquad\quad \rangle O \\ CH_2-C(=O) \end{array} + H_2O \qquad ...(3.1)$$

We cannot compare the rate constants for these two reactions directly because they are expressed in different units. The intramolecular reaction [equation (3.1)] is kinetically first order, whereas the intermolecular reaction [equation (3.2)] is second order. But suppose the two molecules

$$\begin{array}{lcl} CH_3CO_2H & & \\ & \longrightarrow & (CH_3CO)_2O + H_2O \\ CH_3CO_2H & & \end{array} \quad ...(3.2)$$

of acetic acid that enter into reaction (3.2) are labeled isotopically to make them distinguishable and that one type of molecule is present in great excess over the other. The process is then kinetically first order in the concentration of the limiting reactant.

To make the rate constant the same as for reaction the more abundant species has to be present at a concentration of 3×10^5 M! This is far above any concentration that can actually be obtained. This example from organic chemistry shows that tying two reactants together can have an enormous effect on the rate of a reaction.

The effect is due largely to differences between the entropy changes that accompany the inter-and intramolecular reactions. The formation of the transition state requires a larger decrease of translational and rotational entropy in the intermolecular reaction than it does in the intramolecular reaction. In the intramolecular reaction much of this entropy decrease has already occurred during the preparation of the reactant. Enzymes that catalyze intermolecular reactions take advantage of the proximity effect by binding the reactants close together at the active site so that the reactive groups are oriented appropriately for the reaction.

Effect of pH

Chemical bonds are formed by electrons, and formation or breakage of bonds requires the migration of electrons. In broad terms, reactive chemical groups function either as *electrophiles or as nucleophiles. Electrophiles are electron*-deficient substances that react with electron-rich substances; nucleophiles are electron–rich substances that react with electron–deficient substances. *The task of a catalyst often is to make a potentially reactive group more reactive by increasing its electrophilic or nucleophilic character. In many cases the simplest way to do this is to add or remove a proton.* As an consider the hydrolysis of an ester.

Because the electronegativity of the oxygen atom in the C = O group is greater than that of the carbon, the oxygen has a fractional negative charge δ^-, and the carbon has a fractional positive charge δ^+. Hydrolysis of an ester in neutral aqueous solution can occur if the oxygen atom of H_2O, acting as a nucleophile, reacts with the carbonyl carbon. The initial product is an intermediate in which the carbon atom has four substituents in a tetrahedral arrangement. The reaction is completed by the breakdown of the tetrahedral intermediate to release the alcohol. However, water is a comparatively weak nucleophile, and its reaction with esters in the absence of a catalyst is very slow.

Hydrolysis of esters occurs much more rapidly at high pH, when the negatively charged hydroxide ion replaces water as the reactive nucleophile (Fig. 3.1 *a*). The nucleophilic character of water itself also can be increased by interaction with a basic group other than OH^- (Fig. 3.1 bt). By offering electrons to one of the protons of the water, the base increases the electron density on the oxygen. The term "*general base*" is used to describe any substance that is capable of binding a proton in aqueous solution. Enzymes use a variety of functional groups in this role.

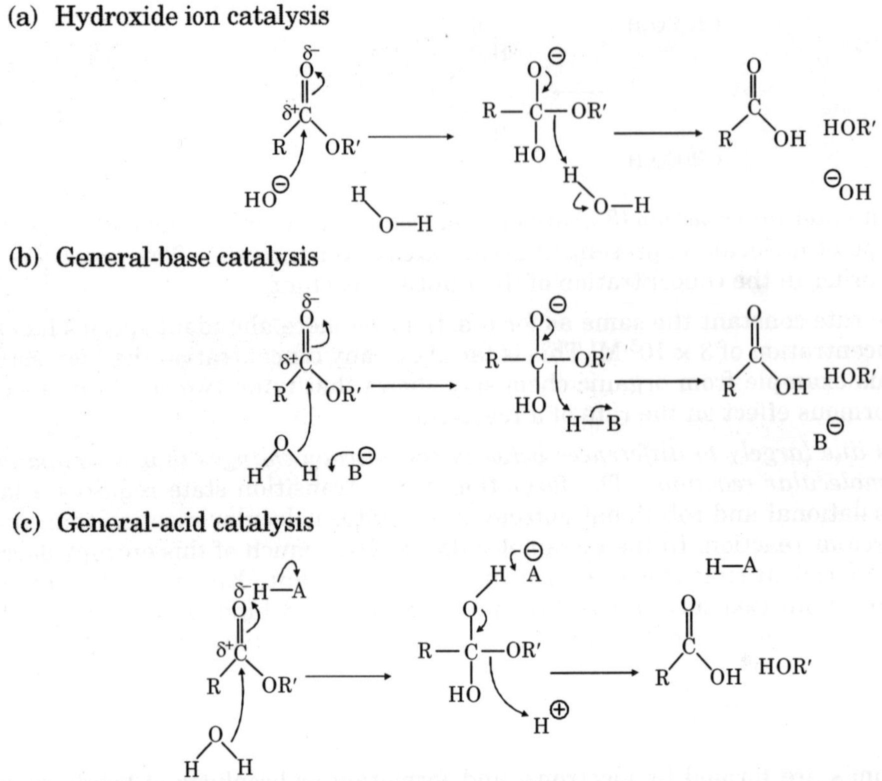

(a) Hydroxide ion catalysis

(b) General-base catalysis

(c) General-acid catalysis

Fig. 3.1. Several ways that the hydrolysis of an ester can occur. A red, curved arrow represents the movement of an electron pair from an electron donor to an acceptor, (a) Catalysis by free hydroxide ion. *(b)* General-base catalysis, (c) General-acid catalysis.

Two factors make free OH^- ions themselves unsuitable for enzymatic catalysis and thus favor the use of a general base. First, the low concentration of OH^- limits its availability at physiological pH. In contrast, proteins contain numerous functional groups that can serve as general bases at moderate pH or even under mildly acidic conditions.

The only requirement is that the base can exist at least partly in its unprotonated form at the ambient pH. This condition can easily be met by selecting a basic group from among the ionizable or polar amino acid side chains from an amino-terminal — NH_2 group or a carboxy-terminal carboxylate (see table 3.3). The second advantage of using a general base instead of OH^- is that a basic group provided by the protein can be positioned precisely with respect to the substrate in the active site, allowing the proximity effect to come into play. Free OH^- ions are much more mobile. In exceptional cases in which OH^- does act as a nucleophile in an enzymatic reaction, it usually is tightly bound to a metal cation. The hydrolysis of an ester also can be catalyzed by an acid (Fig. 3.1)

The acid donates a proton to the carbonyl oxygen, increasing the positive charge on the carbon. The term *'general acid'* is used to refer to any substance capable of releasing a proton in solution, and again enzymes almost always use such proton donors in preference to free H^+ or H_3O^+ ions, presumably because a general acid can operate at moderate pH and is easy to fix in position. An important point to note in Fig. 3.1 is that the same general acid or base that catalyzes the formation of the tetrahedral intermediate also can participate in the decomposition of the intermediate. When a general acid (HA) donates a proton to the ester oxygen it becomes a base (A^-), which can retrieve the proton as the intermediate breaks down. When a general base (B^-) removes a proton from water it becomes an acid (BH), which can provide a proton to the alcohol. Note also that general-acid and general-base catalysis are not mutually exclusive; they can both occur in a concerted manner in the same step of a reaction.

Electrostatic Interactions

The frequent use of general acids and general bases in enzymes illustrates the underlying principle that enzymes act by stabilizing the distribution of electrical charge in transition states. In the enzymatic hydrolysis of an ester, the key transition state probably resembles the tetrahedral intermediates shown in Fig. 3.1. To form such an intermediate, electrons must move from the nucleophile through the carbon atom of the C = O group to the oxygen. There is thus a net movement of negative charge from the nucleophile to the substrate.

In the absence of a general acid or base, a charge approaching + 1 appears on the nucleophile, and a charge approaching – 1 appears on the C = O oxygen. A general base can stabilize this new distribution of charge by offering electrons to the nucleophile so that some of the. positive charge moves to the base. By providing a proton to the C = O oxygen, a general acid can delocalize the negative charge here. But enzymes have other ways to achieve a similar stabilization. Suppose that the active site included a positively charged amino acid side chain, such as that of lysine or arginine, located near the oxygen atom of the C = O group. A positive charge in this region favors the formation of the tetrahedral intermediate.

A negative charge in the region of the nucleophile has a similar effect. The interactions of such charges are termed electrostatic effects. Electrostatic interactions can ber significant even between groups whose net formal change is zero. This is because charge distributions within molecular groups are not uniform but rather vary from atom to atom. We alluded to this previously when discussing the partial charges on the oxygen and carbon atoms of an ester. Similar considerations apply to other functional groups.

The electron distributions around the nuclei leave each atom with a small net positive or negative charge. In an alcoholic —OH group, for example, the oxygen atom has a negative charge of approximately –0.4 atomic charge units, and the hydrogen has a charge of about +0.4.

As a *reacting substrate* is transformed into a transition state, the changing charges on its atoms interact with the charges on atoms of the surrounding protein and any nearby water molecules. The energy difference between the initial state and the transition state thus depends critically on the details of the protein structure. We see illustrations of this in the three enzymes discussed later on.

Enzymatic Functional Groups

Another strategy for catalyzing the hydrolysis of an ester or an amide is to replace water by a stronger nucleophilic group that is part of the enzyme's active site. The $HOCH_2$ — group of a serine residue is often used in this way. In such cases, the reaction of the serine with the substrate splits the overall reaction into a two-step process. Instead of immediately yielding the free carboxylic acid, the breakdown of the initial tetrahedral intermediate yields an intermediate ester that is attached covalently to the enzyme.

$$\mathrm{R-\overset{\overset{\displaystyle O}{\|}}{C}-NHR' + HOCH_2-Enzyme \longrightarrow R-\overset{\overset{\displaystyle O}{\|}}{C}-OCH_2-Enzyme + R'NH_2} \qquad \text{...(3.3)}$$

The *acyl-enzyme* ester intermediate must be hydrolyzed by a second reaction, in which water becomes the nucleophile.

$$\mathrm{R-\overset{\overset{\displaystyle O}{\|}}{C}-OCH_2-Enzyme + H_2O \longrightarrow R-\overset{\overset{\displaystyle O}{\|}}{C}-OH + HOCH_2-Enzyme} \qquad \text{...(3.4)}$$

The proteolytic enzymes trypsin, chymotrypsin, and elastase, discussed later on, all work in this way.

Nucleophilic groups on enzymes participate in a variety of other types of reactions in addition to hydrolytic reactions. An example is acetoacetic acid decarboxylase, which catalyzes the reaction

$$\mathrm{CH_3-\overset{\overset{\displaystyle O}{\|}}{C}-CH_2-CO_2H \longrightarrow CH_3-\overset{\overset{\displaystyle O}{\|}}{C}-CH_3 + CO_2} \qquad \text{...(3.5)}$$

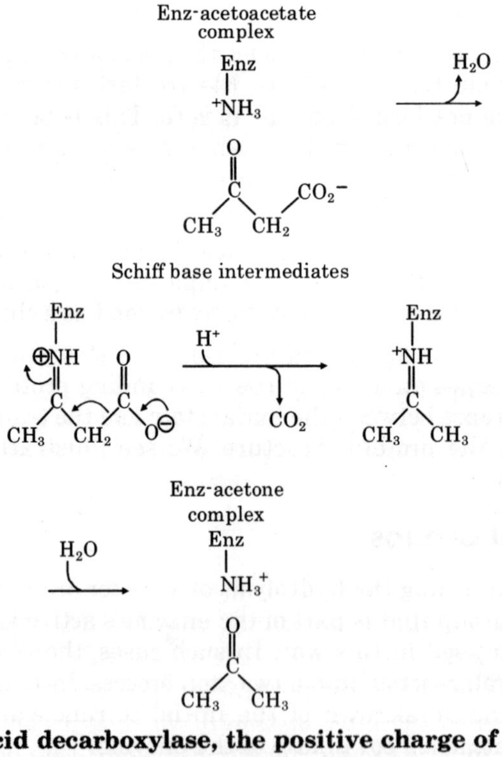

Enz-acetoacetate complex

Enz–$^+NH_3$; $CH_3C(=O)CH_2CO_2^-$ → (H_2O)

Schiff base intermediates

Enz–$\oplus NH=C(CH_3)CH_2C(=O)O^\ominus$ → (H^+, CO_2) Enz–$^+NH=C(CH_3)CH_3$

Enz-acetone complex

(H_2O) → Enz–NH_3^+ ; $CH_3C(=O)CH_3$

Fig. 3.2. In acetoacetic acid decarboxylase, the positive charge of a protonated Schiff base intermediate pulls electrons from a nearby carbon-carbon bond, thereby releasing CO_2.

The reaction proceeds through an intermediate called a Schiff base, in which the substrate is covalently attached to the e-amino group of a lysine residue at the enzyme's active site.

$$CH_3 - \overset{\overset{\displaystyle O}{\|}}{C} - CH_2 - CO_2H + \text{Enzyme} - NH_2 \longrightarrow CH_3 - \overset{\overset{\displaystyle N-\text{Enzyme}}{\|}}{C} - CH_2CO_2H - H_2O \quad ...(3.6)$$

Protonation of the nitrogen atom of the Schiff base introduces a positive charge that pulls electrons from the nearby carbon-carbon bond, causing decarboxylation. The mechanisms outlined in equations 3.3 – 3.6 illustrate a basic feature:

Nucleophilic catalysis by enzymes involves the formation of an intermediate state in which the substrate is covalently attached to a nucleophilic group of the enzyme. In addition to the $—CH_2OH$ group of serine and the ε-amino of lysine, the $—CH_2SH$ of cysteine is often used as a nucleophile. The carboxylate of aspartate or glutamate and the imidazole group of histidine can play a similar role. Some enzymes take advantage of bound coenzymes, such as thiamine, biotin, pyridoxamine, or tetrahydrofolate, to obtain additional nucleophilic reagents. There also are numerous enzymes that use bound metal ions to form complexes with substrates. In these enzymes, the metal ion usually serves is an electrophilic functinal group rather than as a nucleophile.

Carbonic anhydrase, for example, contains a Zn^{+2} ion that binds one of the substrates, hydroxide ion, as a ligand. The bound OH^- reacts with the other substrate, CO_2. In alcohol dehydrogenase, and in the proteolytic enzymes thermolysin and carboxypeptidase A, a Zn^{+2} ion forms a complex with a carbonyl oxygen atom of the substrate. The withdrawal of electrons by the Zn^{+2} increases the partial positive charge on the carbonyl carbon and thus promotes reaction with a nucleophile.

Structural Flexibility

Although precise positioning of the reactants is a fundamental aspect of enzyme catalysis, most enzymes undergo some change in their structure when they bind substrates. A particularly dramatic example is hexokinase, which catalyzes the transfer of a phosphate group from adenosine triphosphate (ATP) to glucose.

$$\text{ATP} + \text{glucose} \rightarrow \text{ADP} + \text{glucose—6—phosphate} \quad ...(3.7)$$

When hexokinase binds glucose, its structure changes in a that brings together the elements of the active site.

The enzyme literally closes like a set of jaws around the substrate! Such a structural change is often referred to as an induced fit. Enfolding a substrate in this way can serve to maximize the favorable entropy change associated with removing a hydrophobic substrate molecule from water. It also allows the enzyme to control the electrostatic effects that promote formation of the transition state. *The substrate is forced to respond to the directed electrostatic fields from the enzyme's functional groups, instead of the disordered fields from the solvent.* Structural cnangesalso contribute to the high specificity of some enzymatic reactions.

In hexokinase, the structural change induced by glucose promotes the binding of the other substrate, ATP. ATP does not bind to the enzyme properly unless glucose is already present in the catalytic site. If ATP were to bind in the absence of glucose, the enzyme might have a tendency to catalyze the transfer of phosphate from ATP to water, resulting in a wasteful loss of ATP.

$$ATP + H_2O \rightarrow ADP + HPO_4^{2-} + H^+ \quad ...(3.8)$$

Hexokinase does not catalyze this side reaction; it waits for glucose to bind first. The structural changes that occur in hexokinase bring home the point that enzyme crystal structures give static snapshots of molecules that actually are highly flexible.

In solution, the structure of an enzyme undergoes fluctuations that vary widely in amplitude and frequency from place to place in the protein. Vibrations and rotations involving only a few atoms occur on time scales of 10^{-13} -10^{-11} s. Some-what larger motions, such as the flipping of the aromatic ring of a tyrosine, typically occur on scales of $10^{-9}-10^{-8}$ s. Major reorganizations may take $10^{-6}-10^{-3}$ s. All of these types of motions can be important in catalysis.

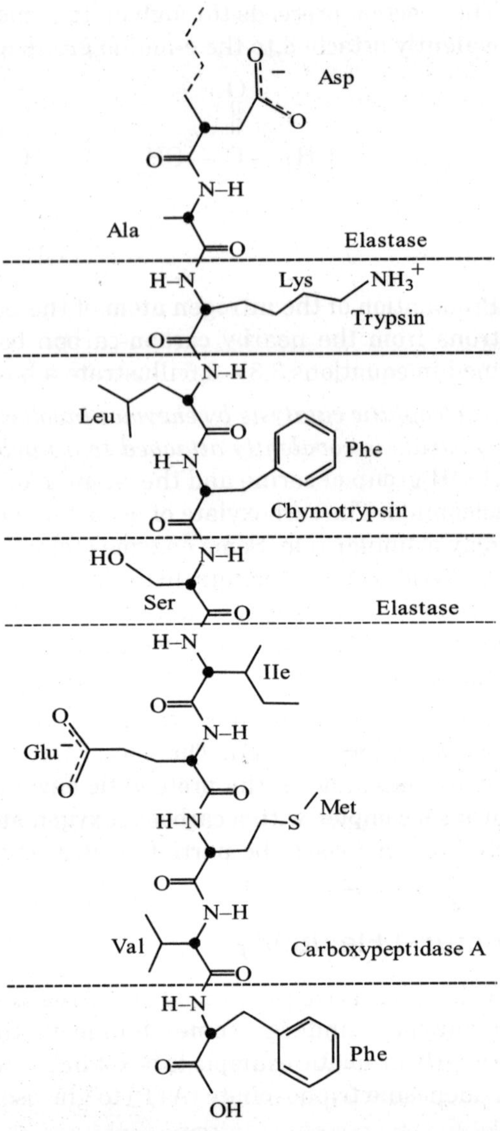

Catalytic Activities

In the preceding sections, we discussed five themes that occur frequently in enzyme reaction mechanisms. We now examine several representative enzymes in finer detail. We focus on three enzymes for which crystal structures have been obtained, because the most decisive advances in our understanding of enzyme reaction mechanisms have come by inspecting such structures.

Nucleophilic Catalysis

The *serine proteases* are a large family of proteolytic enzymes that use the reaction mechanism for nucleophilic catalysis outlined in equations (3.3) and (3.4), with a serine residue as the reactive nucleophile. The best known members of the family are three closely

Fig. 3.4. Trypsin, chymotrypsin, and elastase—three members of the serine protease family—catalyze the hydrolysis of proteins at internal peptide bonds adjacent to different types of amino acids. Trypsin prefers lysine or arginine residues; chymotrypsin, aromatic side chains; and elastase, small, nonpolar residues. Carboxypeptidases A and B, which are not serine proteases, cut the peptide bond at the carboxyl-terminal end of the chain. Carboxypeptidase A preferentially removes aromatic residues; Carboxypeptidase B, basic residues.

related digestive enzymes: *trypsin, chymotrypsin*, and *elastase*. These enymes are synthesized in the mammalian pancreas as inactive precursors termed *zymogens*. They are secreted into the small intestine, where they are activated by proteolytic cleavage. In the digestive system trypsin, chymotrypsin, and elastase work as a team. They are all *endopeptidases*, which means that they cleave protein chains at internal peptide bonds, but each preferentially hydrolyses bonds adjacent to a particular type of amino acid residue.

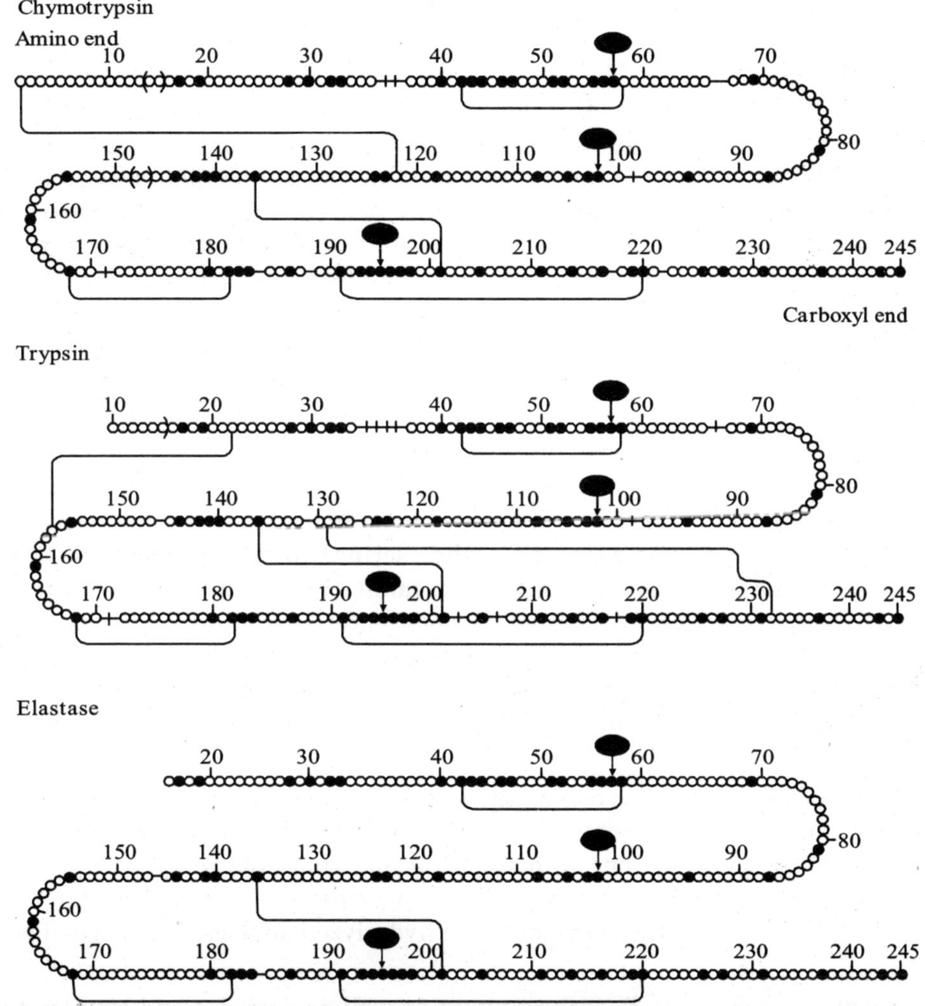

Fig. 3.5. Schematic diagrams of the amino acid sequences of chymotrypsin, trypsin, and elastase. Each circle represents one amino acid. Amino acid residues that are identical in all three proteins are in solid color. The three proteins are of different lengths but have been aligned to maximize the correspondence of the amino acid sequences. All of the sequences are numbered according to the sequence in chymotrypsin. Long connections between nonadjacent residues represent disulfide bonds. Locations of the catalytically important histidine, aspartate, and serine residues are marked. The links that are cleaved to transform the inactive zymogens to the active enzymes are indicated by parenthesis marks. After chymotrypsinogen is cut between residues 15 and 16 by trypsin and is thus transformed into an active protease, it proceeds to digest itself at the additional sites that are indicated; these secondary cuts have only minor effects on the enzymes's catalytic activity.

Trypsin cuts just next to basic residues (lysine or arginina); chymotrypsin cuts next to aromatic residues (phenylalanine, tyrosine, or tryptophan); elastase is less discriminating but prefers small, hydrophobic residues such as alanine. About half of the amino acid residues or trypsin are identical to the corresponding residues in chymotrypsin, and about a quarter of the residues are conserved in all three of the pancreatic endopeptidases. The structural similarities of trypsin, chymotrypsin, and elastase are even more evident in the crystal structures. As is shown in figure 8.6*a* the folding of the polypeptide chain is essentially the same in all three enzymes. These enzymes are classical illustrations of diverging evolution from a common ancestor. The structures of some of the bacterial serine proteasss also are homologous to those of the mammalian enzymes.

On the other hand, subtilisin, a serine protease obtained from *Bacillus subtilis,* has an amino acid sequence that seems totally unrelated to the mammalian sequences. Its three-dimensional structure also is very different from those of the mammalian enzymes. Subtilisin probably joined the serine protease family by convergent evolution. Remarkably, a small set of critical amino acid residues comes together in the folded structure to form the essential elements of the active site in all of these proteases. Using the chymotrypsin numbering system, these are His 57, Asp 102 and Ser 195. That Ser 195 played an important role in the catalytic mechanism was known from early studies on the enzyme inhibitor diisopropylfluorophosphate. This inhibitor reacts irreversibly with chymotrypsin or trypsin to form an inactive derivative in which the diisopropylphosphate group is covalently attached to the serine residue.

A variety of other inhibitors react in a parallel manner with this same serine residue. As a rule, these inhibitors do not react with other serines in the enzyme or with serines in enzymes that are not part of the serine protease family. The reactivity of Ser 195 thus is not a property of serine residues in general but depends on the special surroundings of this residue in the protein. We will see shortly that it is the juxtaposition with His 57 and Asp 102 that makes the serine especially reactive. Although His 57 is far removed from Ser 195 in the primary sequence, studies with affinity labels showed that it must be near the serine residue in the active site.

A derivative of phenylalanine containing a reactive chloromethylketone group was found to inhibit chymotrypsin irreversibly by reacting with the histidine. The inhibition can be prevented by the presence of other aromatic molecules that bind competitively at the active site. The initial evidence for the formation of an acylenzyme ester intermediate came from studies of the kinetics with which chymotrypsin hydrolyzed analogs of its normal polypeptide substrates.

The enzyme turned out to hydrolyze esters as well as peptides and simpler amides. Of particular interest was the reaction with the ester *p*-nitrophenyl acetate. This substrate is well suited for kinetic studies because one of the products of its hydrolysis, *p*-nitrophenol, has a yellow color in aqueous solution, whereas *p*-nitrophenyl acetate itself is colorless. The change in the absorption spectrum makes it easy to follow the progress of the reaction. When rapid-mixing techniques are used to add the substrate to the enzyme, an initial burst of *p*-nitrophenol is detected within the first few seconds, before the reaction settles down to a constant rate.

The amount of p-nitrophenol released in the burst is approximately/equal to the amount of enzyme present in the solution. These observations suggest that the overall enzymatic reaction occurs in two distinct steps.

In the first step, p-nitrophenol is released and the acetyl group is transferred to the enzyme, forming an acyl-enzyme intermediate. In the second step, the intermediate is hydrolyzed, and acetate is released. The finding that diisopropylfluorophosphate prevents the initial burst of p-nitrophenol suggested that the enzymatic group that forms the ester intermediate is Ser 195. When the crystal structures of trypsin, chymotrypsin, and elastase with bound substrate analogs were determined, the substrate analogs were indeed located close to Ser 195. Histidine 57 sits nearby, in an orientation suggesting that the OH group of Ser 195 forms a hydrogen bond to the imidazole side chain of the histidine. Aspartic acid 102 sits on the opposite edge of the imidazole ring, where its negatively charged carboxylate group can interact with the proton on the other nitrogen of the ring.

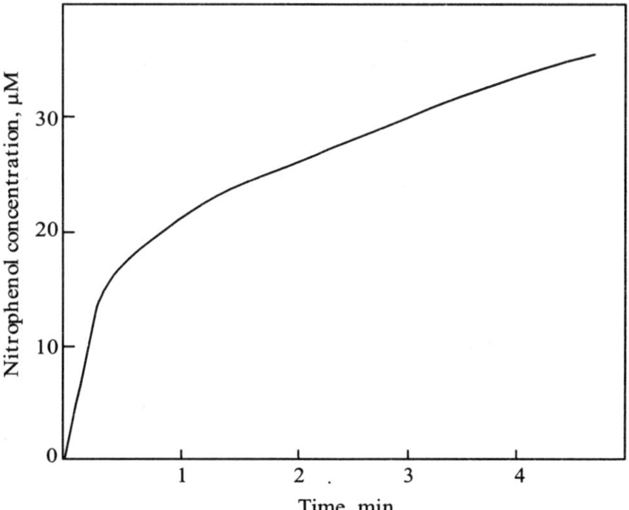

Fig. 3.6. p-Nitrophenol formation as a function of time during the hydrolysis of p-nitrophenyl acetate by chymotrypsin. A rapid initial burst of p-nitrophenol is followed by a slower, steady-state reaction. The amount of p-nitrophenol released in the burst is approximately equal to the amount of enzyme present.

The side chains of the aspartate and histidine residues thus appear to be oriented so as to facilitate removal of the proton from the serine's OH group. Because withdrawing the proton would increase the nucleophilic character of the oxygen, this arrangement explains why Ser 195 is exceptionally reactive.

```
 O          \
 ||          C=C
             /   \
—C—O⁻ ---- HN     N ---- HOCH2—
             \   //
               C
```

The dependence of the enzyme kinetics on pH agrees with this picture. Enzymatic activity depends on the presence of a basic group with a pK_a of about 6.8, which is in the range consistent with a histidine side chain. The k_{cat} decreases abruptly if this group is protonated. Protonating His 57 prevents the histidine from forming a hydrogen bond to Ser 195, thereby decreasing the nucleophilic reactivity of the serine. Tne crystal structures also provided an explanation for the different substrate specificities of trypsin, chymotrypsin, and elastase. In both trypsin and chymotrypsin, the side chain of the substrate fits snugly into a pocket. At the far end of the pocket in trypsin is an aspartic acid residue. The negative charge of the aspartate carboxylate favors the binding of the positively charged side chain of lysine or arginine.

In chymotrypsin the aspartic acid residue is replaced by serine, creating a less polar environment that suits the side chain of tyrosine, phenylalanine, or tryptophan. In elastase, the binding pocket is padded with the bulky side chains of valine and threonine residues, so that it can accommodate only small substrates such as alanine. Another important feature of the substrate-binding site in serine proteases is that the carbonyl oxygen atom of the peptide bond that is cleaved is hydrogen-bonded to one or more NH groups.

In trypsin, chymotrypsin, and elastase, these are the amide NH groups of Ser 195 and Gly 193. The interaction with the two protons favors an increase in the negative charge on the oxygen, facilitating formation of a tetrahedral intermediate state, as we discussed previously in connection with general-acid catalysis and electrostatic effects. The overall reaction mechanism of chymotrypsin is sketched in Fig. 3.11. Part *a* shows the enzyme-substrate complex, with an aromatic side chain of the substrate seated in the binding pocket and the carbonyl oxygen atom hydrogen-bonded to the amide NH hydrogens of Gly 193 and Ser 195. Aspartate 102 and His 57 are aligned as described above, with Ser 195 forming a hydrogen bond with the histidine. Part *b* shows a tetrahedral intermediate state. Serine 195 has released its proton to His 57 and launched a nucleophilic attack on the carbonyl carbon of the substrate.

The histidine residue thus acts as a general-base catalyst in the formation of the intermediate. The movement of negative charge to the carbonyl oxygen creates an "oxyanion," which is stabilized largely by electrostatic interactions with the amide protons of Ser 195 and Gly 193. The tetrahedral oxyanion intermediate is not stable enough to be isolated, and the evidence for its existence is inferential. In

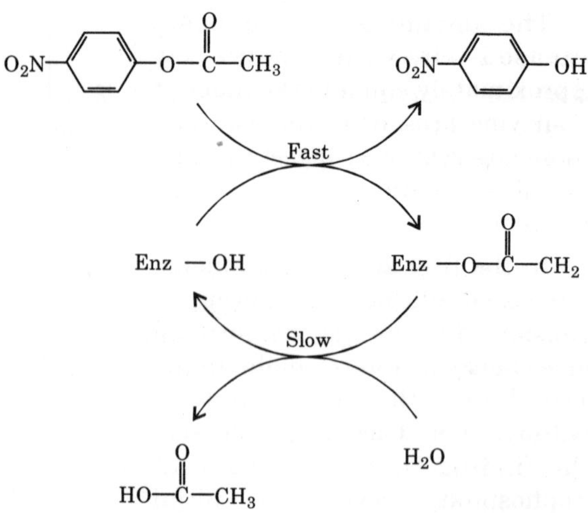

Fig. 3.7. Steps in the hydrolysis of *p*-nitrophenyl acetate by chymotrypsin. In the hydrolysis of this and most other esters, the breakdown of the acyl-enzyme intermediate is the rate-determining step. In the hydrolysis of peptides and amides, the rate-determining step usually is the formation of the acyl-enzyme intermediate. This makes the transient formation of the intermediate more difficult to study because the intermediate breaks down as rapidly as it forms.

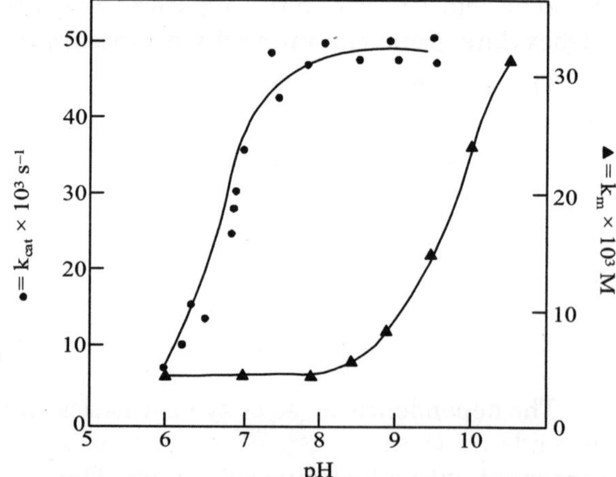

Fig. 3.8. The turnover number (k_{cat}) and the Michaelis constant (K_m) as a function of pH for the hydrolysis of *N*-acetyl-L-tryptophanamide by chymotrypsin at 25°C. The decrease in k_{cat} as the pH is lowered between 8 and 6 probably reflects the protonation of His 57. The increase in K_m above pH 9 probably reflects the deprotonation of Ile 16, which results in the rotation of Gly 193 out of the substrate-binding site.

part *c*, this intermediate has decomposed to form the more stable acyl-enzyme intermediate, in which the serine is linked as an ester to the carboxylic part of the substrate and the amine product has been released. Histidine 57 probably facilitates this decomposition by acting as a general acid. To complete the reaction, the breakdown ot me acyl-enzyme probably occurs as shown in parts *d, e,* and *f* of Fig. 3.11. The steps here are essentially a reversal of the steps through parts *a, b,* and *c*, except that water replaces the amino part of the substrate.

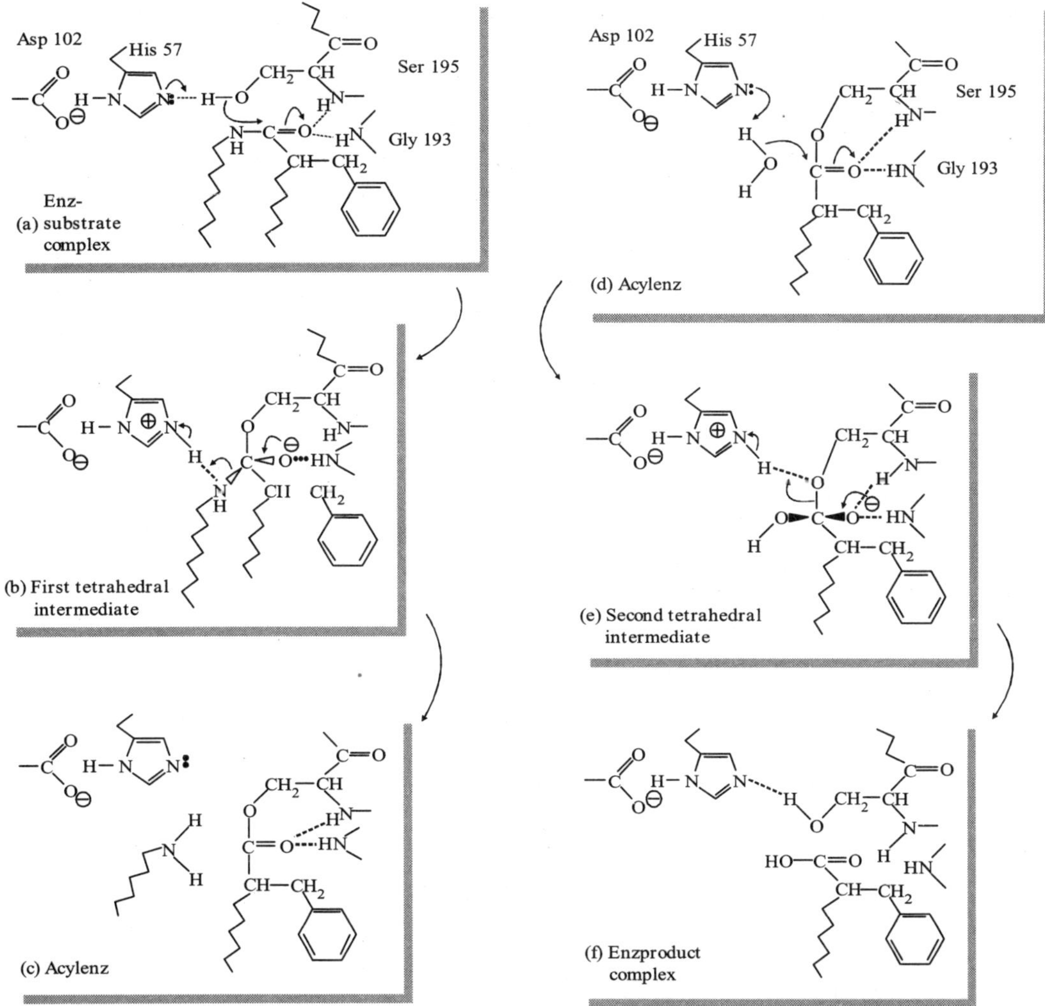

Fig. 3.9. The probable mechanism of action of chymotrypsin. The six panels show the initial enzyme-substrate complex *(a)*, the first tetrahedral (oxyanion) intermediate *(b)*, the acyl-enzyme (ester) intermediate with the amine product departing *(c)*, the same acyl-enzyme intermediate with water entering *(d)*, the second tetrahedral (oxyanion) intermediate *(e)*, and the final enzyme-product complex *(f)*. In the transition states between these intermediates, there probably is a more even distribution of negative charge between the different oxygen atoms attached to the substrate's central carbon atom.

The reaction probably proceeds by way of a tetrahedral intermediate similar to one shown in part *b* except for this substitution (part *e*). In Fig. 3.11, note that Asp 102 doesnot actually remove a proton from His 57 in the formation of either of the tetrahedral intermediates. The effect of the aspartate's negatively charged carboxyl group probably is best viewed as an electrostatic effect that favors movement of positive charge from Ser 195 to His 57. The stabilization of the oxyanion intermediate by the amide protons of Gly 193 and Ser 195 can be described similarly as a favorable electro-static interaction of the negatively charged oxygen with nearby atoms that have positive charges, rather than as an actual transfer of a proton to the oxygen.

The carboxyl group of Asp 102 also helps to align the imdazole ring of His 57 in the proper orientation for removing the proton from Ser 195. One way to test a scheme like that shown in Fig. (3.1l) is to investigate the effects of modifying the amino acid residues that play important roles. The inhibitory effect of the reaction of Ser 195 with diisopropylfluorophosphate was mentioned earlier. To examine the importance of His 57, this residue can be modified by meth ylation, which disrupts the interaction with Asp 102. This process decreases the activity of chymotrypsin by a factor of more than 10^3. The importance of Asp 102 was tested by site-directed mutagenesis of trypsin and subtilisin. Replacing the aspartate by asparagine reduces k_{cal} by a factor of about 10^4.

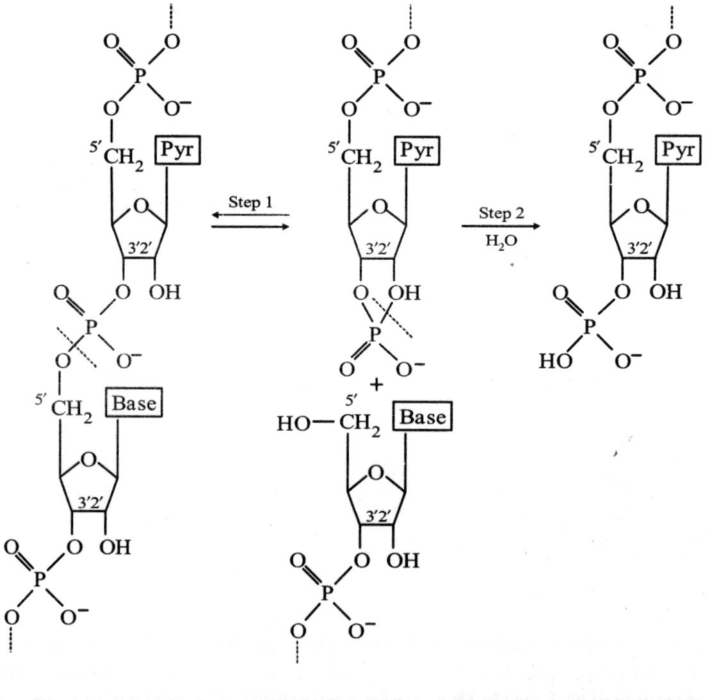

Fig. 3.10. A portion of an RNA chain, indicating points of cleavage by pancreatic ribonuclease. "Pyr" refers (o a pyrimidine; "Base" can be either a purine or a pyrimidine. The *2'*, 3', and 5' carbon atoms are labeled. The enzymatic reaction proceeds in two steps, with a cyclic 2',3'-phosphate diester as an intermediate.

Ribonuclease

Ribonucleases are a widely distributed family of enzymes that hydrolyze RNA by cutting the P—O ester bond attached to a ribose 5′ carbon. A good representative of the family is the pancreatic *enzyme ribonuclease* A. (RNase A), which is specific for a pyrimidine base (uracil or cytosine) on the 3′ side of the phosphate bond that is cleaved. When the amino acid sequence of bo-vine RNase A was determined in 1960 by Stanford Moore and William Stein, it was the first *enzyme* and only the second protein to be sequenced. RNase A thus played an important role in the development of ideas about enzymatic catalysis. It was one of the first enzymes to have its three-dimensional structure elucidated by x-ray diffraction and was also the first to be synthesized completely from its amino acids.

The synthetic protein proved to be enzymatically indistinguishable from the native enzyme. An early discovery by Frederick Richards that turned out to be useful was that the protein could be cleaved between residues 20 and 21 by the bacterial serine protease, subtilisin. The resulting two polypeptides were separated and purified. They were enzymatically inactive individually, but regained the activity of the native enzyme when they were recombined. This work shows that strong, noncovalent interactions occur that can hold protein chains together even when one of the peptide links is cut.

It also makes it possible to modify specific amino acid residues of the two polypeptide chains independently and to explore how each residue contributes to the reassembly of the protein and the recovery of enzymatic activity. The hydrolysis of RNA catalyzed by RNase A occurs in two distinct steps, with a 2′, 3′-phosphate cyclic diester intermediate (see Fig. 3.12). The intermediate can be identified relatively easily, because its breakdown is much slower than its formation. Ribonulceases do not hydrolyze DNA, which lacks the 2′-hydroxyl group needed for the formation of the cyclic intermediate. RNase A is completely inhibited if either of two histdine residues (His 12 or His 119) is modified by carboxymethylation with iodoacetate suggesting that these histidines play important roles in the active site.

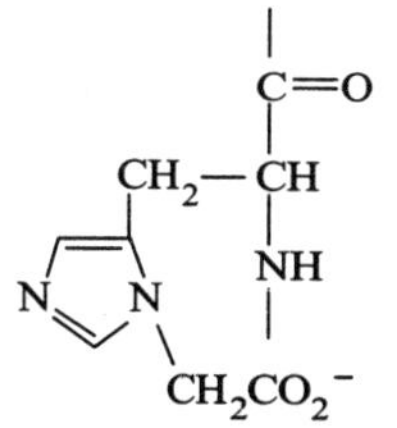

C=O
CH_2—CH
N N NH
$CH_2CO_2^-$

1-carboxymethyl-His 119

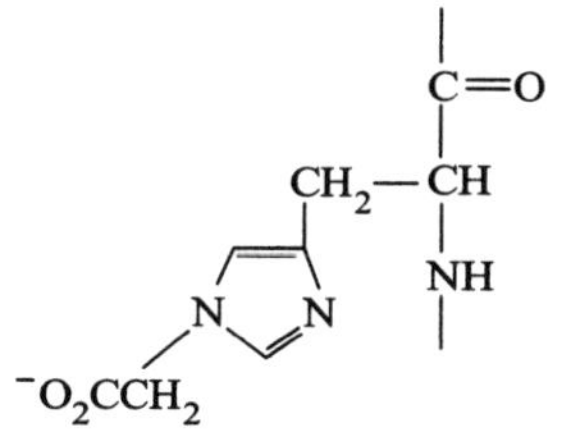

C=O
CH_2—CH
N N NH
$^-O_2CCH_2$

3-carboxymethyl-His 12

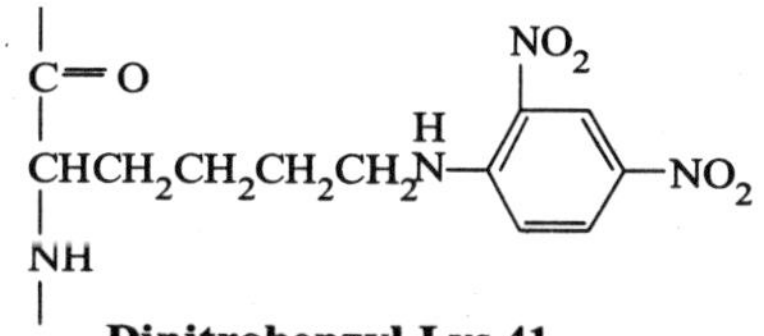

C=O
H
$CHCH_2CH_2CH_2CH_2N$— NO_2 NO_2
NH

Dinitrobenzyl-Lys 41

Fig. 3.11. When RNase A is treated with iodoacetate ($ICH_2CO_2^-$), the two major products obtained are carboxymethylated derivatives of His 12 and His 119. Both of these enzymes are severely inhibited, which suggests that both His 12 and His 119 are important in the active site. The enzyme also is completely inhibited by the reaction of Lys 41 with fluorodinitrobenzene.

In support of this conclusion, the reaction of iodoacetate with His 12 or His 119 is inhibited by cytidine-3′-phosphate and other small molecules that bind at the active site. Lysine 41 has been implicated similarly in the active site by the observation that enzymatic activity is destroyed by the reaction of the ε-amtno group of this residue with fluorodinitrobenzene. A second lysine (Lys 7) also appears to be important for substrate binding, although the enzyme retains some activity when this residue ismodified. The enzymatic activity of RNase A shows a bell–shaped dependence on pH, with an optimum near pH 7. The operation of the enzyme appears to require that a dissociable group with a pK_a of about 6.3 be in the protonated form, and that a group with a pK_a of about 8 be unprotonated.

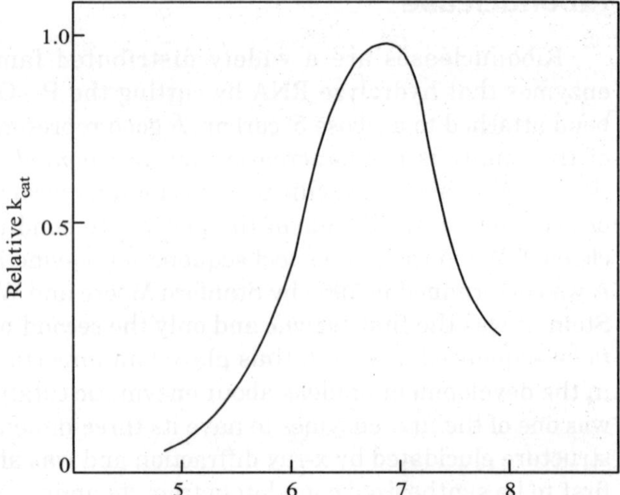

Fig. 3.12. The dependence of k_{cat}, of RNase A on pH. The bell-shaped curve suggests that one histidine residue must be in the protonated state and another must be unprotonated. Similar pH dependences are found for the hydrolysis of either RNA or pyrimidine nucleoside-2′,3′-phosphate cyclic diesters.

These groups have been identified as His 12 and His 1 19 by measuring the NMR spectrum of the enzyme as a function of pH. Spectral peaks due to histidine residues undergo characteristic shifts as the imidazole ring is protonated. The pH dependence of the kinetics, taken with the information that histidines 12 and 119 are essential for enzy-matic activity, suggests that the two histidines participate in the reaction by concerted general-base and general-acid ca-talysis. This supposition is supported by the crystal structure of the enzyme. Fig. 3.15 shows a model of the crystal structure of RNase S (the active enzyme obtained by cutting ribonuclease with subtilisin and recombining the two polypeptides) with a bound competitive inhibitor, "UpCH$_2$A." UpCH$_2$A resembles the dinucleotide substrate UpA but cannot be hydrolyzed because it has a methylene carbon in place of the oxygen between the P and the 5′ — CH$_2$ of the ribose of adenosine.

Similar crystal structures have been obtained with other substrate analogs. In all of the structures, His 12 and His 119 sit on opposite sides of the substrate's phosphate group, with His 12 positioned close to the 2′ OH group. Lysines 7, 14, and 66 also are located nearby, where their positively charged amino groups have favorable electrostatic interactions with the negatively charged phosphate. Near His 119 is the carboxyl group of an acidic residue, Asp 121. The electrostatic effect of Asp 121 favors the transfer of a proton from a molecule of water to His 119, much as Asp 102 of chymotrypsin induces a proton to move from Ser 195 to His 57. The pyrimidine ring on the 3′ side of the substrate fits snugly into a groove on RNase A. Valine 43 is located on one side of the pyrimidine ring, and Phe 120 on the other. The specificity of the enzyme for pyrimidine nucleotides appears to result largely from hydrogen bonding to Thr 45.

The threonine side chain can form a pair of complementary hydrogen bonds with either uracil or cytosine Crystallographic studies have shown that if the pyrimidine is replaced by a purine, the compound can still bind to the enzyme, but the distance between His 12 and the

2′ OH of the ribose increases by about 1.5 Å. This evidently prevents the catalytic reaction from occurring. The probable catalytic mechanism of RNase A is shown in Fig. 3.18. In the formation of the 2′,3′-cyclic ester, His 12 acts as a general base to remove the proton of the 2′ hydroxyl group, which increases the nucleophilic character of the oxygen atom.

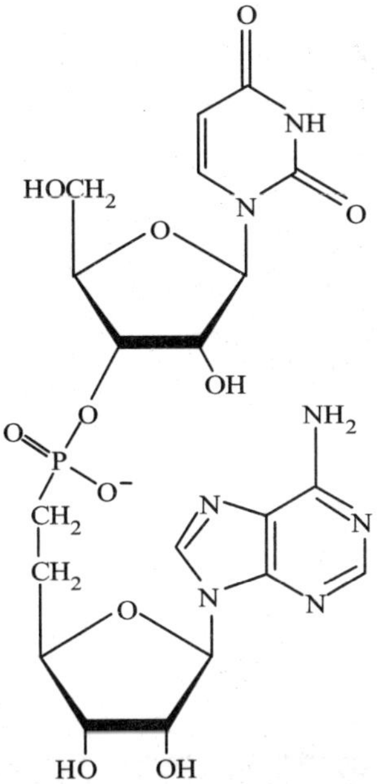

Fig. 3.13. Structure of UpCH$_2$A, a competitive inhibitor of RNase A. UpCH$_2$A differs from the dunucleotide substrate UpA in having a methylene group instead of oxygen between the phosphorus and C-5′ of the adenosine.

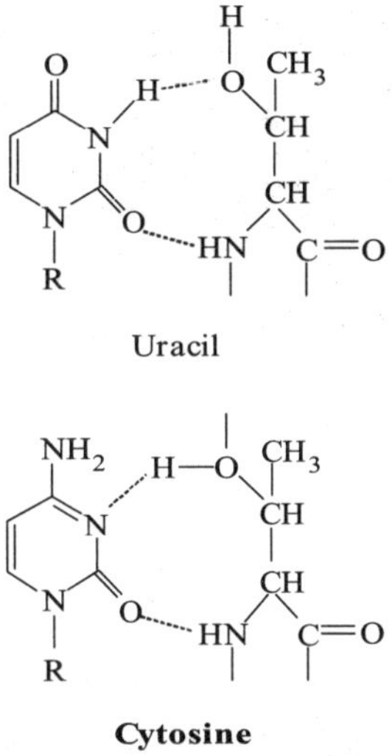

Fig. 3.14. The side chain and amide NH group of Thr 45 in RNase A can form a pair of complementary hydrogen bonds with either uracil or cytosine. These probably account for the specificity of RNase A for pyrimidine nucleotides.

Reaction of the hydroxyl group with the phosphate creates a transient intermediate state in which thephosphorus atom is pentavalent. The increase in negative charge on the phosphate's oxygen atomes is favored by electrostatic interactions with Lys 7 and Lys 41. The formation of the pentavalent intermediate also is promoted by His 119, which acts as a general acid to protonate one of the phosphate oxygens. His 119 then probably participates in transferring a proton to the 5′ oxygen atoms.

Expulsion of the protonated 5′ OH group converts the transient pentavalent intermediate into the more stable 2′,3′-cyclic ester, in which the substituents of the phosphorus have returned to a tetrahedral geometry. In the breakdown of the 2′,3′-cyclic ester, the roles of the two hisidines are reversed.

Histidine 119 now acts as general base to remove a proton from H_2O, and His 12 acts as a general acidto protonate the substrate's 2′ oxygen. Again, the reaction probably proceeds by way of pentavalent intermediate. The pentavalent phosphoryl intermediates that figures in the reactions of RNase A probably have the geometry of a trigonal bipyramid in which three of the phosphorus atom's substituents lie in a plane. The entering andleaving oxygen atoms are on either side of the plane, forming a straight line with the phosphorus. One observation that supports the formation of such pentavalent intermediates is that RNase A is inhibited strongly by uridine vanadate, in which the vanadium atom is surtaunded by five oxygens with a similar geometry. Fig. 3.19 shows the structure of the complex of the enzyme with the inhibitor. The disposition of histidines 12 and 119 on either side of the vanadate group in the crystal structure is consistent with the linear arrangements of the entering and leaving oxygen atoms on either side of the phosphorus atom in Fig. 3.18.

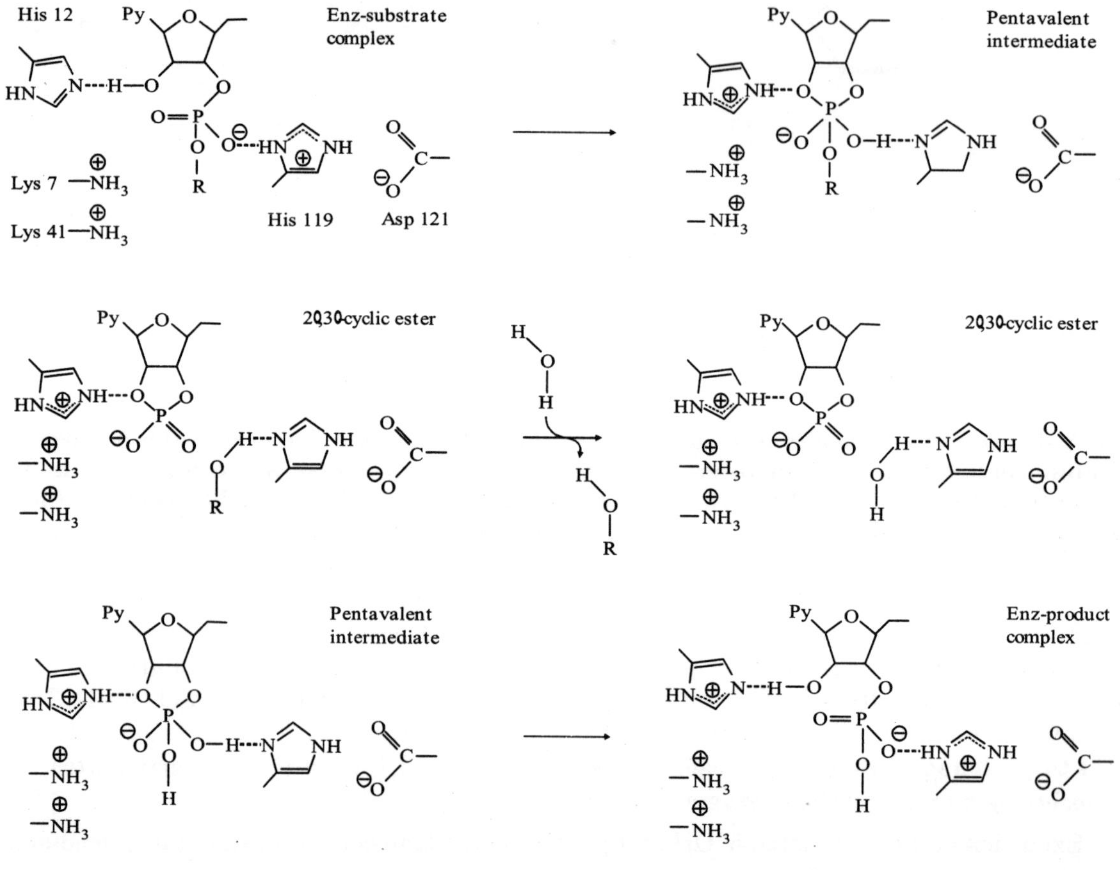

Fig. 3.15. The probable catalytic mechanism of RNase A. In the formation of the 2′, 3′ cyclic ester, His 12 acts as a general base, and His 119 acts as a general acid. The increase in negative charge on the phosphate's oxygen atoms in the transition state is favored by electrostatic interactions with Lys 7 and Lys 41. In the breakdown of the cyclic ester, His 119 acts as a general base and His 12 acts as a general acid. The reaction proceeds through two short-lived intermediates in which the phosphorus atom is pentavalent, in addition to the more stable cyclic ester intermediate. Py = pyrimidine.

Triosephosphate Isomerase

Triosephosphate isomerase catalyzes the interconversion of dihydroxyacetone phosphate and 3-phosphoglyceraldehyde.

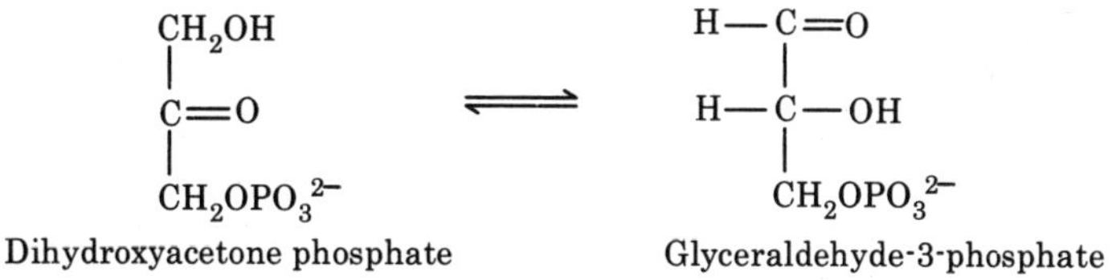

$$\underset{\text{Dihydroxyacetone phosphate}}{CH_2OH-C(=O)-CH_2OPO_3^{2-}} \rightleftharpoons \underset{\text{Glyceraldehyde-3-phosphate}}{H-C(=O)-H-C(OH)-CH_2OPO_3^{2-}} \quad \text{...(3.10)}$$

The interconversion of the two triosephosphates is an essential step in the catabolism of carbohydrates. Examining the catalytic mechanism of triosephosphate isomerase is instructive. It is among the enzymes that approach evolutionary perfection in the sense that its specificity constant k_{cat}/K_m is greater than 10^8 s^{-1} M^{-1}, which is near the limit set by the rates of diffusion of the substrate and protein. The enzyme achieves this high value for the specificity constant by having both a relatively high k_{cat} and a relatively low K_m. Crystal structures have been obtained for triosephos-phate isomerase purified both from yeast and from chicken muscle.

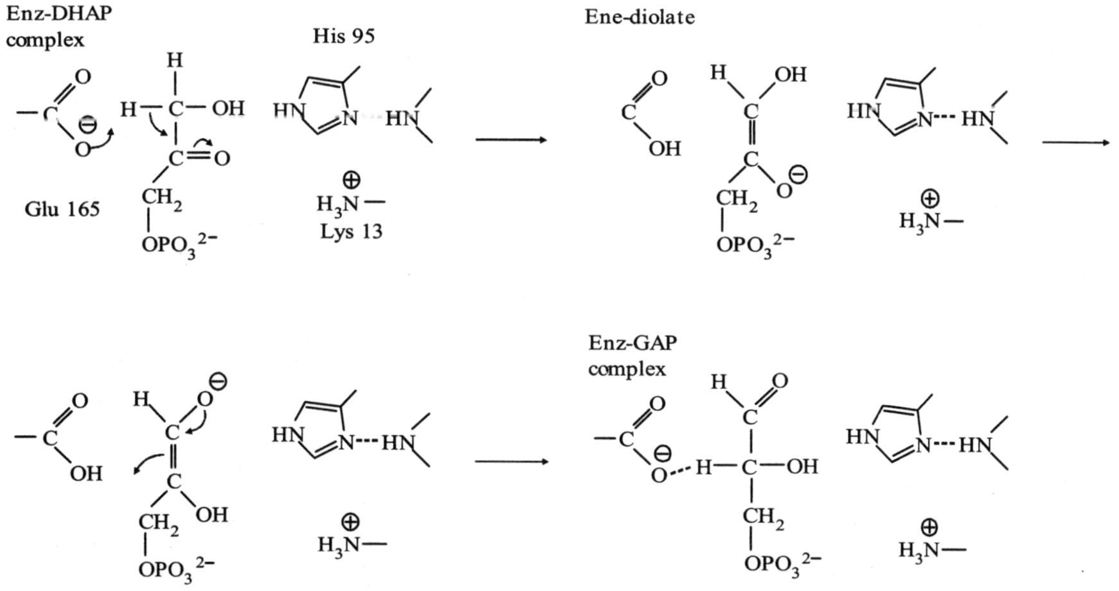

Fig. 3.16. The probable reaction mechanism of trosephosphate isomerase. The γ-carboxylate group of Glu 165 acts as a general base to remove a proton from C-1 of the substrate, dihydroxyacetone phosphate (DHAP). This generates a planar ene-diolate intermediate that has two tautomeric forms. After an interconversion of ene-diolate tautomers, the protonated glutamic acid returns a proton to C-2. The electrostatic effects of His 95 and Lys 12 facilitate the formation of the negatively charged intermediate. GAP = glyceraldehyde-3-P.

The proteins from the two organisms have identical residues at about half the positions in their amino acid sequences, and their three-dimensional structures are very similar. The overall structure is a β barrel composed of eight parallel β sheets linked by eight α helices. The active site is near the center of the molecule and contains a glutamic acid (Glu 165) with nearby histidine and lysine residues. Glu 165 was first implicated in the active site by the observation that it reacts irreversibly with chloroacetone phosphate, an inhibitor that is a close structural analog of the substrate dihydroxyacetone phosphate. (In chloroacetone phosphate, a chlorine atom replaces the hydroxyl group on C–1.) In crystal structures of the enzyme with bound dihydroxyacetone phosphate, the carboxyl group of Glu 165 is located close to C–1 of the substrates.

Changing Glu 165 to aspartate by site directed mutagenesis decreases the rate of the isomerization reaction catalyzed by the enzyme approximately 1,000-fold. This is a dramatic effect considering that the substitution of Asp for Glu only entails shortening the amino acid side chain by one —CH_2-group.

The rate of the isomerase reaction decreases if a basic group is protonated by lowering the pH below 6.5. It seems clear that this is the carboxyl group of Glu 165 because lowering the pH also prevents the reaction of this residue with chloroacetone phosphate. This suggests that, in the enzymatic reaction, Glu 165 acts as a general base to remove a proton from C-l of dihydroxyacetone phosphate, converting the substrate into an ene-diolate intermediate as shown in Fig. 3.21.

To complete the reaction, the protonated glutamic acid residue can return a proton to C-2 of the ene-diolate instead of to C–1. Additional support tor this mechanism comes from experiments in which the enzymatic reaction is run in D_2O. Deuterium is incorporated stereospecifically onto carbon 2 of the product.

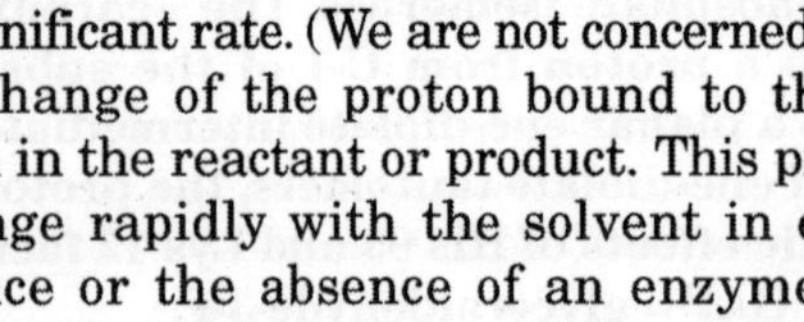

$$\begin{array}{c} CH_2OH \\ | \\ C{=}O \\ | \\ CH_2OPO_3^{2-} \end{array} \xrightarrow{D_2O} \begin{array}{c} H{-}C{=}O \\ | \\ D{-}C{-}OH \\ | \\ CH_2OPO_3^{2-} \end{array} \quad ...(3.11)$$

In the absence of the enzyme, the carbon-bound hydrogen atom on 3-phosphoglyceraldehyde does not exchange with deuterium atoms of the solvent at any significant rate. (We are not concerned here with an exchange of the proton bound to the alcohol oxygen in the reactant or product. This proton does exchange rapidly with the solvent in either the presence or the absence of an enzyme, but the deuteron that is incorporated here can be removed immediately by putting the product back in H_2O.)

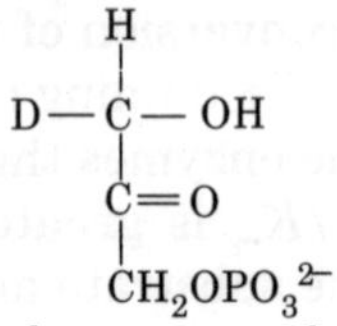

$$\begin{array}{c} H \\ | \\ D{-}C{-}OH \\ | \\ C{=}O \\ | \\ CH_2OPO_3^{2-} \end{array}$$

[1(R)-^{2}H]-dihydroxyacetone phosphate

(a)

$$\begin{array}{c} H\diagdown\;\diagup\!\!\!\!{}^{\,}O \\ C \\ | \\ D{-}C{-}OH \\ | \\ CH_2OPO_3^{2-} \end{array}$$

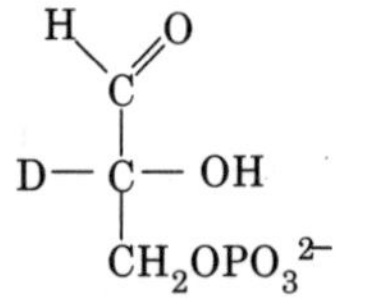

[2-^{2}H]-glyceraldehyde-3-phosphate

(b)

Fig. 3.17. The substrates of triosephosphate isomerase can be labeled specifically with deuterium. The rate of the isomerase reaction with [1(R)-^{2}H]-dihydroxyacetone phosphate (a) as substrate is slowed by about a factor of 3 relative to the rate with ordinary dihydroxyacetone phosphate. This isotope effect indicates that breaking the H—C bond to form the ene-diolate intermediate is the rate-limiting step in conversion of dihydroxyacetone phosphate to 3-phosphoglyceraldehyde. Little or no isotope effect is obtained when the reaction is run in the opposite direction starting with [2-^{2}H]-3-phosphoglyceraldehyde (*b*), indicating that the breakdown of the ene-diolate must be rate-limiting in this direction. Note that although the two hydrogen atoms on C–1 of dihydroxyacetone phosphate are chemically identical, they are stereochemically distinguishable. The enzyme always removes the one that has been replaced by deuterium in (a).

The incorporation of deuterium can be explained as follows: After a proton is transferred from C – 1 of the substrate to the carboxyl group of Glu 165, it has an opportunity to escape into the solution and to be replaced by a deuteron. A proton attached to a carboxylic acid oxygen usually exchanges rapidly with the solvent. The deuteron on Glu 165 then can be transferred to C–2 of the product. The hydrogen-bonding pattern in the crystal and studies of the pH dependence of the reaction suggest that the histidine and lysine residues in the active site do not act as general acids in the catalytic pathway but rather serve to stabilize the negative charge on an ene-diolate intermediate by electrostatic effects. Assuming that the mechanism shown in Fig. 3.21 is generally correct, how can we determine whether the formation or the breakdown of the ene-diolate intermediate is the rate-limiting step in the overall reaction? One approach is to study the kinetics using substrates that are specifically labeled with deuterium.

With [1(R)–^{2}H]-dihydroxyacetone phosphate k_{cat} is about three times smaller than the k_{cat} obtained with the ^{1}H analog. An isotope effect of this magnitude is consistent with the view that the bond holding the H or D atom to C-1 must be broken in the rate-limiting step. This is consistent with the view that the formation of the ene-diolate is rate-limiting. Similar measurements with [2-^{2}H]-3-phosphoglyceraldehyde show that there is little or no isotope effect on k_{cat} for the reverse direction, from 3-phosphoglyceraldehyde to dihydroxyacetone phosphate. In this direction, the rate-limiting step must be the breakdown of the ene-diolate intermediate. By measuring the kinetics of the isotopic exchange reactions catalyzed by the enzyme, along with the overall kinetics of the isomerization reactions in both directions, Jeremy Knowles and his colleagues obtained rate constants for the individual steps of the catalytic mechanism.

Fig. 3.18 shows the pathway in the form of a free energy profile. The free energy changes for binding of the reactant rar product to the enzyme are calculated on the basis or typical concentrations of dihydroxyacetone phosphate and 3-phosphoglyceraldehyde found in muscle cells (40 µM). At this low concentration of substrate, when the overall rate of the reaction is described by the specificity constant k_{cat}/K_m, the rate-limiting step appears to be the formation of the enzyme-substrate complex. This step occurs with a rate constant of about 10^8 $M^{-1}S^{-1}$ and evidently is set by the rate of diffusion of the enzyme and substrate. It can be slowed down by increasing the viscosity of the solution. Because the activation

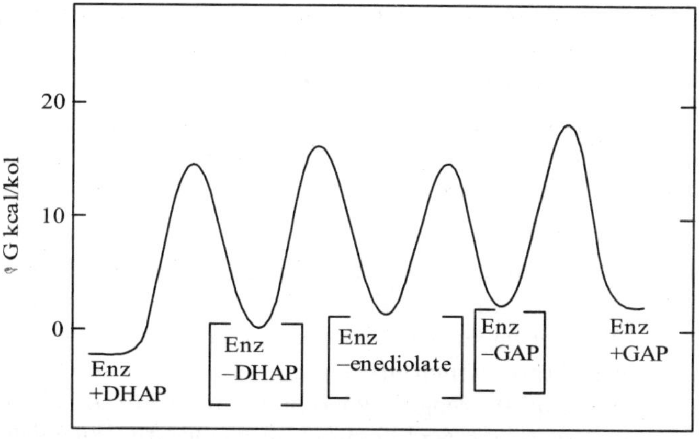

Fig. 3.18. The calculated free energy profile of the reaction catalyzed by triosephosphate isomerase. (Enz = enzyme; DHAP = dihydroxyacetone phosphate; GAP = glyceraldehyde-3-phosphate.) The free energy changes associated with binding of DHAP and GAP to the enzyme are calculated on the assumption that DHAP and GAP are present at concentrations of 40 µM.

free energies for the additional steps in the interconversion of the enzyme-substrate and enzyme–product complexes are low enough so that these steps do not limit the overall rate of the reaction, there is no evolutionary pressure to increase the rate constants for these steps any further.

Evolutionary changes that led to tighter binding of the substrate might actually be harmful, because lowering the free energy of the enzyme-substrate complex can increase the activation free energy for one or more of the steps in the conversion of this complex into the product. Given the fixed, low concentration of its substrate, triosephosphate isomeras appears to have reached perfection!

SUMMARY

Like ordinary chemical catalysts, enzymes interact with the reacting species in a manner that lowers the free energy of the transition state.

1. All known enzymatic reaction mechanisms depend on one or more of the following five themes:
   (*a*) proximity effects (enzymes hold the reactants close together in an appropriate orientation)
   (*b*) general-acid or general-base catalysis (acidic or basic groups of the enzyme donate or remove protons and often do first one and then the other)
   (*c*) electrostatic effects (charged or polar groups of the enzyme favor the redistribution of electric charges that must occur to convert the substrate into the transition state)
   (*d*) nucleophilic or electrophilic catalysis (nucleophilic or electrophilic functional groups of the enzyme or a cofactor react with complementary groups of the substrate to form covalently linked intermediates
   (*e*) structural flexibility (changes in the protein structure can increase the specificity of enzymatic reactions by ensuring that substrates bind or react in an obligatory order and by sequestering bound substrates in pockets that are protected from the solvent).
2. The serine proteases trypsin, chymotrypsin, and elastase are very similar in structure but have substrate-binding pockets that are tailored for different amino acid side chains. The active site of each enzyme contains three critical residues: serine, histidine, and aspartate, which are positioned so that the serine hydroxyl group becomes a strong nucleophilic reagent for reaction with the substrate's peptide carbon atom. This reaction generates an acyl-enzyme intermediate. Formation and hydrolysis of the acyl-enzyme intermediate probably are promoted by electrostatic interactions that stabilize tetrahedral transition states and by general-acid and general-base catalysis.
3. Ribonuclease A hydrolyzes RNA adjacent to pyrimidine bases. The reaction proceeds through a 2′,3′-phcsphate cyclic diester intermediate. Formation and breakdown of the cyclic diester appear to be promoted by concerted general-base and general-acid catalysis by two histidine residues, and by electrostatic interactions with two lysines. These reactions proceed through pentavalent phosphoryl intermediates. The geometry of these intermediates resembles the geometry of vanadate compounds that act as inhibitors of the enzyme.

Proteases

A landmark in the development of any field of study is the appearance of a sound system of nomenclature and classification for the objects with which it deals. The introduction of the Linnaean system for naming and classifying organisms in the eighteenth century and the invention of a system of nomenclature for enzymes in the 1950s were such key events, and their value has been obvious. Both nomenclature and classification are vitally important for information-handling, allowing people to communicate efficiently, knowing that they are talking about the same thing, and to store and retrieve information unambiguously.

A good system will also serve to highlight important questions, and thus prompt new discoveries. The nomenclature and classification become increasingly necessary as a field of study grows, and the field of proteolytic enzymes is clearly in a stage of rapid expansion. This is seen as the accelerating rate of discovery of new proteolytic enzymes and the increase in what is known about them. These factors undoubtedly increase the need for sound nomenclature and classification, but happily they can also help to provide a basis for it. The purpose of this chapter is first to reviewthe terminology that is used in relation to proteolytic enzymes, and then to look at two systems that exist for their classification, the EC system of the IUBMB and the MEROPS system of peptidase clans and families. We shall consider what these systems do and how they can help the student or research worker.

TERMINOLOGY AND NOMENCLATURE

Peptidase and related terms

The term *protease* appeared in the German literature of physiological chemistry in the latter part of the nineteenth century in reference to proteolytic enzymes, and was used as a general term embracing all the hydrolases that act on proteins, or further degrade the fragments of them. Around 1930, the need to distinguish separate kinds of protease activity began to be recognized and two sets of terms were independently proposed, one in Germany and one in the United States. In Germany, Grassmann and Dyckerhoff started to call the enzymes that act on proteins 'proteinases', and those that act preferentially on oligopeptides '*peptidases*'. The reason for the ability of some proteolytic enzymes to act preferentially on small peptides became much clearer when peptide substrates with and without terminal blocking groups could be synthesized.

It was found that the free end-groups were not required, and frequently were not tolerated, in the specificity sites of the proteinases, so that the enzymes acted well on long chains, away from the ends, but acted much less well on the oligopeptide products.

In contrast, the enzymes that acted well on oligopeptides generally required at least one free terminus close to the scissile bond, so that they had little action on intact proteins, in which exceedingly few of the peptide bonds are close to a terminus. Bergmann and Ross then used the term *peptidase* as a general term for peptide bond hydrolase, not in the restricted sense that had been proposed by Grassmann, and extended it to form the terms 'endopeptidase' and 'exopeptidase'. The use of the term peptidase in both the narrow and the broad senses has caused confusion, and I favour restricting it to the broad sense, as is recommended by IUBMB. For the narrower meaning of a peptide bond-hydrolysing enzyme that acts only or preferentially on 'peptide' substrates (i.e. oligopeptides, or small polypeptides), the best terms are exopeptidase if the specificity is a requirement for free terminal groups or oligopeptidase if the specificity is substrate size-dependent. We return to these and related terms, including aminopeptidase and carboxypeptidase which are defined in the EC system.

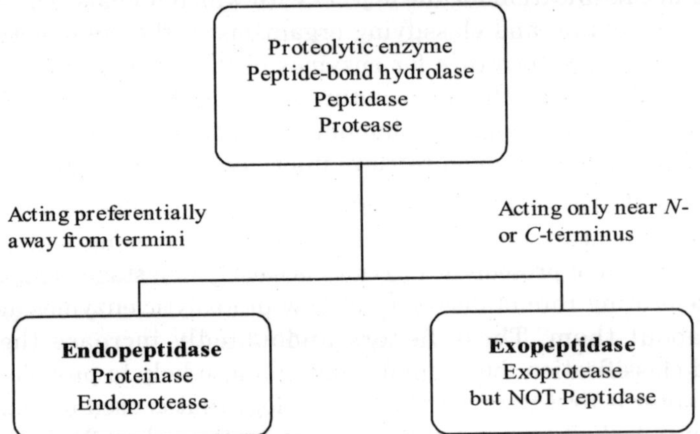

Fig. 4.1. The synonymous terms for some kinds of proteolytic enzymes. The terms within each box presently have identical meanings. The terms peptidase, endopeptidase and exopeptidase (printed in bold type) are those that are recommended by IUBMB and they have a clear logic. The other synonyms are not particularly troublesome except for peptidase when used as a synonym for exopeptidase, since this is the general term for proteolytic enzymes.

Specificity-subsite Terminology

Crystallographic structures show that the active site of a peptidase is commonly located in a groove on the surface of the molecule between adjacent structural domains, and the substrate specificity is dictated by the properties of binding sites arranged along the groove on one or both sides of the catalytic site that is responsible for hydrolysis of the peptide bond. Accordingly, the specificity of a peptidase is described by use of a conceptual model in which each specificity subsite is able to accommodate the side-chain of a single amino acid residue. The sites are numbered from the catalytic site, S_1, S_2...S_n towards the N-terminus of the substrate, and S'_1, S'_2 ...S_n' towards the C-terminus. The amino acids they accommodate are numbered P_1, P_2...P_n. and P_1', P_2'...P'_n, respectively, as follows :

$$\text{Substrate: } - P_3 - P_2 - P_1 + P_1' - P_2' - P_3' -$$
$$\text{Enzyme: } - S_3 - S_2 - S_1 * S_1' - S_2 - S_3' -$$

Here, the catalytic site of the enzyme is marked *, and the peptide bond that is cleaved (the scissile bond) is indicated by the symbol '+'. This system is based on that of Berger and Schechter, but for typographical convenience, the original system is modified in that the subsite numbers are not subscripted, and are followed rather than preceded by the prime characters, when needed.

Catalytic type

It was first pointed out by Hartley that proteolytic enzymes could be grouped according to the chemical nature of the catalytic site. Hartley recognized four distinct types of catalytic site, and minor refinements of the names for these in the light of subsequent discoveries have led to the present recognition of serine, cysteine, aspartic and metallo-types of peptidases. Most recently, the group of threonine peptidases has been discovered. These-groupings contribute to both current systems of peptidase nomenclature and classification, the EC system and the *MEROPS* system.

Homology

In discussing the evolutionary and structural relationships of peptidases, the concept of homology is important. As is the case with other proteins, a peptidase is correctly described as being homologous with another when the two amino acid sequences are so similar that the resemblance cannot be attributed to chance; this is most readily explained by a common evolutionary origin. The concept of homology is thus absolute, not relative. The effect of this is that one may correctly say that two proteins are homologous, or that they are not homologous, but it is not correct to say that two proteins are more homologous than two others.

THE EC CLASSIFICATION OF PEPTIDASES

What is the EC system?

The internationally recognized scheme for the classification and nomenclature of all enzymes, including peptidases, is that of the IUBMB. The Nomenclature Committee of the IUBMB is the successor to the Enzyme Commission, and the index numbers that it assigns to enzymes are still described as 'EC' numbers. The latest full publication of the EC list is in *Enzyme Nomenclature 1992,* but this has been updated by several supplements relating specifically to peptidases. In the EC system all enzymes are divided between six classes: (1) as Oxido-reductases. (2) Transferes. (3) Hydrolases, (4) Lyases, (5) Isomerases. (6) Ligases. The hydrolases of peptide bonds, the peptidases, form subclass 3.4. There is an important difference between the way in which the great majority of enzymes are treated in the EC list and the way in which peptidases are dealt with, however.

Throughout most of the EC list, an 'enzyme' is defined by the fact that it catalyses a single reaction. This has the important implications that (*i*) several quite different proteins are all described as the same enzyme, under a single EC number, if they catalyse the same reaction;

(*ii*) conversely, a single protein that catalyses more than one reaction is treated as more than one enzyme, with more than one EC number; and (*iii*) a satisfactory name for any enzyme (catalysing a single reaction) can be formed directly from the description of the reaction it catalyses. However, a decision has been taken to treat peptidases differently from the other enzymes. The are several reasons for this, (*i*) The reaction catalysed is not a satisfactory basis for the classification of most peptidases.

For all of the peptidases, the basic reaction is the same, hydrolysis of a peptide bond. The specificities differ, but for the endopeptidases, and for many of the exopeptidases, the specificity is extremely difficult to determine in any rigorous way and, if determined, is almost impossible to describe simply, or to use as the basis of a name, (*ii*) The value of the classification of peptidases by catalytic type was quickly recognized, and accordingly, this system was introduced into the EC list in 1972. (*iii*) It was felt that it would be useful to scientists working in the field of proteolytic enzymes to be free to regard different proteins as different peptidases, even when they express similar or identical peptidase activitles.

An example would be the distinction of pancreatic elastase and leukocyte elastase. It happens that these enzymes have slightly different specificities, but even if they did not, the genetic and biological differences between them would mean that they needed to be distinguished in biomedical science. The decision not to classify peptidases purely on the basis of the reactions they catalyse created the need for special systems for both classifying and naming them. The peptidases in the EC list are divided between 13 sub-subclasses. Two major sets of these sub-subclasses are recognized, comprising the exopeptidases and the endopeptidases. The *exopeptidases* act only near the ends of polypeptide chains. Those acting at a free N-terminus liberate a single amino acid residue (*aminopeptidases*, 3.4.11) or a dipeptide or a tripeptide (dipeptidyl-peptidases and tripeptidyl-peptidases). Those acting at a free C-terminus liberate a single residue *carboxypeptidases* or a dipeptide (peptidyl-dipeptidases.

The carboxypeptidases are allocated to three groups on the basis of catalytic groups: serine-type carboxypeptidases, metallocarboxypeptidases and cysteine-type carboxypeptidases. Other exopeptidases are specific for dipeptides (dipeptidases, or remove terminal residues that are substituted, cyclized or linked by isopeptide bonds {peptide linkages other than those of α–carboxyl to α-amino groups) (omega peptidases). The terms '*dipeptidyl-peptidase*' and '*peptidyl-dipeptidase*', for exopeptidases liberating dipeptides from N- or C-terminus, respectively, deserve a word of explanation.

Historically, names involving aminopeptidase and carboxypeptidase have sometimes been used for these peptidases, but it is important to maintain the distinction between the enzymes that release single amino acids, and those that release dipeptides. The systematic basis for the term dipeptidyl-peptidase is that the substrate is thought of as a dipeptide linked to the N-terminus of another peptide (*i.e.*, a dipeptidyl-peptide), and it is the linking peptide bond that is hydrolysed, so the enzyme is a dipeptidyl-peptide hydrolase or dipeptidylpeptidase (the hyphen indicating the scissile bond). Similar reasoning gives rise to the term peptidyldipeptidase.

Endopeptidases cleave internal bonds in polypeptide chains. In the EC list, the endopeptidases are divided into sub-subclasses on the basis of catalytic mechanism, and the specificity is used only to identify individual enzymes within the groups. The enzymes of sub-subclass (serine endopeptidases) have an active centre serine involved in the catalytic process, those of 3.4.22 (cysteine endopeptidases) have a cysteine in the active centre, those of 3.4.23

depend on an aspartic acid residue (or commonly two) for their catalytic activity (aspartic endopeptidases), and those of 3.4.24 use a metal ion (often, but not always, Zn^{2+}) in the catalytic mechanism (metalloendopeptidases). Recently, sub-subclass has been added for threonine peptidases.

The initial assignment of an endopeptidase to a catalytic type commonly depends upon the results of tests with inhibitors, as is described elsewhere in the present. Generally, the polypeptide chain hydrolysed by an endopeptidase may be of any length, so that proteins are among the substrates, but a subset of endo-peptidases is recognized that are restricted to acting on substrates smaller than proteins; these are termed 'oligopeptidases'. Among the first oligopeptidases to be clearly recognized as such was thimet oligopeptidase. This endopeptidase hydrolyses oligopeptides and polypeptides of up to 17 amino acids, but has no action on larger polypeptides or on proteins. Prior to the general awareness of the existence of the oligopeptidase subset of endopeptidases, a pitfall was repeatedly encountered in studies designed to identify endopeptidases specific for hydrolysis of particular peptide bonds in protein substrates, such as the 3-secretase cleaving the Alzheimer's amyloid precursor protein.

For experimental convenience, oligopeptide substrates were synthesized with sequences spanning the scissile bond of interest. These model substrates were then used in the search for the enzyme that cleaves the protein, but on several occasions it was actually an irrelevant oligopeptidase that was discovered. Oligopeptidases include prolyl oligopeptidase, oligopeptidase A, oligopeptidase B, pitrilysin, insulysin, neprilysin and others.

What Information does the EC List Contain?

The value of the EC list for peptidases lies in providing standard terminology for the various types of peptidase activity, and especially in the assignment of a unique identification number and a recommended name to each peptidase. The special strengths of the EC system are thus in the area of nomenclature rather than classification. The EC entry for a peptidase contains the 'EC number', the "Recommended name', a description of the 'Reaction', and a list of 'Other names' that have been used.

There are also 'Comments', providing a brief summary of what is known about the enzyme (including the peptidase family number in the *MEROPS* system, for all new entries), links to other databases, and up to five 'References', complete with titles and inclusive page numbers.

When and how can a Newly-Discovered Pepetidase be added to the EC list?

In view of the valuable role that the EC list has to play in the terminology of peptidases, it is important that newly discovered peptidases are added to the list as soon as they have been characterized in sufficient detail. The present section summarizes the requirements that normally apply, and describes the procedure by which a new entry in the EC list may be proposed.

A Distinctive Peptidase

Clearly, a peptidase must be shown to be distinct from any already in the EC list before a new entry can be created for it. In the EC list, as a whole, the requirement would be simply that the type of reaction or its specificity are unlike those of any enzyme already included.

These criteria are also sufficient for a peptidase, but it follows from what has been said above about the special handling of peptidases in the EC list that other factors also can justify the new entry. One such factor would be a difference in catalytic type, so that two peptidases that might have indistinguishable specificity would merit separate entries if they were of different catalytic types. Biological differences, such as encoding by separate genes in the same organism, or expression in different cell types, have also justified new entries. The over-riding consideration is that it is of practical usefulness to a significant number of scientists to be able to make unambiguous reference to the peptidase as a distinct entity.

Purification and Characterization

The properties of a peptidase cannot be determined unambiguously without the use of a homogeneous preparation of the enzyme. Methods for the purification of naturally occurring forms of the enzymes are constantly improving as are the methods for cloning and expression, which may lead to simpler purification. A proposal for addition of a peptidase to the EC list should normally include a reference to a refereed publication describing the isolation to homogeneity, commonly shown as a single zone of protein in gel electrophoresis or column chromatography which is associated with the enzymatic activity.

Evidence of Catalytic Type

In view of the importance that is placed on catalytic type in the EC classification of peptidases, it is expected, with rare exceptions, that the catalytic type will have been determined for any peptidase proposed for addition to the list. As was explained above, the catalytic types are the serine, cysteine, threonine, aspartic and metallo-peptidases, and the sub-subclasses of carboxypeptidases and endopeptidases depend directly upon catalytic type. This gives rise to just seven sub-subclasses of these enzymes, since there are no known aspartic-type carboxypeptidases (or indeed exopeptidases of any kind). The type of catalytic sitels usually revealed most clearly by the effects of class-specific inhibitors.

The use of these is described elsewhere but I should like to emphasize the usefulness of analysing any data showing partial inhibition by taking account of the effect of time on the degree of inhibition, in conjunction with the type of inhibition (irreversible or reversible) that is expected. For example, if using phenylmethane sulfonyl fluoride as a diagnostic inhibitor of many serine peptidases, one might note that since this is a slow-reacting, irreversible inhibitor, the expectation is that there will be time-dependent inhibition progression towards completion, with a rate that is first-order in inhibitor concentration. Partial inhibition that cannot be shown to have the expected time-dependence should be regarded with suspicion.

Conversely, a reversible inhibitor such as soybean trypsin inhibitor will generally give apparently instantaneous inhibition under practical conditions, and when inhibition is partial, it will most probably depend not on time, but on inhibitor concentration that is in accord with a K_i value. The point is simply that it is known for all the catalytic-type-specific inhibitors that are commonly used what kind of inhibition is to be expected, and any partial inhibition obtained should be examined for dependence upon time of incubation and concentration of inhibitor that is consistent with this. Deviations from expectation may indicate that the enzyme preparation is inhomogeneous. Many published sets of data clearly have not been examined in this way and may be suspeet. If the amino acid sequence of a peptidase is known and allows the peptidase to be placed in one of the structure-based families then it can safely be assumed that the catalytic type will be that already known for the family, and biochemical tests may be unnecessary.

Amino acid Sequence Data

It is now uncommon for a peptidase to be added to the EC list without its amino acid sequence being known. Although knowledge of the amino acid sequence is not a formal requirement it is clearly of great value, not least in allowing the compilers of the list to know that the enzyme being put forward really is distinct from all those already in the list. It usually allows the peptidase to be assigned to one of the known families and confirms its catalytic type. The number of the family to which a peptidase belongs is included in the Comments section of new entries to the EC list. If the protein has not been cloned and sequenced, but has been purified to homogeneity, then limited chemical microsequencing of N-terminus and internal fragments may well yield sufficient data to allow at least tentative assignment to a family.

A Satisfactory Name

One of the most important parts of the entry for a peptidase in the EC list is the Recommended name. Those compiling the list see it as part of their responsibility to help ensure that satisfactory names are available for peptidases, and it is not uncommon for the inclusion of a peptidase to be delayed because none of the names in use can be recommended. Even if a peptidase were not to be included in the EC list, it would be important that it had a satisfactory name. The literature contains many sound reports of peptidases described only by a term such as 'beef brain acid proteinase'; such a term is not easily remembered or stored in indexes, so the report tends to sink into obscurity, to the disadvantage of science and the scientist concerned.

It is therefore wise to put forward a distinctive name for any newly discovered peptidase at an early stage. If it should happen that the enzyme proves to be one for which a name already exists, that is not difficult to deal with later. Although a new name will initially sound strange, it is surprising how soon it becomes a natural and useful part of one's thinking and communication. Creating a name for a newly-discovered peptidase is seldom easy. For most enzymes a name can be constructed from the reaction catalysed, but that is valid only because those enzymes are defined purely by the reaction catalysed so that the enzyme name does not have to distinguish a unique protein. This is not the case for the names of peptidases.

Do not Mistake a Descriptive Term for a Name

The difficulty of inventing a good name has often been avoided by using a descriptive term in its place. The difference between a descriptive term and a true name is very important, but is not widely recognized. To take an example from the existing nomenclature of peptidases, the serine endopeptidase found in the acrosome of the sperm, for which the recommended name is acrosin, could equally well have been termed acrosomal neutral proteinase. I should describe acrosin as a true name, whereas acrosomal neutral proteinase is a descriptive phrase.

The name, while often implying some information about the enzyme, does not explicitly state it, and is therefore more robust in surviving further discoveries about the enzyme. If acrosin is found to occur elsewhere than in the acrosome, the name will still be valid, at least reflecting the history of its discovery, but the descriptive designation acrosomal neutral proteinase would become untenable: clearly one could not say that one was working on acrosomal neutral proteinase from the ovary! Similarly, if another neutral proteinase were discovered in the acrosome, it could create a problem for the descriptive term, but not for the name, since it would not be acrosin. Much the same considerations apply to naming on the basis of a source organism.

One should assume that any peptidase will be found to occur in more than one organism. Imagine that a cysteine endopeptidase has been isolated from a fruit fly. The discoverer might consider calling the enzyme either drosophilain or *Drosophila melanogaster* cysteine proteinase. If species variants of the same enzyme are found to exist in other flies, the name drosophilain would still be satisfactory (reflecting the history of discovery of the enzyme), but the descriptive phrase would not. True names are also more convenient than descriptive phrases, because they are shorter and can contain unique strings of characters that will be valuable in computer searches and indexing.

Do not Try to Use the Name of its Gene as the Name for a Peptidase

Gene names are species-specific and, since we assume that any peptidase is likely to occur in more than one species, it will be encoded by genes of more than one name. Thus, *kex2* protease would not be a good designation for the processing enzyme from *Saccharomyces cerevisiae,* because genes with other names encode the same enzyme, even in other yeasts. However, the name kexin works well.

Base the Name of an Exopeptidase on the Type of Reaction

A look at the existing EC list will show that the names of exopeptidases almost invariably have been based on the type of activity—aminopeptidase, dipeptidase, etc.—qualified by a specificity descriptor (*e.g.*, leucyl aminopeptidase) or a serial identifier (*e.g.*, carboxypeptidase A). This is generally proving satisfactory.

For an Endopeptidase, Consider Inventing a Novel, One-word

The majority of endopeptidases have one-word names, good examples of which would include trypsin, pepsin, acrosin and stromelysin. Biological similarities have sometimes led to the establishing of a series of names with a single root, such as the cathepsins and the granzymes. It is uncommon for the specificity of an endopeptidase to be so distinctive and so easily described as to form a satisfactory basis for a name. However, a term such 'glutamyl endopeptidase' has been found to need the addition of further qualifiers (*e.g.*, glutamyl endopeptidase II) as other enzymes with similar specificity are discovered.

For the Name of an Endopeptidase, try to use a Meaningful Ending

The existing names of many cysteine peptidases end in '-ain', and it makes sense to try to follow this convention. Similarly, the ending '-epsin' is good for aspartic endopeptidases and the ending 'lysin' works well for metallo-endopeptidases. Take care with the ending '-ase', though: this is recommended to be reserved for the construction of names from those of substrates (*e.g.* collagenase), but even then only with caution, since it may tend to suggest a greater degree of specificity, or a primary biological function, that the enzyme does not prove to show.

Avoid Using a Set of Capital Letters as a Name

Descriptive phrases are much easier to coin than real names, but apart from their other disadvantages described above, they are usually too long for regular use. As an attempt to get around this problem, they have sometimes been condensed to sets of initial capital letters. A string of capital letters does not make a good name, however, because it looks like an abbreviation and seems to need to be defined whenever it is used. Try changing it to lower case, and adding a few letters, as necessary, to make it into a novel and pronounceable name.

Do not Make the Name for your new Peptidase into an Advertisement

Occasionally, scientists have chosen to name a peptidase after their institute, for example. This may have pleased the chairman but generally it is a bad idea since it tends to delay acceptance by other scientists.

Some Other Points

It is important to invent a name that is easy to use, in speech, in writing and at a computer keyboard. A one-word name is usually best and, if the sequence of letters is unique, it will work well in a computer search. Avoid Greek letters and typographical complications such as the bar characters that are used in the notation of the complement system. It is important to be absolutely sure that the new name really is novel and not already in use, so an essential step is to search the literature databases for it. This will also show how easy or difficult such a search will be for others later. Even if the planned new name is not yet in the literature, it may be on its way.

Serial names (a root name with a letter or number, such as cathepsin K) are particularly likely to be duplicated. At any given time, many scientists feel that they know what the 'next' letter or number should be, but they have no way of knowing whether someone else is also adopting that same designation for their own new peptidase. When the publications come out it will be too late to avoid a muddle, so it is worth making enquiries and perhaps using a special interest news group on the Internet to announce the intention of using the name at an early stage.

Procedure for Proposing a New or Amended EC Entry

When the description of the isolation and specificity of the new peptidase has been accepted for publication in a refereed journal, and the sequence has been deposited, it is time to propose the EC entry. Proposals for inclusion of new entries in *Enzyme Nomenclature*, as well as for amendment of existing entries, should be sent to Prof. K. F. Tipton, Trinity College, Dublin.

CLASSIFICATION OF PEPTIDASES

As we have seen, the EC system places each of the peptidases in one of just 13 sub-subclasses, and only five of the sub-subclasses are available for the very numerous endopeptidases (comprising 204 of the total of 281 entries). This inevitably lumps together peptidases that are very different.

Most importantly, it takes no account of structural groupings of peptidases that would reflect evolutionary relationships. About 1992, N, D. Rawlings and the present author started to develop a new form of classification of peptidases that was designed to group them in a way that would reflect essential structural features and evolutionary relationships, taking advantage of the wealth of amino acid sequence data that was becoming available.

The beginnings of the system were described in 1993; it was then further developed as the organizational basis for collections of articles in *Methods in Enzymology* and in 1996 the system was published on the WWW as the *MEROPS* database. The name *MEROPS* has no special significance, being simply the name of a bird whose social behaviour prompted our adoption of the term 'clan', but it now forms a convenient label for reference to the system of peptidase families and clans wheter it be in the form of the World Wide Web database or printed versions. Cross-references to the *MEROPS* system will be found in the SWISS-PROT databse.

MEROPS is a three-layered system in which we allocate the peptidases to families and group these as far as possible into clans. The summary of the resulting classification is presented in Table 4.1. We shall now look at each of the three levels of *MEROPS* in a little more detail.

Table 4.1. The three levels of the *MEROPS* classification. The identifiers used for clans, families and peptidases in the *MEROPS* classification each start with a letter indicating the catalytic type of peptidases in the group. These letters are: A, aspartic; C, cysteine; M, metallo; S, serine; T, threonine, or U, unclassified; the letter P is also used, to identify two clans that contain peptidases of more than one of the types C, S and T. At the time of writing, no clan is yet recognised that contains peptidases only of type T or of type U. The identifier of a clan is completed with a second capital letter, assigned sequentially (e.g. CD), and the identifier for a family is completed with a number *(e.g.* C14). The *MEROPS* identifier for each individual peptidase starts with the family identifier and is completed with a decimal number *(e.g.* C14.001). All of the identifiers are stable, being changed only under very special circumstances, and never re-used

Level	*Description*	*Catalytic types recognized in forming identifiers*
Clan	A clan is a set of families in which all the peptidases have evolved from a single ancestor. Families in the same clan are best recognised by the similar protein folds of their peptidases.	A, C M, P, S
Family	A family contains peptidases that can be proved to be homologous by comparison of amino acid sequences.	A, C, M, S, T, U
Peptidase	An individual peptidase is distinguishable from all others by differences in activity, structure or genetics.	A, C, M, S, T, U

Families

The families of peptidases are constructed by comparisons of ammo acid sequences, with the use of strict statistical criteria such that there is no realistic possibility of two peptidases being placed in the same family when they are not truly homologous. Each family is built around a 'type example' peptidase, and every member of the family shows a statistically significant relationship in amino acid sequence to at least one other member of the family in the part of the molecule that is responsible for peptidase activity. The point that we require that the relationship be demonstrable in the catalytically active part of the molecule is an important one, necessitated by the existence of peptidases that are chimeric proteins.

For example, procollagen I N-endopeptidase is a chimeric protein that contains a catalytic domain related to that of astacin, but also contains segments that are clearly homologous with non-catalytic parts of the complement components C1r and C1s, which are in the chymotrypsin family the procollagen endopeptidase is placed in the family of astacin (M12), and not that of chymotrypsin (SI).

Table. 4.2. Clans and families of peptidases. The clans and families of peptidases, together with the type example one example from each family. The clan PA contains protein nucleophile peptidases of more than one catalytic type, previously assigned to clans SA and CB, as described in the text. The miscellaneous groups of families that cannot yet be assigned to clans are listed as if in clans SX, CX, AX, MX and UX for convenience.

Clan	*Family*	*Example*
(a) Serine, threonine and cysteine peptidases		
PA	S1	Trypsin
	S2	Streptogrisin A
	S3	Togavirin
	S6	IgA1-Specific serine endopeptidase
	S7	Flavivirin
	S29	Hepatitis C virus NS3 polyprotein peptidase
	S30	Potyvirus P1 proteinase
	S31	Pestivirus polyprotein peptidase *p*80
	S32	Equine arteritis virus serine endopeptidase
	S35	Apple stem grooving virus protease
	C3	Poliovirus picornain 3C
	C4	Tobacco etch virus Nla endopeptidase
	C24	Feline calicivirus endopeptidase
	C30	Mouse hepatitis coronavirus picornain 3C-like endopeptidase
	C37	Southampton virus processing peptidase
SB	S8	Subtilisin
SC	S9	Prolyl oligopeptidase
	S10	Carboxypeptidase C
	S15	X-Pro dipeptidyl-peptidase
	S28	Pro-X carboxypeptidase
	S33	Prolyl aminopeptidase
	S37	PS-10 peptidase *(Streptomyces lividans)*
SE	S11	D-Ala-D-Ala carboxypeptidase A
	S12	D-Ala-D-Ala carboxypeptidase B
	S13	D-Ala-D-Ala peptidase C

Clan	*Family*	*Example*
SF	S24	Represser LexA
	S26	Signal peptidase 1
	S41	Tail-specific protease
SH	S21	Assemblln
TA	T1	Proteasome
CA	C1	Papaln
	C2	Calpain
	C6	Tobacco etch virus HC-proteinase
	C7	Chestnut blight virus p29 endopeptidase
	C8	Chestnut blight virus p48 endopeptidase
	C9	Sindbis virus nsP2 endopeptidase
	C10	Streptopaln
	C12	Deubiqultinating peptidase Yuhl
	C16	Mouse hepatitis virus endopeptidase
	C19	Isopeptidase T
	C21	Turnip yellow mosaic virus endopeptidase
	C23	Blueberry scorch cariavirus endopeptidase
	C27	Rubella rubivirus endopeptidase
	C28	Foot-and-mouth disease virus L proteinase
	C29	Mouse hepatitis coronavirus papain-like endopeptidase 2
	C31	Porcine respiratory and reproductive syndrome arterivirus α-endopeptidase
	C32	Equine arteritis virus PCP α-endopeptidase
	C36	Beet necrotic yellow vein furovirus rapair-like endopeptidase
CD	C11	Clostripain
	C13	Legumain
	C14	Caspase-1
	C25	Gingipain R
`CE	C5	Adenovirus endopeptidase
CX	S14	Endopeptidase Clp
	S16	Endopeptidase La
	S18	Omptin
	S19	Chymotrypsirvlike protease (*Coccidioides*)
	S34	HflA protease
	S38	Chymotrypsinlike protease (*Treponema denticota*)
CX	C15	Pyroglutamyl peptidase I
	C26	-y-Glutamyl hydrolase
	C33	Equine arterivirus Nsp2 endopeptidase
	C40	Dipeptidyl-peptidase VI
	C41	Hepatitis E cysteine proteinase
Aspartic peptidases		
AA	A1	Pepsin
	A2	HIV 1 retropepsin
	A3	Cauliflower mosaic virus endopeptidase
	A9	Simian foamy virus potyprotein peptidase
	A10	*Schizosaccharomyces* retropepsin-like transposon
	A15	Rice tungro bacilliform virus protease
AB	A6	Nodavirus endopeptidase

Clan	***Family***	***Example***
AX	A4	Scytalidopepsin B
	A5	Thermopsin
	A7	Pseudomonapepsin
	A8	Signal peptidase II
	A11	*Drosophila* transposon copia peptidase
	A12	Maize transposon bsl peptidase
Metallopeptidases		
MA	M1	Membrane alanyl aminopeptidase
	M2	Peptidyl-peptidase A
	M4	Thermolysin
	M5	Mycolysin
	M6	Immune inhibitor A
	M7	*Streptomyces* small neutral endopeptidase
	M8	Leishmanolysin
	M9	*Vibrio* collagenase
	M10	Interstitial collagenase
	M11	Gametolysin
	M12	Astacin
	M13	Neprilysin
	M30	Hyicolysin
	M36	Fungalysin
MC	M14	Carboxypeptidase A
MD	M15	Zinc D-Ala-D-Ala Carboxypeptidase
ME	M16	Pitrilysin
MF	M17	Leucyl aminopeptidase
MG	M24	Methionyl aminopeptidase
MH	M18	Aminopeptidase I
	M20	Glutamate Carboxypeptidase
	M25	X-His dipeptidase
	M28	Aminopeptidase Y
	M40	Carboxypeptidase *(Sulfolobus sulfataricus)*
	M42	Glutamyl aminopeptidase *(Lactococcus)*
MX	M3	Thimet oligopeptidase
	M19	Membrane dipeptidase
	M22	O-Sialoglycoprotein endopeptidase
	M23	β-Lytic endopeptidase
	M26	IgA-speciflc metalloendopeptidase
	M27	Tentoxilysin
	M29	Aminopeptidase T
	M32	Carboxypeptidase Taq
	M34	Anthrax lethal factor
	M35	Penicillolysin
	M37	Lysostaphin
	M38	β-Aspartyl dipeptidase
	M41	FtsH endopeptidase

Clan	*Family*	*Example*
	M45	D-Ala-D-Ala dipeptidase *(Enterococcus)*
Unclassified peptidases		
UX	U3	Endopeptidase gpr
	U4	Sporulation sigma E factor processing peptidase
	U6	Murein endopeptidase MepA
	U7	Protease IV
	U9	Prohead proteinase (bacteriophage T4)
	U12	Prepilin type IV signal peptidase
	U26	vanY D-Ala-D-Ala Carboxypeptidase
	U27	ATP-dependent protease *(Lactococcus)*
	U28	Aspartyl dipeptidase
	U29	Cardiovirus endopeptidase 2A
	U32	Microbial collagenase *(Porphyromonas)*
	U39	Hepatitis C virus NS2-3 protease
	U40	Protein P5 murein endopeptidase (bacteriophage phi6)
	U43	Infectious pancreatic necrosis birnavirus endopeptidase

A further aspect of the way families are assembled in *MEROPS* that has sometimes given rise to confusion is the use of 'linking sequences'. The essential characteristic of peptidases within a single family is that they are evolutionarily related. If peptidase A is related to peptidase B, and peptidase B is related to peptidase C, then peptidase A must obviously be related to peptidase C, even if they have rather different structures. In these circumstances it is the existence of peptidase B that proves the relationship of peptidases A and C, and it can therefore be termed a 'linking sequence'. As more sequence information is obtained, we are discovering new linking sequences and from time to time these allow families to be merged.

The new, enlarged families may, however, contain individual peptidases that differ greatly from each other. At the time of writing, there are about 140 families in the *MEROPS* database. A few of the families contain two or more rather distinct groups of peptidases, as shown by a deep divergence in the dendrogram. For these, subfamilies are recognized. The naming of the families follows the system introduced by Rawlings and Barrett in which each family is named with a letter denoting the catalytic type (S, C, T, A, M or U, for serine, cysteine, threonine, aspartic, metallo- or unknown, respectively), followed by a sequentially assigned number.

The threonine-type peptidases are the most recently discovered catalytic type. When a family disappears, usually because it is merged with another, the family name is not reused. For this reason, there are interruptions in the numerical sequences of families that are of no current significance.

Clans

Although the families are the largest groupings of peptidases that can be proven rigorously to be homologous, there are persuasive lines of evidence that many of the families do share common ancestry with others. That is to say, there are sets of families in which all of the proteins have diverged from a single ancestral protein, but they have diverged so far that we can no longer prove their relation-ship by comparison of the primary structures. The term clan is used to describe such a group of families.

The best kind of evidence to support the formation of a clan is similarity in three-dimensional structures when the data are available, but the arrangement of catalytic residues in the polypeptide chains and limited similarities in amino acid sequence around the catalytic amino acids are also taken into account, so that the assignment of a family to a clan normally depends upon several lines of evidence. Sometimes, the assignment can be made with a high level of confidence but on other occasions it has to be considered provisional. The name of each clan is formed from the letter for the catalytic type of the peptidases it contains (as for families) followed by an arbitrary second capital letter.

As with families, if a clan disappears, the name is not reused. A clan that contains families of more than one of the C, S and T types is treated as being of 'P' type. An example is clan PA that contains trypsin and many other serine peptidases but also contains cysteine peptidases from RNA viruses (27) (see also Table 3.2). About 32 clans are recognized at the time of writing.

Individual Peptidases

The *MEROPS* system— in the form of version 2 on the WWW, and the databanks tables in the Handbook *of Proteolytic Enzymes*— includes data for all the peptidases that are recognized in the EC system and for which sequence data are available. A good many other peptidases are included that are not yet in the EC system. All of the available amino acid sequences of peptidases are included, after removal of 'redundant' items that are essentially identical. Each individual peptidase is given an identifier that acts much like an accession number. This is formed by concatenation of the three-character identifier of the family to which the peptidase belongs, a point, and a three-figure number.

Uses of the *MEROPS* System

The literature on peptidases contains many sweeping statements about the properties of peptidases of serine, cysteine or metallo-type that should perhaps be restricted to the enzymes of particular families. The 30 and more clans that can already be recognized indicate that there are at least this number of quite distinct lines of peptidase that have had different evolutionary origins and have very different structures and chemistry. These merit separate consideration, and the classification in *MEROPS* provides for this. The existence of distinct 'families' of proteins is well recognized, but the exact meaning of a family has had no single, clear definition, so the concept has been a vague one.

Within the *MEROPS* system there is a rigorous definition of family that is applied across the hundreds of known peptidases and, most importantly, it provides a way of dealing with the problem of mosaic proteins by restricting the comparison of amino acid sequences to the part of each enzyme that is responsible for its catalytic activity. Until now, there has been no unambiguous way to name families of peptidases, and a family has generally been referred to by the name of one of its members.

For example, one might speak of the 'prolyl oligopeptidase family' or the 'dipeptidyl-peptidase IV family, it happens that both of these would be references to the same family, S9 in the *MEROPS* system, since both peptidases named are in this family. One of the useful features of *MEROPS* is therefore that it provides an identifier that can be used as a concise and unambiguous way to refer to any family or clan of peptidases. As the *MEROPS* family names have been increasingly introduced into the records of the sequence databases and the EC list, they have become effective ways of searching for information about the family as a whole. The WWW version of the *MEROPS* database forms a convenient mode of access to a great deal of the information about peptidases that is available on the WWW.

Caspase-1

MEROPS *classification*			
Clan: CD	Family: C14	Subfamily:	MEROPS ID: C14.001
IUBMB classification			
3.4.22 Cysteine endopeptidases		EC Number: EC 3.4.22.36	
Nomenclature			
Recommended Name	Caspase-1		
Other names	Interleukin 1-beta-converting enzyme		
Other information			
For a review, see *Handbook of Proteolytic Enzymes* chapter 248			
Catalytic type	Cysteine		
Links:	Functional relevance	3D Image	

Human Genetics			
Gene names	*Chromosome location*	*GDB*	*OMIM*
1L 1BC; 1L 1BCE; CASP 1	11q22.2-q22.3		147678

Protein Sequence Data		
Species (and isoform)	*Swiss Prof*	*PIR*
Homo sapiens	P29466	A42677
Mus musculus	P29452	A46495
Rattus norvegicus	P43527	I53300

Nucleic Acid Sequence Data				
	Comment	*EMBL*	*Gen Bank*	*CDS*
Gallus gallus				
	cDNA	AF031351	AF031351	G2642241
Homo sapiens				
	cDNA	M87507	M87507	

Tertiary Structure Data					
	Comment	*Resolution*	*PDB*	*DSSP*	*RASMOl*
Homo sapiens					
	mature peptidase	2.6A	1ICE	1ICE	1ICE

Fig. 4.2. One View of the *MEROPS* database as it appears on the World Wide Web. This view of the output for an individual peptidase is just a sample of what is contained in the database. The figure shows only the format of the data for caspase 1: many sequence identifiers have been omitted to save space. The clickable links to other parts of *MEROPS* and to other database can be seen.

At the family level, there is access to the SCOP, MIPS, HSSP and PROSITE databases. These provide sequence alignments, secondary structure predictions, dendrograms, and conserved sequence motifs. *MEROPS* itself contains family alignments, family trees and search facilities. At the level of individual peptidases there is access to the EC list. At the level of sequences there is access to the annotated amino acid and nucleic acid sequence data of SWISS-PROT, PIR and EMBL, as well as the data on human genetics in OMIM. Data for three-dimensional structures are available from the Brookhaven Protein Databank (PDB), and *MEROPS* provides for viewing and manipulation of the three dimensional images by use of RADMOL (29). Bibliographic data are accessible through SWISS-PROT and EMBI.

DISCOVERING A NEW PEPTIDASE

Clearly, the systems of nomenclature (EC list) and classification (*MEROPS*) for peptidases can work well only if they are kept up to date. It is therefore important that newly discovered peptidases are included when the time is right. I whould therefore urge anyone who feels that they have discovered a new peptidase to consider putting it forward for an EC entry, as described above. Once the sequence has been deposited in the databanks, it should find its way into the *MEROPS* system automatically. However, the mass of data is such that this can take longer than it should, and the curator of the WWW version of *MEROPS* (see below) would be happy to be informed of new peptidases that should be included.

Inhibition of Proteolytic Enzymes

The two most frequent questions asked by those not working on proteases every day are: 'How do I prevent proteolysis in my particular system?' and 'How do I find out which protease is cutting my protein?' There are no 'bullet-proof, recipes to help with these questions, possibly because of the many varied or even overlapping specificities and catalytic strategies adopted by proteases. Nevertheless, there are a number of general issues and specific techniques that we and others use to address these and other related questions. This usually involves employing natural and synthetic protease inhibitors in complex biological environments such as tissue fluids or cells in culture. In keeping with the scope of this book, we concentrate primarily on synthetic inhibitors in the characterization and prevention of proteolysis.

With a few notable exceptions, the synthetic inhibitors generally afford more practical uses in experimental biology. Inhibitors of proteases come in many different forms and are often grouped on the basis of their reaction mechanism, origin or structural similarity. We divide them here into specificity groups on the assumption that readers will be more interested in how to inhibit given proteases than how to classify inhibitors. Three classes are designated: firstly, those that react with more than one class of protease; secondly, those that are ostensibly specific for one of the classes; and finally, those that show high selectivity for a single protease.

IRREVERSIBLE AND REVERSIBLE INHIBITORS

General Structure of Synthetic Inhibitors

Proteases bind their substrates in a cleft on the enzyme surface, with the scissile bond positioned over the catalytic machinery, and with side–chains of the substrate-filling specific pockets in the cleft. Most design strategies for synthetic protease inhibitors have used this property to attach a group ('warhead') reactive with components of the catalytic machinery to a short stretch of peptide chain, as shown in Fig. 5.1 The peptide side-chains (P) deliver a certain degree of specificity, and the N-blocking group is frequently added to prevent digestion by aminopeptidases and increase binding since many endopeptidases will not accept a free a-amine in their specificity pockets. Usually, the peptide is designed to fit the unprimed (S) protease subsites, but sometimes it fits the primed (S′) ones, as in the case of several effective hydroxamate peptidyl inhibitors of metalloproteases.

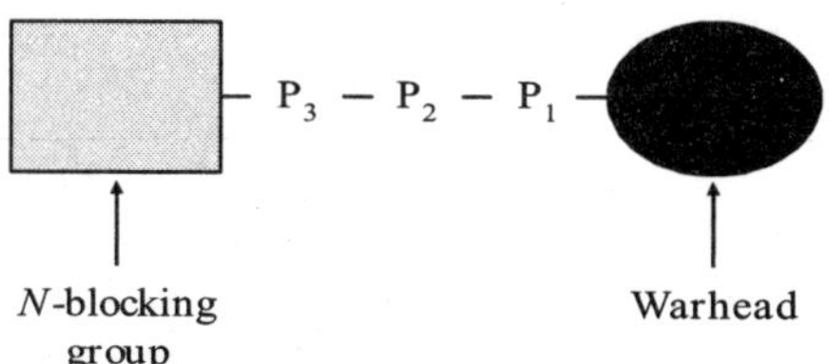

Fig. 5.1. General structure of a tripeptidyl protease inhibitor. Sometimes a single residue (P_1) is sufficient for directing inhibition, but extending the peptide chain satisfies additional extended subsite preferences and imparts selectivity to the inhibitor. The 'warhead' may be one of a number of molities that are reactive with specific protease catalytic components.

How to Live with Crossreactivity

A new inhibitor is usually intended to be directed against a specific target enzyme; it is then tested against a number of other related and unrelated proteases to gauge the specificity and selectivity of the compound. The inhibitors in this chapter have been invaluable in characterizing proteases, but ultimately there is often a later study that demonstrates unexpected crossreactivity with other proteases, and sometimes even other non-proteolytic proteins. Crossreactivity is particularly noticeable between closely related proteases, but occasionally occurs even between distinct catalytic classes of serine, cysteine and threonine (proteasome).

Members of these classes tend to stabilize the transition state by similar mechanisms, and so transition-state inhibitors such as aldehydes can inhibit all three of these classes. 'All contain powerful catalytic nucleophiles that participate in covalent catalysis, so good electrophilic warheads such as chloromethyl ketones can demonstrate cross-mechanism reactivity. These issues do not pose a problem if all that is needed is to abolish general proteolytic activity, but if specificity is desirable they must be considered.

Use of a single inhibitor to demonstrate a single protease involvement in a biological process can, therefore, lead to misinterpretations. Happily, there are solutions to this problem. So, whenever using synthetic inhibitors to characterize or discover proteases or biological functions, it is always best to consider crossreactivity and use a concerted approach of several of the inhibitors available.

Most inhibitors in this chapter are commercially available, and those that are not can sometimes be obtained from the original source. Surveys of the literature and experience have led the authors to recommend the procedures listed here and, where space permits, reasons are given for particular choices made. Where it was felt necessary to point out errors resulting from popular (but often inappropriate) methods, we have tried to replace these with more valid procedures.

INHIBITION CONSTANTS

The importance of reaction kinetics as a basis for using inhibitors to investigate proteolysis. If one wishes to suppress the activity of a given protease and knows the kinetic constants for the reaction of this protease with an inhibitor, one can determine how much inhibitor to add and how much time to allow for inhibition. Although more complex reaction pathways can be written for all protease-inhibitor interactions, for most practical applications the following simplifications apply, considered in the generic case of a two-step irreversible inhibitor with the following definitive kinetic quantities. If the inhibitor is reversible, one can simply truncate the mechanism by omitting the second step that is, by omitting the rate governing the formation of the EI* complex (k_2).

Reaction pathways $$E + I \underset{k_{-1}}{\overset{k_1}{\rightleftharpoons}} EI \xrightarrow{k_2} EI^*$$

Definitions $$k_i = \frac{k_{-1}}{k_1} \quad k_{ass} = \frac{k_2}{K_i}$$

Fig. 5.2. Protease inhibitor reaction pathway and definitions. K_i and k_{ass} are constant for a particular enzyme-inhibitor reaction under specified environmental conditions. To make relevant comparisons of enzymes or inhibitors, and derive useful predictions, relationships between the constants and the total inhibitor concentration are used as described in the text.

Irreversible Inhibitors

If an inhibitor operates by an essentially irreversible mechanism, both the rate of complex formation (the second order reaction rate, k_2/K_i) and the equilibrium concentration of the first step (K_i) are important in principle. For most practical purposes k_{ass} takes into account both steps in the reaction and so the magnitude of the K_i step can be ignored. This is because most good inhibitors have a K_i well below the common working concentration of the inhibitor, and the first step in the reaction pathway (see above) is therefore mostly saturated. Examples of such inhibitors are α_1-protease inhibitor, TICK, PMSF, and Z-VAD-FMK, which form covalent bonds with the enzyme that often remain when the enzyme is denatured.

The most important quantity to consider when using irreversible protease inhibitors is the time required for the free enzyme concentration to decrease by 50% (half life, $t_{1/2}$). During the reaction of a protease with an irreversible inhibitor is then given by:

$$t_{1/2} = \ln 2/k_{ass}[I] \qquad ...(5.1)$$

One may use this relationship to compare the inactivation of an enzyme by a variety of inhibitors. It is used later in this chapter to compare the efficacy of different inhibitors for a variety of proteases. The relationship, as discussed by Bieth, can be very powerful in revealing the potential role of protease inhibitors in physiological processes, and is widely applicable. It is also extremely useful in deciding how long a synthetic inhibitor should be allowed to incubate with a protease solution to achieve adequate inhibition. The time required for an irreversible inhibitor to diminish the activity of a protease by 99.9% (the delay time) is 10 half-lives of the reaction.

Reversible Inhibitors

An equilibrium between the complexed form of an enzyme and free enzyme characterizes reversible inhibition. Sometimes, a reversible inhibitor may bind so tightly that it is not possible to measure a binding constant. Under this condition the inhibitor is sometimes classified as 'pseudo-irreversible' and it is then appropriate to analyse it as an irreversible inhibitor.

For freely reversible inhibitors, provided that [I] > [E] × 10, the fraction of free enzyme existing in equilibrium with a reversible inhibitor is given by the relationship.

$$[EI]/[E] = [I] \qquad ...(5.2)$$

If [I] > lOOOK$_i$. and if [I] > 10[E], then less than 0.1% of the total enzyme is free at any time. Only when [I] approaches K_i should one anticipate significant substrate hydrolysis. Therefore the ratio [I]/K_i describes the efficiency of a reversible inhibitor.

The concentration of a reversible inhibitor required to diminish the activity of a protease by 99.9% (the stability concentration) is $K_i \times 1000$. Thus the stability concentration is an extremely useful quantity to use when adequate inhibition of a protease by an reversible inhibitor is required. The effect of substrate on the delay time or stability concentration is not considered, as this is irrelevant to comparisons of inhibitor efficacy in vitro and in vivo.

NON-SPECIFIC INHIBITORS

α-Macroglobulins

The most noteworthy example of an inhibitor that fails to discriminate between protease classes is the human protein α_2M, a member of a group of high molecular weight proteins known collectively as the α-macroglobulins that bind and inhibit most endopeptidases. The inhibitory capacity of α_2M results from a conformational change following endopeptidase reaction that appears physically to entrap the reacting molecule(s). Thus, α_2M is mechanistically different from the active site-directed protein inhibitors of proteases. Binding of a protease to α_2M is irreversible; however, the complexed enzyme retains substantial activity against small peptides, but is almost completely inactive against larger ones (proteins larger that 25 kDa).

This unusual reactivity is a result of steric hindrance placed on the protease by enclosure within the large α_2M molecule.

Frequently, inhibited endopeptidases may also become covalently attached via displacement of an internal α_2M thiol-ester. The covalent attachment phenomenon can be extremely useful in determining whether a protein of interest is an endopeptidase. The first step in the reaction of an endopeptidase with α_2M is cleavage of the 'bait region', a stretch of about 20 residues that is the most sensitive and non-specific of all known protein substrates.

After bait region cleavage, the protease is frequently covalently attacned to α_2M, in a reaction that depends partly on the lysine content of the protease. Thus, if the protein of interest (purified or recombinant) is a protease, there is a very good chance that it will become covalently linked when added to α_2M. This can be monitored by a 90-200 kDa shift in the apparent size of the catalytic chain of the putative protease upon SDS-PAGE. This reaction will only be seen with endopeptidases since exopeptidases do not cleave the bait region. Not all endopeptidases will cleave the bait region, so absence of a gel shift does not mean the candidate is not an active endopeptidase. However, an obvious gel shift is very characteristic and will almost certainly mean endopeptidase activity.

Caution must be exercised to ensure that the samples for SDS-PAGE are boiled for 5 min in 1% SDS and at least 10 mM DTT to overcome artefactual, non-specific binding. The lack of specificity or selectivity of α_2M usually argues against its use in most instances, since it binds almost all endopeptidases. Moreover, because proteases in complex with α_2M may retain significant activity against protein substrates, especially disordered ones such as casein, the use of α_2M as an additive to prevent proteolysis in sensitive samples is discouraged.

Peptide Aldehydes

Peptide aldehydes, such as leupeptin and antipain, are di-, tri- or tetra-peptides containing an aldehyde moiety in place of the usual α-carboxylate. They are reversible inhibitors and are often referred to as transition-state analogues because they act by mimicking the tetrahedral intermediate formed during peptide bond hydrolysis. Leupeptin and antipain in particular are sometimes injudiciously used to classify the catalytic mechanism of a newly identified protease. Such use is inadvisable as these react equally well with serine and cysteine proteases; for example, leupeptin reacts equally well with trypsin. and cathepsin B, and antipain equally well with trypsin and calpain I.

Similarly, some aldehydes that are used to inhibit calpain also inhibit the proteasome, for example Ac-Leu-Leu-nVal-CHO. Many peptide aldehyde derivatives have been synthesized in attempts to improve specificity and, largely because of their low toxicity, they are used frequently in animal and cell culture models of proteolysis. However, two major drawbacks limit their usefulness as selective or specific inhibitors. Firstly, the lack of selectivity allows for the inhibition of proteases other than the one(s) under investigation. This is especially true in cell culture and animal experiments.

Secondly, in aqueous solution most aldehydes exist primarily as inactive hydrates, and in the case of arginals as cyclic carbinolamines. Accordingly, the rate-limiting step for inhibition can be formation of the aldehyde rather than encounter with a suitable target enzyme.

Peptide Chloromethyl Ketones

These irreversible inhibitors represent one of the first systematic approaches in the study of active-site-directed protease-modifying reagents. The two best known members of the group are TLCK and TPCK. The mechanism of inactivation of serine proteases has been studied extensively, and it is clear that inhibition results from the protease binding the inhibitor in a substrate-like manner followed by alkylation of the active-site histidine by the chloromethyl moiety.

Chloromethyl ketones also inactivate cysteine proteases and the mechanism of inactivation. is thought to be similar to that for serine proteases, although this remains unproven. The design of selective inhibitors has reached a high art in these compounds. However, given the poor specificity of some of them we caution against their use in analysing catalytic mechanisms. Nevertheless, peptide chloromethyl ketones are considered by many to be invaluable in preventing proteolysis in well-defined systems. Their use in cell culture and animal studies is limited by toxicity resulting from alkylation of cellular components and interaction with intracellular glutathione.

Metal Chelators

Metalloproteases contain a zinc ion that participates in catalysis by polarizing a water molecule to attack the substrate-peptide bond. This differentiates them from proteases such as the calpains and some trypsin-like serine proteases, whose activities are stabilized by, but not necessarily dependent on, the presence of calcium. Both the zinc-dependent metalloproteases and some calcium stabilized proteases from other classes can be inactivated by chelating agents such as EDTA and EGTA.

Care should therefore be exercised when using these to characterize new proteases. 1.10-Phenanthroline is preferred as an inactivator of metalloproteases, as it has a much higher stability constant for zinc (2.5×10^6/M) than for calcium (3.2/M). Thus, 1 mM 1,10-phenanthroline will inactivate a metallo-protease even in the presence of 10 mM calcium, whereas this concentration of 1,10-phenanthroline will not remove calcium from a calcium-binding protein. Therefore, this chelator is usually diagnostic for a zinc metalloprotease. The related, non-chelating compound 1,7-phenanthroline is generally used as a control to rule out non-specific effects of the relatively hydrophobic heterocyclic frame of the chelator. Confirmation of a protease as metallo-enzyme is accomplished by removal of zinc. This is followed by titration with zinc (0.1-100 µM) to restore full activity).

Removal of zinc, followed by re-addition, if desired, can also be used to quantify the contribution of metalloproteases to proteolysis in a complex mixture. High concentrations of zinc (in the mM range) often inhibit metalloproteases. Inhibition by zinc results from the formation of zinc monohydroxide which bridges the catalytic zinc ion to a side-chain in the active site of the enzyme. This inhibition is therefore competitive with substrate. The non-competitive inhibition by other heavy metal ions is attributed to binding of the ion to a site distinct from the active.

CLASS-SPECIFIC INHIBITORS

Serine Proteases

Organophosphates

Diisopropylphosphofluoridate (DFP), historically one of the most widely used protease inhibitors, is a member of the organophosphorus compounds that irreversibly inactivate serine proteases. DFP is a toxic compound by virtue of its inactivation of acetylcholine esterase, and has lost favour as a general inhibitor of serine proteases. On the whole, DFP and its relatives react less rapidly with trypsin-like serine proteases than with chymotrypsin-like ones.

The toxicity of DFP, its relatively low reaction rates and its instability in aqueous solutions limit its practical use as a general inhibitor of serine proteases. Radioactive derivatives of DFP are however, of considerable use as active site labels for serine proteases and are therefore suitable for autoradiography.

Sulphonyl Fluorides

Phenylmethylsulphonyl fluoride (PMSF), another well-known irreversible inhibitor of serine proteases, reacts relatively slowly with common proteases although it also reacts reversibly with some cysteine proteases, so care should be exercised in interpreting results. Attempts to design faster-acting, more selective derivatives have met with some success, but the derivatives tend not to discriminate very well between closely related proteases. Thus, 2-(benzyl-oxycarbonylaminoethyl carbonyl)-benzene sulphonyl fluoride reacts about 100-fold faster with chymotrypsin-like enzymes than does PMSF, but more slowly with elastases.

Similarly, APMSF reacts much more rapidly with trypsin-like proteases than does PMSF, but very slowly with chymotrypsin. These data point to the potential use of sulphonyl fluorides as inhibitors of general types of serine proteases. PMSF, the least selective and therefore most useful as a general inhibitor of serine proteases, decays relatively rapidly in aqueous solutions ($t_{½}$ 25°C, pH 7.5, 55 min) and this must be taken into account when using this particular inhibitor. Spontaneous hydrolysis in aqueous solutions is a characteristic of the sulphonyl fluorides.

A recently popular alternative to PMSF is AEBSF (frequently known by its trade name Pefabloc), which shares a broad range of serine protease targets, but is reportedly more soluble and stable in aqueous solutions. AEBSF also is reported to inhibit NADH oxidase activation through a non-proteolytic route, so clearly some non-protease targets must be expected. In all instances, one should assume that the sulphonyl fluorides are unstable in aqueous solutions, and repeated additions are required for each fractionation step during protein purification.

Coumarins

Research into the design of inhibitors of serine proteases based on coumarin and related heterocyclic compounds has been reviewed before. Interest in the design of heterocyclic protease inhibitors takes advantage of their capacity to act as mechanism-based inactivators, and the more pragmatic fact that most successful drugs are heterocyclics. Considerable current synthetic inhibitor research is geared towards the production of potentially therapeutic compounds to inhibit proteolysis in vivo, and the coumarins show promise in this area: Only one coumarin-based inhibitor 3,4-DCI, is commercially available and this compound shows good reactivity with a large number of serine proteases.

Table 5.1. Relative efficacies of general serine protease inhibitors.

	Half-life (s) for Inhibition at 1 mM Inhibitor				
	Trypsin	*Chymotrypsin*	*Neutrophll elastase*	*Cathepsin G*	*S. aureus protases*
DFP	43	2.8	—	—	>300
PMSF	237	2.8	33	49	Nl
3,4-DCI	3.5	1.2	0.08	25	0.25

[a]Data is expressed as half-time for inhibition, therefore a lower number means faster inhibition. Note that 3, 4-DCl is by far the most effective inhibitor for all proteases listed.

[b]*Staphylococcus aureus* serine protease is sometimes called *S. aureus* V8 protease and is frequently used for protein digestion. Its only other known inhibitor is α_2M (J. Potempa, personal communication).

[c]Ni, not inhibited.

It is relatively non-toxic, does not react at appreciable rates with acetylcholine esterase, papain, leucine aminopeptidase or β-lactamase and is faster-reacting with serine proteases than DFP or PMSF. It does, however, inactivate glycogen phosphorylase *b*, so clearly some non-protease targets must be expected. 3,4-DCI is not diagnostic for serine proteases since it also inhibits caspase-1 (J. Powers, personal communication) and the proteasome. Nevertheless, its use as a general inhibitor of serine proteases is strongly recommended. Although kinetically irreversible, activity of some proteases acylated by 3,4-DCI can be recovered following treatment with hydroxylamine, a property that could be of use in mechanism studies of protease-inhibitor interactions.

Peptide Boronic Acids

The extended boronic acid peptides shown in Table 5.2 are among the most potent reversible protease inhibitors yet developed. The boronic acids are similar to the peptide aldehydes in their apparent function as transition state analogues, but they appear to be much more selective for serine proteases. Relatively few studies on design of peptide boronic acids for protease selectivity have been reported but it is clear from the current data that these derivatives have a future. The compounds show little toxicity at millimolar concentration in cell-culture experiments, although it is not known whether they penetrate cells.

Protein Inhibitors of Serine Proteases

The protein inhibitors of serine proteases fall into several groups according to various schemes. The avian ovomucoids, the Kunitz-type trypsin inhibitors (such as aprotinin) and soybean trypsin inhibitor all form very tight, but reversible, complexes with a number of proteases. The high-molecular-weight inhibitors of serine proteases from the plasma of mammals, the serpins (serine protease inhibitors), form such tight complexes that reversibility is difficult to demonstrate. In most instances, the protein inhibitors show little selectivity for proteases in vitro, although they probably show effective selectivity in *vivo*.

The most extensively studied group of protein inhibitors of proteases, the avian ovomucoids, demonstrate hypervariability of the amino acid residues that are in contact with proteases and this has probably arisen from attempts to fit the specificity of, as yet undefined, proteases during the evolution of avian species. The serpins, although less well conserved in primary structure than the ovomucoids, also show hypervariability of their presumed reactivesite loops. Given these points, it is surprising that the members of these two groups have not achieved higher degrees of selectivity, although this may be due to the relative invariability of extended substrate binding sites of proteases, or because our undemanding of the requirements for protease selectivity are far from complete.

Attempts to alter selectivity of inhibitors using site-directed mutation and recombinant DNA strategies have met with partial success, but it has seldom been possible to achieve the hoped for degree of selectivity. Indeed, these approaches have unfortunately raised more questions about inhibitory selectivity than they have answered. This may change as our understanding improves.

Inhibitor Selectivity for Serine Proteases

The main goal of inhibitor design is to maximize selectivity for closely related proteases. Thus, it should be possible to suppress the activity of a single protease in a mixture and determine the result of this on any given biological phenomenon. Such a selective inhibitor should have a K_i for the protease of 1000-fold less than for any other protease in the mixture, or a k_{ass} of 1000-fold greater than for any other protease. This ideal has been very difficult to achieve and one can only speculate about the types of inhibitor that will provide the required degree of selectivity.

Valuable lessons have, however, been learned from the chloromethyl ketones. Most of the work on chloromethyl ketones has focused on exploring selectivity for plasma serine proteases, and on discriminating between the proteases from neutrophils. Examples of these successes are shown in Table 5.2. Because chloromethyl ketones react well with cysteine proteases, one should be wary of inferring too much regarding the role of a protease in a very complicated mixture (such as a cell or tissue extract or culture medium) from inhibitory data such. It is also likely that other, as yet unidentifid proteases reacting well with supposedly selective inhibitors are at work.

Nevertheless, with the judicious manipulation of experimental conditions and use of cysteine protease inhibitors such as E-64 (for papain family proteases) or fluoromethyl ketones and acyloxymethyl ketones (see below), one can track down the likely causes of specific proteolysis. Despite the selectivity engineered into chloromethyl ketones, it may transpire that the peptide boronic acids will ultimately be of greater use in cell culture experiments. They are generally deemed less toxic than chloromethyl ketones.

Further, the principle of using a reversible inhibitor in a relatively longterm experiment is sounder than that of using an irreversible one. This is because an irreversible inhibitor will eventually inhibit proteases that it reacts with very slowly, whereas a reversible one will never decrease the activity of an untargeted protease when that inhibitor is at a concentration above K_i for the reaction. For example, for the inhibitors shown in table 5.2 10 μM Glu-Phe-Arg-CH_2Cl will inhibit 99.9% of free plasmin in 30 s and 99.9% of free t-PA in just under 3 h. This is an excellent degree of selectivity, but consider that at 20 μM MeO-Suc-Ala-Ala-Pro-Boro-Phe, cathepsin G will be 99.9% inhibited but elastase will never be more than 25% inhibited, even after an incubation lasting several hours. These points should be taken into account when choosing inhibitors for long-term experiments.

Table 5.2. Selectivity in serine protease inhibiors

	Half-life (s) for Inhibition at 10 μM Inhibitor				
	t-PA	*uPA*	*Plasmin*	*Neutrophil elastase*	*Cathepsin G*
D-Val-Gly-Arg-CH_2Cl	13	140	260	–	–
Glu-Gly-Arg-CH_2Cl	172	9	653	–	–
Glu-Phe-Arg-CH_2Cl	1000	3800	3	–	–
Z-Gly-Leu-Phe-CH_2Cl	–	–	–	N.I.	1360
MeO-Suc-Ala-Pro-Val-CH_2Cl	–	–	–	44	N.I.
K_f(nm)					
Meo-Suc-Ala-Ala-Pro-Boro-Val				0.57	359
Meo-Suc-Ala-Ala-Pro-Boro-Phe				74000	359

Cysteine Proteases

Popular class-specific inhibitors of cysteine proteases fall into five categories: the synthetic peptide diazomethanes, the peptide epoxides, the fluoromethyl ketones, the acyloxymethyl ketones and the natural protein inhibitors known as the cystatins.

Peptide Diazomethanes

These are a group of oligopeptide derivatives in which the OH of the terminal carboxylate is replaced with a diazomethyl moiety, CH-N = N. Inhibition is irreversible and is thought to proceed, via a thiohemiketal intermediate, to the formation of an alkylated active site cysteine existing as a thio-ether to the methyl group of the inhibitor. These inhibitors are almost completely specific for cysteine proteases, although one report indicates the Z-Phe-Arg-CHN_2 can also inhibit the serine protease plasma kallikrein.

Although most strategies have focused on inhibition of endopeptidases, the compound Gly-Phe-CHN_2 is a useful inhibitor of dipeptidyl peptidase I, the activator of several white blood cell associated serine proteases.

Peptide Epoxides

The peptide epoxides are a group of irreversible inhibitors based on compound E-64 isolated by Hanada *et al.* from an extract of *Aspergillus japonicus*. Several derivatives of E-64 have been synthesized in attempts to explore this group as class-specific inhibitors of cysteine proteases. Inhibition of papain family proteases by E-64 and its relatives results from occupancy of the enzyme subsites followed by alkylation of the catalytic cysteine by the trans epoxide group. Interestingly, most structures show the inhibitor peptide backbone binding the unprimed subsites in the opposite direction to substrate, with the exception of CA074 which binds in the same direction as substrate, but occupies the primes subsites. The inhibition of several papain family proteases by E-64 and its relatives is quantified in Tables 5.3. The degree of selectivity is minimal, with the exception of CA074 , which shows remarkable specificity for cathepsin B compared with other members of the family. The compounds are relatively nontoxic and appear to be completely specific for papain family proteases under commonly used conditions.

Table 5.3. Selectivity in papain family inhibitors

	Half-life (s) for Inhibition at 10 μM Inhibitor				
	Papain	*Cathepain B*	*Cathepsin H*	*Cathepsin L*	*Calpain*
E-64(L)	0.1	0.8	17	0.7	9.2
Ep-479	0.08	0.2	33	0.5	14
Ep-460	0.15	0.4	89	0.4	3
CA074	>5000	0.6	>5000	>5000	–
Z-Phe-Phe-CHN_2	23	374	–	0.4	3
Z-Phe-Ala-CHN_2	2	57	–	1.0	–
Z-Phe-Tyr(*O*-but)-CHN_2	–	7000	–	0.3	–
Z-(Iodo-Tyr)-Ala-CHN_2	–	2.5	–	0.06	>7000
Iodoacetic acid	30	156	593	–	–
K_i(nm)					
Human cystain A	0.019	8.2	0.31	1.3	NI
Human cystain B	0.13	73	0.58	0.23	NI
Human cystain C	0.005	0.25	0.28	0.005	NI
Human Kininogen	0.015	600	1.2	0.017	1.0
Chicken ovocystatin	0.005	1.7	0.06	0.019	NI

At very high concentration (> 0.1 mM) E-64 may inhibit trypsin or the cysteine protease R-gingipain, although binding is weak and reversible (J. Potempa, personal communication).

Since their rate of reaction with protein or low molecular weight thiols other than the active cysteine of a protease is negligible, peptide epoxides are excellent inhibitors in the presence of reducing agents used to activate cysteine proteases, this is an additional advantage over sulphydryl alkylating agents such as iodoacetate.

Fluoromethyl Ketones

The fluoromethyl ketones (CH_2F, frequently abbreviated FMK) are useful specific cysteine protease inhibitors, and should not be confused with serine protease specific trifluoromethyl ketones (CF_3). Originally Z-Phe-Ala-CH_2F was used to target the lysosomal cysteine protease cathepsin B. CH_2F compounds have the advantage of being less reactive with serine proteases and far more stable in the presence of thiols than CH_2Cl equivalent compounds. They are cardiotoxic, limiting their usefulness *in vivo,* but experiments in cell culture and *in vitro* have been extremely informative.

Table 5.4. Selectivity in caspase inhibitors

	Half-life (s) for Inhibition at 1 μM Inhibitor								
Inhibitor	*Casp-1*	*Casp-2*	*Casp-3*	*Caap-4*	*Casp-5*	*Casp-6*	*Casp-7*	*Casp-8*	*Casp-9*
Z-VAD-FMK	**2.5**	2400	43	130	5.3	98	39	2.5	3.9
K_i(nM)									
Ac-DEVI>CHO	18	1710	0.23	132	205	31	1.6	0.92	60
AC-IETD-CHO	<6	9400	195	400	223	5.6	3280	1.05	108
Ac-YVAD-CHO	0.76	Nl	Nl	362	163	Nl	Nl	352	970

The good specificity of CH_2F compounds has been exploited particularly well with another family of cysteine proteases, completely unrelated to the cathepsins, namely the caspases. The caspase-specific inhibitor Z-Val-Ala-Asp-CH_2F has been used in many studies to establish the role of these enzymes in apoptosis. Unfortunately, this compound does not discriminate well between caspases, and, as demonstrated in Table 5.4 extreme caution must be exercised in interpreting results because of the non-discriminatory nature of this frequently used compound.

Acyloxymethyl Ketones

The AOMK unit (–$COCH_2OCOR$), containing a weak leaving group, functions as an irreversible affinity label. Inhibitors with this unit are remarkably selective for cysteine proteases. Extensive work on a number of leaving groups within the frame Z-Phe-Ala-CH_2OCOR demonstrate that R = 2.6–$(CF_3)_2$-Ph provides an extremely powerful and rapid cathepsin B inhibitor, primarily because this unit is such a good leaving group. As with the fluoromethyl ketones (see above) the AOMKs react with members of the papain family and the caspases, depending on specificity dictates in the substrate subsites, and so do not distinguish between these separate families. The general order of effectiveness is similar to the FMKs, and from a practical viewpoint there is presently little to choose between them.

Cystatins

The cystatin superfamily of cysteine protease inhibitors has been divided into three families based on primary amino acid sequence homology and domain structure; the cystatins are tight-binding reversible inhibitors of papain family cysteine proteases and are not known to inhibit other families of cysteine proteases. The structure of cystatin B in complex with papain reveals three non-contiguous loops of the inhibitor, creating a three dimensional wedge that blocks the substrate cleft of the enzyme, and provide a pleasing explanation of why cystatins do not inhibit other protease families.

Inhibitor Selectivity for Cysteine Proteases

Neither the cystatins nor the peptide epoxides show a significant degree of selectivity for individual proteases with the notable exception of the cathepsin B-inhibitor CA074, and the possible exception of human plasma kininogen. This protein contains three cystatin-like tandem repeats of which the third one, which can be released by limited digestion with trypsin, has a very high affinity for cathepsin L and a low affinity for cathepsin B but does not inhibit calpains. Papain famlly proteases tend to lack a well-defined primary specificity.

This has hampered attempts to design inhibitors with selectivity for proteases, with the result that random approaches have proved more effective than directed ones. Several of these proteases seem to have a favourable interaction with hydrophobic amino acids in the P_2 position, a finding that aided Shaw and Co-workers in their design of peptide diazomethanes. The most selective diazomethane inhibitors favour cathepsin L. The problem of differentiating the activities of cathepsin B and L in cell culture may be overcome with the aid of compounds such as Z-(iodo-Tyr)-Ala-CHN_2, which shows promising selectivity for cathepsin L. Occupancy of the S_1 pocket of caspases by Asp is definitive of this family and in stark contrast to the papain family.

This is exploited by using peptides containing Asp adjacent to the warhead, and the currently most useful reagent is Z-Val-Ala-Asp-FMK, which inhibits most caspases. The Asp-OMe derivative is reportedly cell-penetrating, and has been used in a large number of studies to ablate caspase activity in cell culture. However, caution should be exercised in attributing caspase activity solely on the basis of the use of Z-Val-Ala-Asp-FMK, since this compound can inhibit the proteasome, at least in vitro. Perhaps more useful from the objective of distinguishing between caspases are some of the aldehydes described in Table 5.4.

Here, it is fairly easy to distinguish the pro-inflammatory caspase 1 from the apoptotic ones by using Ac-Tyr-Val-Ala-Asp-CHO since there is over a 100-fold difference in K_i. However, the frequently used Ac-Asp-Glu-Val-Asp-CHO only demonstrates less than 10-fold selectivity between caspases 3, 7, and 8, and is therefore of dubious value in dissecting apoptotic pathways. Presumably the same would be true for outer warheads placed on the -Asp-Glu-Val-Asp- frame.

Proteasome

The proteasome presents a particular problem in inhibitor design owing to its ability to react with several warheads, including aldehydes, chloromethyl ketones, isocoumarins and even diazomethanes (52) which were originally designed for other proteases. Many investigators use tripeptide aldehydes such as Ac-Leu-Leu-nLeu-CHO or Ac-Leu-Leu-Val-CHO in cell culture to demonstrate an involvement of the proteasome, however, these may crossreact with calpain or lysosomal papain family members. Therefore, the use of lactacystin is recommended. This reagent, a microbialnatural product has become widely used. Lactacystin binds some, but not all, of the catalytic subunits of the 20S proteasome and ablates proteolysis.

The mechanism is complex, and varies with individual catalytic subunits, but effectively terminates all proteolytic activity. It is a cell-penetrant compound and is currently thought to be proteasome-specific. Another inhibitor warhead based on peptidyl vinyl sulphones has recently shown promise in proteasome investigations. Although also reactive with members of the papain family of cysteine proteases biotinylated tripeptide or tetrapeptide vinyl sulphones may be used as affinity labels for individual proteasome subunits.

Metalloproteases

Peptidic Inhibitors

The design of class-specific inhibitors of metalloproteases has focused on attempts to chelate the catalytic zinc atom. Synthetic inhibitors, therefore, commonly contain a chelating moiety, such as a carboxyl, a thiol, a phosphorous or a hydroxamic acid group, to which is attached a series of other groups designated to fit the specificity pocket of a particular metalloproteases. Examples are shown in Table 5.5. Some naturally occurring low molecular weight inhibitors such as actinomin, which is an aminopeptidase M inhibitor found in actinomycetes, also contain a hydroxamate moiety. Many hydroxamate inhibitors originally targeted rowards matrix metallo-proteases (MMPs or matrixins) have been designed.

Although selective inhibitors towards particular MMPs have been successfully generated many also inhibit metalloproteases other than MMPs. For example, tumour necrosis factor α-converting enzyme (a member of the ADAM/reprolysin family) is inhibited by hydroxamate inhibitors originally designed for MMPs. Useful distinctions between MMPs and other family members of metalloproteases can be made by the use of TIMPs, which tend to be specific for MMPs.

Protein Inhibitors

The most widely used protein inhibitors of metalloproteases are the TIMPs of which four (TIMPs 1-4) have been described. TIMPs inhibit MMPs, but not metalloproteases belonging to other families and are therefore useful reagents to distinguish MMPs from other families. TIMPs and MMPs form a tight 1:1 complex with subnanomolar K_i values. The crystal structure of the complex of TIMP-1 and the catalytic domain of MMP-3 revealed that the α-aminoacyl group of Cys1 bidentately chelates the catalytic zinc and the side-chain of Thr2 extends into the Large S_1' pocket of MMP-3.

Because of their slow dissociation rates, they are suitable for active site titration of MMPs. However, TIMP-1 fails to inhibit the membrane-bound MTl-MMP (MMP-14) whereas TIMP-2 and TIMP-3 inhibit MT1-MMPr. Other protein inhibitors of metalloproteases are not so widely used as TIMPs, but originate from organisms throughout the biotic kingdoms. For example, oprin is a single-chain glycoprotein of 52 kDa isolated from opossum, which inhibits snake venom metalloprotease atrolysin AB (formally known as Ht-b) from *Crotalus atrox.* The serum of Japanese Habu snake Trimerasurus *flavoviridis* contains a 323-amino acid glycoprotein that inhibits several metalloproteases in Habu venom.

The molecule contains two copies of a cystatin-like domain but it does not inhibit cysteine proteases. A 12-kDa inhibitor from *Streptomyces nigrescens* TK-23 inhibits thermolysin and *Pseudomanas aeruginosa* metalloproteases. Microbial proteinaceous inhibitors of metalloproteases are also found in *Envitria crysanthemi,* in *Serratia marscescens* and in P. *aeruginosa.*

Table 5.5. Potentially useful synthetic metalloprotease inhibitors

Compound	*Target enzyme*	K_i*(nM)*	IC_{50}*(nM)*
(I) Carboxylates			
N-[1-(S) carboxy-3-phenylpropyl]-Leu-Trp	Thermolysin		300
N-[1-(S) carboxy-3-phenylpropyl]-Ala-Pro	ACE	0,05	
(enalapril)			
(II) Thiois			
D-HS-CH_2CH(Me)CO-L-Pro	ACE	1.3	
(captopril)			
HS-CH_2-CH(CH_2ph)-CO-Gly	ACE	140	
(Thiorphan)	Neprilysin	4	
Ac-Trp-Leu-SH	MMP-1		2000
(III) Phosphorus-containing Inhibitors			
Rhamnosyl-PO_2-Leu-Trp	Neprilysin	3	
(Phosphoramidon)	Thermolysin	28	
	Pseudolysin	250	
$C_6H_5CHCH_2$-PO_2-Ala-Pro	ACE	0.5	
Phthaloyl-N-$(CH_2)_4$-PO_2-lle-(β-naphthyl)-Ala-NH-CH_3	MMP-3	7	
Pro-L-Phe (PO_2CH_2)Gly-Pro	Neurolysin	4	
	Thimet oligopeptidase	8100	
(iv) Hydroxamates			
HONHCOCH_2CH(iBu)CO-Trp-NHMe	MMMP1	0.4	
(GM6001)	Thermolysin	20	
	Pseudolysin	200	
Batimastat™	MMP-1	3	
	MMP-2	4	
	MMP-3	20	
	MMP-9	1	
	Atrolysin C	6	
Actinonin	Meprin A	135	
	Astacin	130000	

Unlike its counterpart in mammals, the ovomacroglobulin from chicken eggs (ovostatin) inhibits only metalloproteases well. Ovostatins from other species show broader reactivity with proteases.

Aspartic Proteases

The best characterized aspartic proteases from mammals (pepsin, chymosin, cathepsin D and renin) are all inhibited by pepstatin A, a pentapeptide-like compound secreted by *Streptomyces* species, which contains two residues of the unusual amino acid statine [(3S,4S)-4-amino-3-hydroxy-6-methylheptanoic acid]. K_i values range from 4.5×10^{-11}M for pepsin to 5×10^{-7}M for renin, and the original pepstatin structure has been modified in many ways to increase selectivity for different aspartic proteases.

Renin is an extremely selective enzyme (its only known protein substrate is angiotensinogen) and the other mammalian aspartic proteases encountered during protein isolation are inactive above pH 6. This class of enzymes is unlikely to cause problems during protein isolations. A tremendous amount of work has gone into designing potent, selective, and bioavailable inhibitors of HIV protease, and the topic will not be covered in this chapter.

INHIBITORS AS ACTIVE-SITE TITRANTS

Often, one must determine the concentration of active enzyme in a crude, or pure but partially inactive, preparation of a protease. This is most accurately accomplished using specific 'burst titrants' (substrates that are turned over only once). However, these titrants are available for only certain serine proteases and require relatively large quantities of protein (about 1-10 μM enzyme per assay). If material is precious, or burst titrants are unavailable, then titrations using protease inhibitors may be employed.

To successfully determine the active protease concentration one should use an active-site-directed inhibitor that: (*i*) has defined purity, (*ii*) is relatively stable in aqueous solution, (*iii*) shows no turnover during inhibition, (*iv*) reacts quickly and tightly enough to allow essentially complete inhibition in a short time. The most reliable inhibitors for performing active-site titrations are, in principle, irreversible ones. Not all appropriate inhibitor titrants react sufficiently quickly to be useful in determining small concentrations of proteases.

Affinity titrants can only rarely be used in crude samples because other proteins may sequester the titrant to give an erroneously high estimate of the target protease concentration. For crude samples, consider using a radioactive or biotinylated inhibitor affinity label followed by SDS-PAGE and quantification of bound label.

Cysteine Proteases

Given their broad specificity for papain family members and fast-reacting mechanism the *trans*-epoxysuccinyl peptides are almost ideal active-site titrants. Compound E-64 is used to titrate these cysteine proteases primarily because of its ready availability. The procedure, taken from Barrett *et. al.* is applicable to the lysosomal proteases cathepsins B, H, L and S and the plant proteases papain, chymopapain, facin and bromelain, but not clostripain, it is carried out as described in Protocol 2. Caspases are not inhibited by E-64, and the authors recommend determining the active concentration of this family by using Z-Val-Ala-Asp-CH_2F, which is the broadest currently available caspase inhibitor.

Serine Proteases

The general serine protease-selective inhibitors DFP, PMSF and 3,4–DCI are too unstable in aqueous solutions to be used as active-site titrants. The protein inhibitors can, however, be useful for this purpose. A preparation of inhibitor is standardized using trypsin with known activity, by using the burst-titrant procedure of Chase and Shaw. Trypsin acts as the primary standard upon which other inhibitors may be standardized for secondary titrations. The standardized inhibitor is then used to titrate activity of proteases known to react well with it.

PROTEASE INHIBITORS IN CELL CULTURE

Many of the forgoing procedures can be adapted from purified systems to cell culture. The key is to use inhibitors that are cell-penetrating and stable to culture conditions. Often, these inhibitors are methylated to block acidic moieties and improve cell penetration and, whenever possible, it is preferable to use the methylated derivative. The methylated compounds are frequently inactive *in vitro,* and it is assumed that cellular esterases demethylate them to the active inhibitor.

They are thus technically considered pro-drugs. Success has been reported with most types of proteases, and some examples are included below. Since the extent of cell penetrance by these compounds is not known, it is best to conduct several experiments at different inhibitor concentrations to determine the minimal realistic concentration required to achieve a measurable result.

Table 5.6. Inhibitors for cell culture

Proteases targeted	*Inhibitor*	*Comments*
Papain family	E-64D (methylated derivative of E-64)	Broadly reactive with members of the papain family
	CA074 metriyl ester	Selective cathepsin B pro-inhibitor
	Gly-Phe-CHN_2	Selective for dipeptidyl peptidases over endopeptidases
Caspase family	Z-Val-Ala-(OMe)Asp-FMK	Non-selective, inhibits apoptosis and IL1β processing
Proteasome	Lactacystin	Reputedly, very selective
Calpains	Ac-Leu-Leu-nLeu-CHO	Possible cross-reactivity
	Ac-Leu-Leu-Met-CHO	with proteasome
Serine proteases	CH_2Cl derivatives	Depending on sequence, can be selective, but watch for cross-reaction with cysteine proteases and proteasome
Metallo-proteases	Batimastat	Broadly reactive with MMPs Also inhibits ADAMS

SUPPRESSION OF PROTEOLYSIS

There are many instances in which proteolysis has been identified as a cause of artefacts. The problem usually arises from the disruption of cells or tissues during analytical or purification procedures. Certain procedures are highly prone to adventitious proteolysis, including the isolation or identification of structural proteins and intracellular protein precursors following cell lysis, preparations of immunoprecipitates (IgG preparations are often contaminated by plasma kalli-krein) from cell extracts and preparation of protein samples for SDS-PAGE.

Denatured proteins are much more sensitive to proteolysis than native ones, and several proteases are active in 1% SDS. Classically, biologists have included protease inhibitors to overcome some of these problems, and it is probably a good idea to incorporate this as part of a standard procedure when proteolytic artefacts are suspected. Therefore, the compounds in Table 5.7 are recommended as general protease inhibitors for suppressing.

Table 5.7. Inhibitors for prevention of adventitious proteolysis

	*Inhibitor solution**		
Compound	*Stock solution*	*Working concentration*	*Catalytic class*
EDTA	0.5 M in H_20, pH 7	5.0 mM	Metallo-
1,10-Phenanthroline	0.1 M in DMSO	1.0 mM	Metallo-
3,4-DCI	5.0 mM in DMSO	0.1 mM	Serine/threonine
PMSF	0.1 M in dry PrOH	1.0 mM	Serine
E-64	1.0 mM in H_2O	0.02 mM	Cysteine
lodoacetic acid	1.0 M in H_20 (fresh)	10 mM	Cysteine

EDTA can be omitted if the use of a calcium chelator presents a problem. 3,4-DCI is a much faster inhibitor of serine proteases than PMSF, but not as widely tested, lodoacetic acid modifies cysteine residues on many other proteins. Therefore, these three compounds are used only if failure of the following cocktail is indicated.

General inhibitor cocktail solution from mixture of above stocks:

10 mM 1,10-phenanthroline

2 mM 3,4-DCI

0.4 mM E-64

Make up in H_2O (except for 1,10-phenanthroline) and use within 1 h. Dilute 20-fold into sample. Make up 1,10-phenanthroline in DMSO, as it is not soluble In aqueous solvents above 1 mM.

If samples are prepared from animal cell lysates, caspases may be a problem. Since E-64 does not inhibit caspases, lysates are treated with 20 μM (final concentration) of Z-Val-Ala-Asp-FMK (unmethylated derivative). This only applies when caspases are suspected to be a problem, and when iodoacetic acid cannot be used.

If checking for proteolytic artefacts arising from SDS-reducing gel electrophoresis, first add reducing agent in presence of cocktail and sample. Incubate 10 min, then add SOS. This will eliminate activation of cysteine proteases during simultaneous addition of SDS and reducing agent.

Boehringer supply a general inhibitor reagent which is a mixture of some different compounds. Although we have not compared its effectiveness with the above reagents, it should also prove acceptable when used according to the manufacturers recommendations.

They are mostly irreversible general inhibitors of low molecular weight that can readily be removed without regeneration of proteolytic activity. Occasionally, protein inhibitors such as aprotinin or soybean trypsin inhibitor may be added, but these are far less broad in their specificity than 3,4-DCI or PMSF. Some soybean inhibitor preparations are contaminated with metalloprotease activity, and the addition of the protein inhibitors is discouraged unless one wishes to inhibit a specific protease with which they are known to react well.

Benewishes (an inhibitor of trypsin-like serine proteases) is not recommended as its K_i values (0.01-1 mM) with various enzymes require it to be used at 10-1000 mM to be an effective inhibitor. The peptide aldehydes, such as leupeptin or antipain, should be used in the 1-100 mM range for complete inhibition but, since they are reversible inhibitors, proteolytic activity may reappear upon dilution or dialysis. Aspartic proteases do not present a problem as they are inactive at neutral pH, with the exception of renin which, however, is not a problem because of its highly restricted specificity. Their zymogens can, however, be stable at this pH and can be activated by a decrease in pH to below 5. If a procedure calls for decreasing the pH, and if aspartic proteases may be present, one should include 10 ng/ml pepstatin.

THERAPEUTIC VALUE OF PROTEASE INHIBITORS

A major incentive in inhibitor research is that control of proteolysis is a valid pharmacological principle. Based on successes (and failures) of inhibitor therapy, Schnebli and Braun noted common properties of pharmacologically effective protease inhibitors. First, the target enzyme for therapy must be precisely known. Second, the target enzyme must be in a compartment (plasma, lung lumen or peritoneum) easily accessible by common drug delivery routes. Third, the interaction between inhibitor and protease must be either irreversible or have $K_i < 10^{-10}$M in order to maintain inhibitory activity following administration

Inhibitors have indeed proved useful in controlling pathogenesis in many animal models of proteolysis, and their application to human counterparts of these models shows promise. For example, captopril, a potent inhibitor of angiotensin I-converting enzyme is in therapeutic use for hypertension. A recent advancement is the design of a meta-substituted benzofused macrocyclic lactam that inhibits both ACE and neprilysin with IC_{50} values of 4nM and 8nM, respectively. The design of such dual inhibitors is considered to be a potent antihypertensives resulting from an increase of the circulating levels of atrial natriuretic peptide due to the inhibition of neprilysin and a decrease of angiotensin II due to the inhibition of ACE.

Clinical trials with several synthetic inhibitors of coagulation and fibrinolytic enzymes, and those for MMPs in cancer metastasis and arthritis, are in progress. Successes with AIDS therapy by using inhibitors of HIV protease has led to searches for therapy of other viral diseases by selective targeting of viral processing enzymes. Significantly, the success with HIV protease inhibitors indicates that intracellular compartments are no longer a practical barrier to inhibitor therapy.

Specific Enzymes

We have already considered reactions by which living cells are able to transfer groups from one molecule to another using nucleophilic displacements. We also showed how transfer reactions can be utilized by ligases to join two molecules together with the Gibbs energy of cleavage of ATP or of a related molecule driving the reaction. In this chapter we will examine addition reactions, which provide a simple way of joining two molecules by means of C–O, C–N, C–S, or C–C bonds. Among these are the aldol and Claisentype condensations by which C–C bonds are formed. We will also consider elimination reactions and decarboxylations, which are the reverse of addition and condensation reactions, as well as mechanistically related isomerizations. Many reactions of these types occur in the major pathways of metabolism.

ADDITION OF R–OH, R–NH, AND R–SH

Next to nucleophilic displacement, the commonest mechanistic processes in enzymatic catalysis are addition to double bonds and elimination to form double bonds. These often involve addition of a nucleophile together with a proton to a highly polarized double bond such as C = O or C = N–. In other reactions, which are discussed in Section C,2, the nucleophile attacks one end of a C = C bond that is polarized by conjugation with C = O or C=N. Alcohols, amines, and thiols add readily to the electrophilic carbon of the carbonyl group to form *hemiacetals, carbinolamines, hemiketals, and hemimercaptals.* An example is the formation of ring structures of sugars. Water can also add to carbonyl groups and most aliphatic carbonyl compounds exist in water as an equilibrium mixture with a covalent hydrate (Eq. 6.1).

For example, acetaldehyde in aqueous solution consists of a mixture of about 50% free aldehyde and 50% hydrate in rapidly reversible equilibrium and formaldehyde is over 99.9% hydrated.

$$CH_3-C(=O)H + H_2O \rightleftharpoons CH_3-C(OH)(OH)H \qquad \text{....(6.1)}$$

Addition reactions often occur as parts of more complex reactions. For example, a thiol group of glyceraldehyde-3-phosphate dehydrogenase reacts with the aldehyde substrate to form a hemimercaptal, which is subsequently oxidized to a thioester.

Carbonic Anhydrase

Another simple addition reaction is the hydration of CO_2 to form the bicarbonate ion. Without catalysis the reaction may require several seconds, the apparent first-order rate constant being ~0.03 s^{-1} at 25°C. Cells must often hasten the process. The specific catalyst carbonic anhydrase is widespread in its distribution and is especially active in tissues (*e.g.*,

$$O{=}C{=}O \xrightarrow{H_2O} HO{-}C({=}O){-}OH \xrightarrow{-H^+} HO{-}C({=}O){-}O^- \quad ...(6.2)$$

red blood cells, and lungs) that are involved in respiration. One liter of mammalian blood contains 1–2 g of this enzyme, a monomeric 30–kDa protein containing –260 amino acids and one tightly bound ion of Zn^{2+}. Erythrocytes contain two isoenzymes (I and II) of carbonic anhydrase and the human body contains at least eight distinct isoenzymes (I – VIII). They are found wherever there is a high demand for CO_2 or bicarbonate. Isoenzyme I, II, III, and VII are cytosolic. Carbonic anhydrase I is specific to erythrocytes, while isoenzyme II is present in most cells. Hereditary lack of carbonic anhydrase II is associated with *osteopetrosis* (marble bone disease), a condition involving failure of bone resorption and the calcification of other tissues.

The generation of acidity according to Eq. 6.2 is presumably required for dissolution of bone by the osteoclasts. Isoenzymes III and VII have a more specialized distribution. Carbonic anhydrase III is abundant in adipocytes which use bicarbonate in fatty acid synthesis. Isoenzyme V is present in the mitochondrial matrix and is also abundant in both adipocytes and liver. Isoenzyme IV is a larger membrane–associated form, while VI is secreted into the saliva. Carbonic anhydrase has also been identified in *E. coli.*, in a methanobacterium, and in green plants. A 60–kDa carbonic anhydrase called *nacrein* is found in the organic matrix of the nacreous layer of the pearl oyster, the layer that forms aragonite (orthorhombic calcium carbonate) in the shell and in pearls. X–ray studies of carbonic anhydrases I and II, from human blood, revealed that both have an ellipsoidal shape of dimensions ~4.1 × 4.1 × 4.7 nrn.

The zinc atom in each molecule lies in a deep pocket –1.2 nm from the surface and is surrounded by three histidine side chains and one H_2O or OH^- ion, the four ligands forming a distorted tetrahedron (Fig. 6.1). The coordinating imidazole group from His 119 is hydrogen bonded to a carboxylate group of Glu 117, a feature reminiscent of the charge-relay system of serine protease. This carboxylate group is also bound into a more extended hydrogen–bonded network, part of which is indicated in Fig. 6.1. The other imidazole groups also form hydrogen bonds to protein groups and the zinc–bound H_2O is involved in an extensive hydrogen–bonded network with several other bound water molecules and protein side chains. Most of these structural features are conserved in the other mammalian isoenzymes.

However, from X–ray absorption spectroscopy (EXAFS) it appears that spinach carbonic anhydrase contains one or more sulfur ligands to the zinc, while the enzymes from the archaeon *Methanosarcina thermophila* have a left–handed β helix structure. Nevertheless, this enzyme has the same three-histidine Zn–OH structure found in the mammalian enzymes.

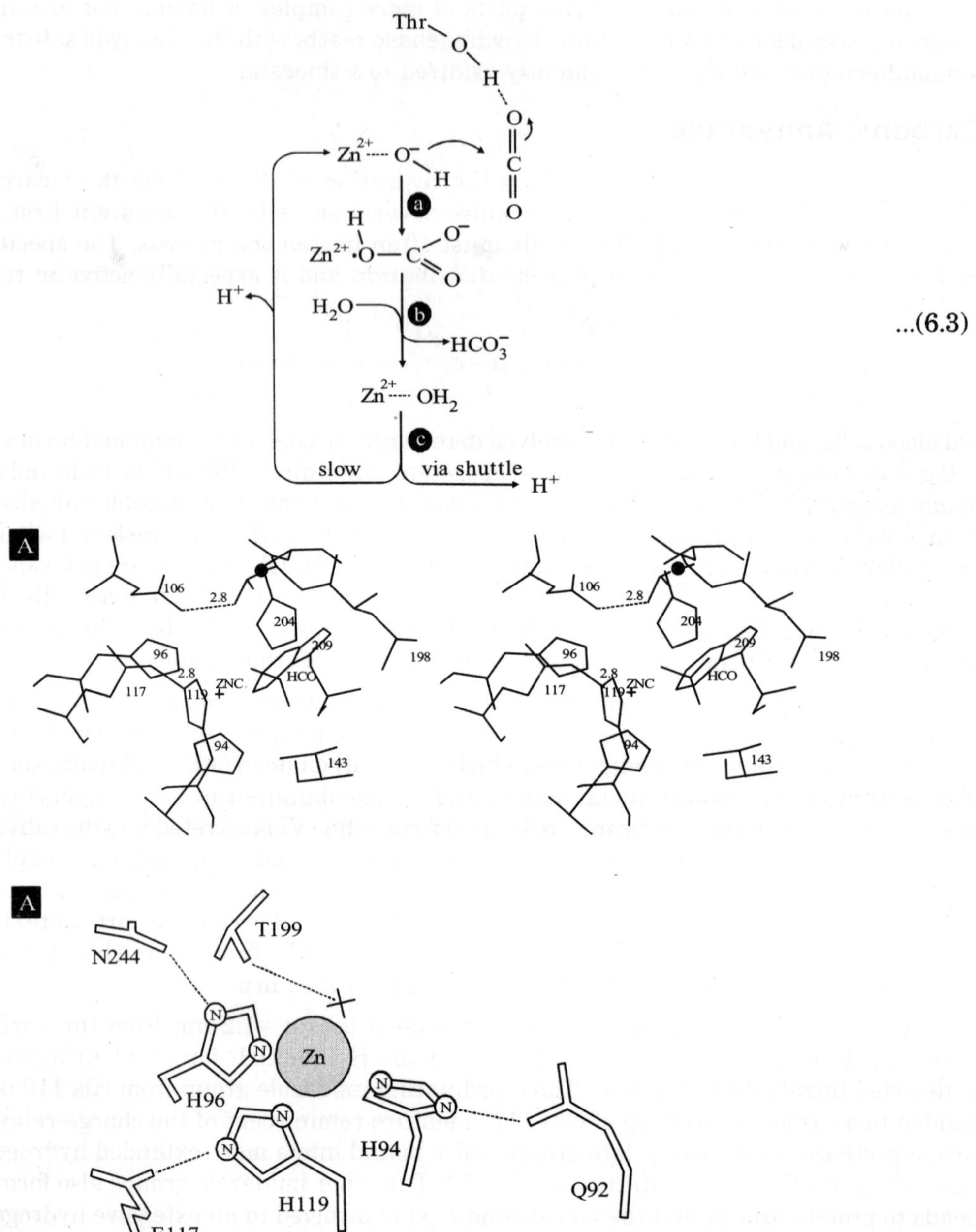

Fig. 6.1. (A) Stereoscopic view showing the binding pocket for HCO_3^- in the active site of human carbonic acid anhydrase I. Also shown are the hydrogen-bonded pairs E117–H119 and E106–T199. ZNC, zinc ion; HCO, bicarbonate. (B) View of the active site of carbonic anhydrase II. The orientation is a little different than in (A). The location of H_2O or – OH bound to the zinc ion is marked X. Hydrogen bonds from the three histidines that form coordinate bonds to the zinc, Q92, E117, the backbone carbonyl of N244, and the hydrogen bond from the zinc-bound hydroxyl to the T199 side chain are shown as dashed lines.

Carbonic anhydrase II is among the most rapid enzymes known, with the turnover number at 25°C for hydration of CO_2 being ~10^6 s^{-1}. The same enzyme catalyzes hydration of acetaldehyde (Eq. 6.1) but at a 1000–fold slower rate. A pK_a of ~7 controls the activity. This appears to represent the loss of H^+ from the Zn^{2+}–OH_2 complex to give Zn^+–OH. The latter is in effect a stabilized hydroxide ion existing at a pH at which OH^- is normally not present in quantity. It is this hydroxide ion that adds to the CO_2 or to the aldehyde substrate (Eq. 6.3, step *a*). In step *b a* water molecule replaces the departing bicarbonate.

A variety of data indicate that proton transfers mediated by the enzyme are essential parts of the carbonic anhydrase mechanism. One proposal is that the nearby imidazole group of His 64 (not shown in Fig. 6.1) deprotonates the bound H_2O via a hydrogen-bonded network of bound water molecules (Eq. 6.3, step *c*). The side chain of Thr 199 may function in a cyclic proton transfer in step *c*. The proton generated in step *c* is released to the solvent in a process that is catalyzed by buffer anions or by amines such as histamine. The latter binds at the edge of the active site and forms an additional hydrogen-bonded pathway to the zinc–bound H_2O. A different proton shuttle pathway has been proposed for the slower carbonic anhydrase III. Related to the reaction catalyzed by carbonic anhydrase is the addition of an amino group to CO_2 form a carbamino group (–NH–COO^-). This reaction is essential to the functioning of hemoglobin, which must carry large amounts of CO_2, as carbamino groups, from tissues to the lungs, and to some enzymes such as ribulose bisphosphate carboxylase.

Imines (Schiff Bases)

As we have seen already, many enzymatic reactions depend upon formation of *imines,* which are commonly called Schiff bases. The two–step formation of Schiff bases consists of addition of an amino group to a carbonyl group to form a carbinolamine followed by elimination of water (Eq. 6.4). One group of *aldolases* (Section D) have, at their active centers,

H_2N—R'

R—C(H)=O···H^+ —(a)→ R—C(H)(OH)—N(H)—R'

Carbinolamine

(b) ↓ — H_2O

R—C(H)=N—R'

Imine of Schiff base ...(6.4)

lysine side chains which form Schiff bases with the ketone substrates prior to the principal reaction of breaking or forming a C – C bond. Similarly, the initial reaction of the aldehyde coenzyme *pyridoxal phosphate* with amino acid substrates is the formation of Schiff bases (Fig. 6.4). Indeed, *the groups C = O and H_2N–are inherently complementary* and their interaction through imine formation is extremely common. Schiff bases often form within a fraction of a second, but one or both steps of Eq. 6.4 may require catalysis to achieve enzymatic velocities.

The reaction is usually completely reversible and formation constants are often low enough that a carbonyl compound present in small amounts will not react extensively with an amine unless the two are brought together on an enzyme surface. If the amino group is hydrogen bonded to the enzyme, the proton may remain on the Schiff base nitrogen, enhancing the electron-accepting properties of the C=N group. The pK, values of Schiff bases with aliphatic aldehydes are usually 1– 2 units lower than those of the corresponding primary amines. However, the local environment may sometimes cause the pK_a to be increased. Most Schiff bases are reduced readily by sodium borohydride or sodium cyanoborohydride to form secondary amines in which the original aldehydes are bound covalently to the original amino groups. This provides a method for locating sites of Schiff base formation in enzymes. An isotopically substituted aldehyde or amine can be employed, or an isotopic label can be introduced, by using ^{2}H-or ^{3}H-containing sodium borohydride in the reduction.

Stereochemistry of Addition to Trigonal Carbon Atoms

Adducts formed by enzymatic addition of nucleophiles to carbonyl groups are usually chiral. For example, addition of an amino group to the "front" side of the carbonyl carbon in Eq. 6.5 creates a carbinolamine of the S configuration. Addition of the amino group from behind the plane of the paper (the *re* face) would lead to a carbinolamine of the R configuration.

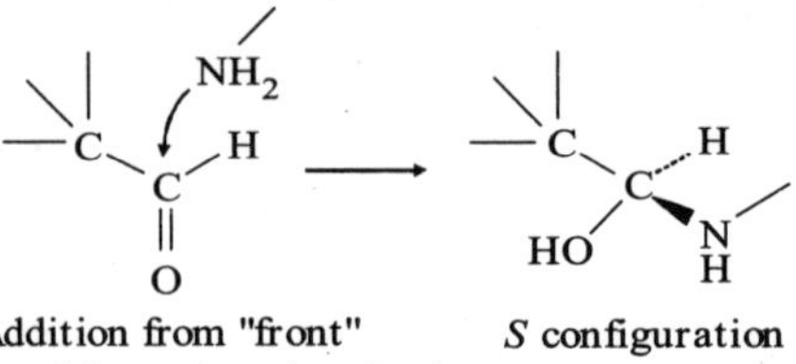

NH_2 H C O → C H HO N H ...(6.5)

Addition from "front" (*re* face of the carbonyl carbon) — *S* configuration

Addition to Carbon–Carbon Double Bonds, Often Reversible Reactions

Most of the reactions that we will consider in this chapter involve addition of a proton to a carbon atom or removal of a proton attached to a carbon atom. A frequent metabolic reaction is addition of water to a carbon–carbon double bond that is conjugated with a carbonyl group. This transmits the polarization of the carbonyl group to a position located two carbon atoms further along the chain.

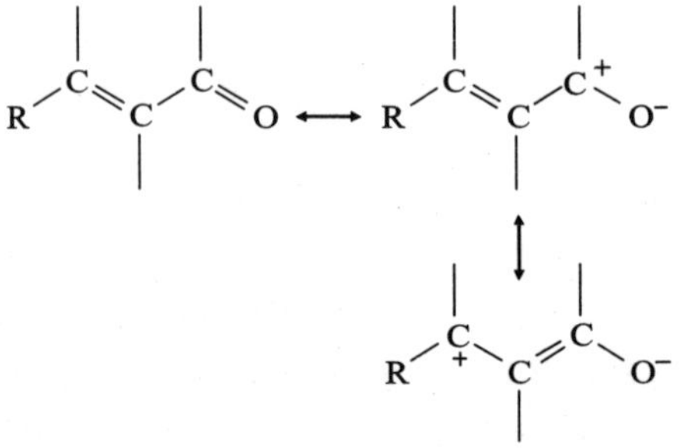

R–C=C–C=O ⟷ R–C=C–C^+–O^- ⟷ R–C^+–C=C–O^-

Because of this effect bases can add to a carbon–carbon double bond in a position β to the carbonyl group as in Eq. 6.6, step *a*, The product of addition of HO^- is the anion of an *enol* in which the negative charge is distributed by resonance between the oxygen of the cabonyl group and the carbon adjacent to the carbonyl. A stable product results if a proton adds to the latter position (Eq. 6.6, step *b*). Many nucleophilic groups (Nu in Eq. 6.6) other that HO^-, may add, the reaction being known as a *Mechael addition.* If a neutral nucleophile Nu adds, the product may lose a proton, or transfer it to the adjacent carbon in Eq. 6.6, step *b*, to give a neutral end product. The reverse of an addition reaction of this type is also known as an *elimination reaction.*

Enzymatic reactions involving addition to a C = C bond adjacent to a carbonyl group (or in which elimination occurs α, β to a carbonyl) are numerous. Except for some enzymes acting by free radical mechanisms, the nucleophilic group always adds at the β position suggesting

$$R-C=C-C=O \;+\; HO^- \,(Nu) \xrightarrow{\;a\;} R-C(HO)(Nu)-C=C-O^-$$

$$\updownarrow$$

$$R-C(HO)-\overset{\ominus}{C}-C=O \quad \text{Enolate anion}$$

$$\xrightarrow{\;b,\;H^+\;} R-C(HO)(Nu)-C(H)-C=O \qquad \text{...(6.6)}$$

that the mechanism protrayed by Eq. 6.7 is probable. It is noteworthy that *frequently in a metabolic sequence a carbonyl group is deliberately introduced to facilitate elimination or addition at adjacent carbon atoms.* The carbonyl may be formed by oxidation of a hydroxyl group or it may be provided by a thioester formed with coenzyme A or with an *acyl carrier protein.*

Enoyl-CoA Kydratase

A specific example of the reaction in Eq. 6.6 is the addition of water to trans-α, β-unsaturated CoA derivatives (Eq. 6.7). It is catalyzed by enoyl–CoA hydratase (crotonase) from mitochondria and is a step in the β oxidation of fatty acids.

$$R-CH=CH-C(=O)-S-CoA \xrightarrow[\;H_2O\;]{\text{Enoyl-CoA hydratase}} R-CH(OH)-CH_2-C(=O)-S-CoA$$

L-β-Hydroxyacyl-CoA ...(6.7)

The enzyme is a hexamer, actually a dimer of trimers made up of 291–residue polypeptide chains. Aceto–acetyl–CoA is a competitive inhibitor which binds into the active site and locates it. From the X–ray structure of the enzyme–inhibitor complex it can be deduced that the carboxylate group of E144 abstracts a proton from a water molecule to provide the hydroxyl ion that binds to the β position (Eq. 6.6, step *a*) and that the E164 carboxyl group donates a proton to the intermediate enolate anion in step *b*. The hydroxyl group and the proton enter from the same side of the double bond in a *syn* addition. The X–ray structure shows that the E144 and E164 side chains are on the same side of the double bond, accounting for this stereochemistry.

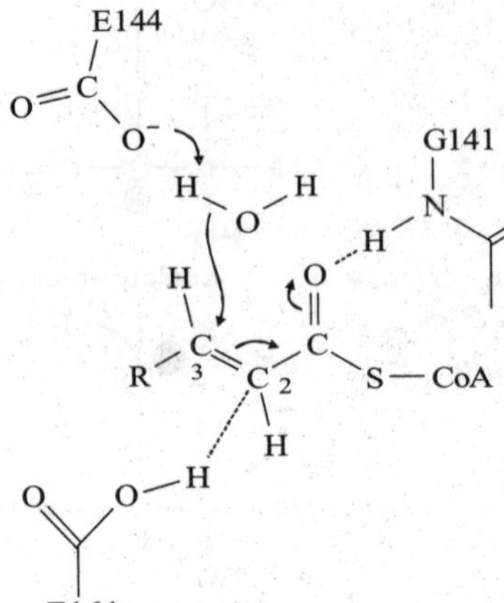

Cinnamoyl–CoA thiol esters (R = phenyl in the foregoing structure) contain a good light–absorbing group (chromophore). Binding to the protein induces distinct shifts in the ultraviolet absorption spectrum and also in ^{13}C NMR and Raman spectra. These shifts suggest that binding induces an enhanced positive charge at C-3 and formation of a strong hydrogen bond to the carboriyl group. The X-ray studies indicate that this bond is to the G141 peptide nitrogen as shown. These results favor an enolate anion intermediate as in Eq. 6.6. However, kinetic isotope effects as well as studies of proton exchange have suggested a concerted mechanism with water adding at the same time that a proton binds at C-2. A closely related *E. coli* protein is a 79-kDa multifunctional enzyme that catalyzes four different reactions of fatty acid oxidation.

The amino-terminal region contains the enoyl hydratase activity. A quite different enzyme catalyzes dehydration of thioesters of (β-hydroxyacids such as 3-hydroxydecanoyl-acyl carrier protein to both form and isomerize enoyl-ACP derivatives during synthesis of urtsaturated fatty acids by *E. coli*. Again, a glutamate side chain is the catalytic base but an imidazole group of histidine has also been implicated. This enzyme is inhibited irreversibly by the N-acetylcysteamine thioester of 3-decynoic acids (Eq. 6.8). This was one of the first enzyme-activated inhibitors to be studied.

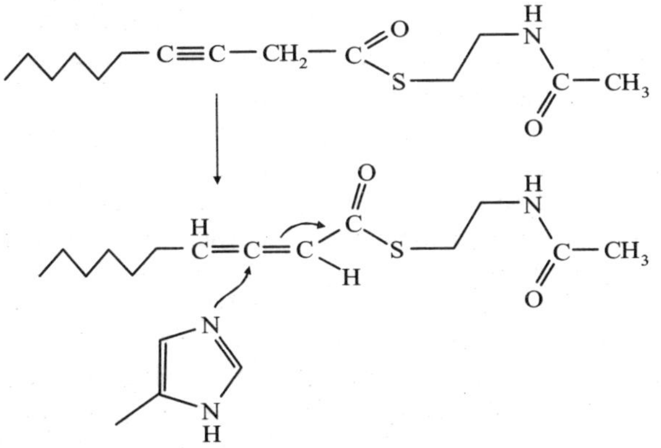

...(6.8)

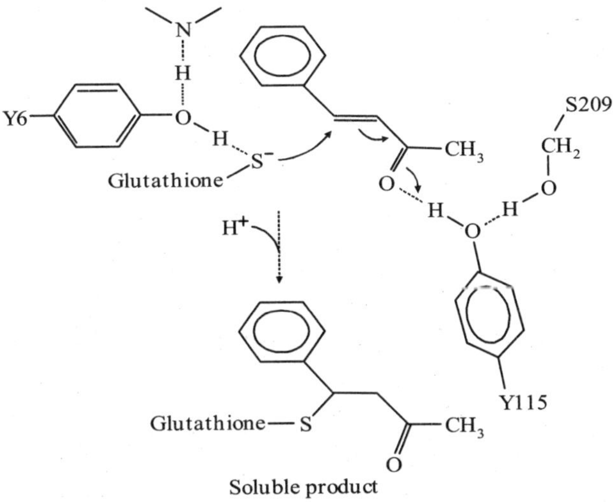

Y6 Glutathione S209 CH_2 CH_3 H^+ Y115 Glutathione—S CH_3

Soluble product

...(6.9)

Glutathione S-transferases

Addition of glutathione to a large variety of different substrates containing electrophilic centers, such as that at the β position in an α, β-unsaturated ketone (Eq. 6.9), is catalyzed by the ubiquitous group of enzymes called glutathione S-transferases. There are six classes of eukayotic glutathione S-transferases-one membrane-associated microsomal and five cytosolic. They play an important role in detoxifying many xenobiotics and other relatively hydrophobic compounds by converting them to more soluble compounds that can be degraded and excreted easily. A pair of tyrosines appear to participate in catalysis, as is indicated in Eq. 6.9. Glutathione S-transferases have attracted attention because of their role in detoxifying anticancer drugs.

Cancer cells can become resistant to drugs as a result of excessive synthesis of these detoxifying enzymes. At the same time glutathione transferases protect human patients from drugs. In plants these enzymes may provide protection from insecticides and herbicides. Glutathione transferases of pathogenic organisms such as schistosomes are appropriate targets for new drugs as well as for vaccines. Nonenzymatic Michael addition of glutathione to such compounds as 4-*hydroxynonenal,* a product of peroxidation of the polyunsaturated arachidonate are also biochemically important.

Chlorobenzoyl-CoA dehalogenase

The enzymatic release of chloride ion from 4-chloroxybenzoyl-CoA would be hard to explain by a simple nucleophilic displacement by HO^-However, *addition* of HO^-, followed by elimination of Cl^-, would be chemically reasonable. Nevertheless, single-turnover studies revealed a more complex mechanism involving formation of a covalent adduct with the enzyme. A carboxylate group adds and Cl^- is eliminated to form of an oxygen ester, which is then hydrolyzed to the final product (Eq. 6.10). Studies of mutant enzymes together with X-ray crystallography support this mechanism. As with enoyl-CoA hydratase, binding of substrate to chlorobenzoyl-CoA dehalogenase causes alterations in ultraviolet, NMR, and Raman spectra that can be interpreted as indicating enhanced polarization of the benzoyl group. For example, changes in the C = O stretching frequency suggest that the bond is elongated, presumably as a result of hydrogen bonding to the N – H of G114, which lies at the N terminus of an α helix and experiences the additional polarizing effect of the helix dipole. This effect can be compared with that mentioned previously for enoyl-CoA hydratase.

...(6.10)

In fact, close sequence and structural homologies show that chlorobenzoyl-CoA dehalogenase, enoyl–CoA hydratase, and a variety of other hydratases, isomerases, synthases, lyases, and hydrolases belong to a large family of related proteins.

Addition to Double Bonds Adjacent to Carboxylate Groups

Biochemical reactions often involve addition to C = C bonds that are not conjugated with a true carbonyl group but with the poorer electron acceptor – COO^-. While held on an enzyme a carboxylate group may be protonated, making it a better electron acceptor. Nevertheless, there has been some doubt as to whether the carbanion mechanism of Eq. 6.6 holds for these enzymes. Some experimental data suggested a quite different mechanism, one that has been established for the nonenzymatic hydration of alkenes. An example is the hydration of ethylene by hot water with dilute sulfuric acid as a catalyst (Eq. 6.11), an industrial method of preparation of ethanol. The electrons of the double bond form the point of attack by a proton, and the resulting carbocation readily abstracts a hydroxyl ion from water. Direct addition of OH^- to ethylene to form a carbanion is not favored, because there is no adjacent carbonyl group to stabilize the negative charge.

$$H_2C{=}CH_2 \xrightarrow{H^+} \underset{\text{Carbocation}}{H_3C{-}\overset{+}{C}H_2} \xrightarrow[H^+]{H_2O} H_3C{-}CH_2OH \qquad \ldots(6.11)$$

Fumarate Hydratase

The most studied enzyme of this group is probably the porcine mitochondrial fumarate hydratase, a tetramer of 48.5–kDa subunits with a turnover number of ~$2 \times 10^3\ s^{-1}$. It accelerates the hydration reaction more than 10^{15}–fold. A similar enzyme, the 467–residue *fumarase C* whose three-dimensional structure is known, is found in cells of *E. coli* when grown aerobically. The product of the fumarate hydratase reaction is L-malate (S-malate). The stereospecificity is extremely high. If the reaction is carried out in ^2H_2O an atom of ^2H is incorporated into the *pro–R* position, *i.e.*, the proton is added strictly from the *re* face of the trigonal carbon (Eq. 6.12).

Hydroxyl added in front (*si* face)

$$\underset{\text{Fumarate}}{^-OOC{-}CH{=}CH{-}COO^-} \xrightarrow[{}^{*}H^+]{^-OH} \underset{\text{L-Malate}}{^-OOC{-}C(H)(OH){-}C(H)(^{*}H){-}COO^-} \qquad \ldots(6.12)$$

Proton added from behind (*re* face)

To obtain L-malate the hydroxyl must have been added from the opposite side of the double bond. Such *anti* (*tram*) addition is much more common in both nonenzymatic and enzymatic reactions than is addition of both H and OH (or – Y) from the same side (*syn, cis,* or adjacent addition). For concerted addition it is a natural result of stereoelectronic control.

Almost *all* enzymatic addition and elimination reactions involving free carboxylic acids are anti with the proton entering from the *re* face, suggesting that there has been conservation throughout evolution of a single basically similar mechanism. The pH dependence of the action of fumarate hydratase indicates participation of both an acidic and a basic group with pK_a values of 5.8 and 7.1 for additional information. However, either anion or carbocation mechanisms might be possible. That the cleavage of the C–H bond is not rate limiting is suggested by the observation that malate containing ^2H in the *pro–R* position is dehydrated at the same rate as ordinary malate. If the anion mechanism (Eq. 6.13) is correct, the ^2H from the *pro-R* position of specifically labeled malate might be removed rapidly, while the loss of OH^- could be slower. If so, the ^2H would be "washed out" of L-malate faster than could happen by conversion to fumarate followed by rehydration to malate. In fact, the opposite was observed.

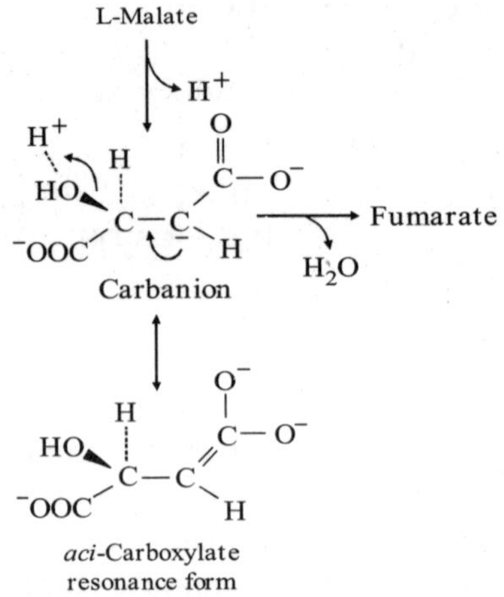

L-Malate
H^+
H^+ HO H O C–O⁻ ⁻OOC C–C H → Fumarate H_2O
Carbanion
HO H O⁻ C–O⁻ ⁻OOC C–C H
aci-Carboxylate resonance form

...(6.13)

The hydroxyl group was lost rapidly and the ^2H more slowly. This result suggested the carbocation ion mechanism of Eq. 614.

While no *primary* isotope rate effect was observed, when the hydrogen at C–2 or the *pro–S* hydrogen at C–3 of malate was replaced by ^2H or ^3H distinct secondary isotope effects were seen. Thus, $k(^1H) / k(^2H) = 1.09$ for both the *pro–S* and the C–3 hydrogen atoms. These findings appeared to support the carbocation mechanism for the reasons considered in and

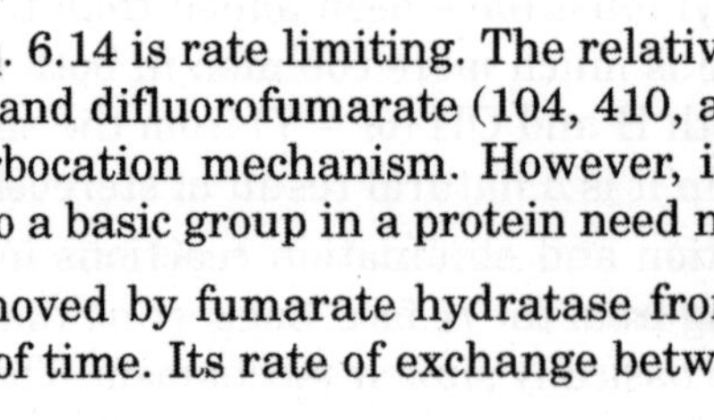

L-Malate
a ^-OH
H COO⁻ C⁺–C ⁻OOC H *H b → Fumarate $*H^+$
Carbocation

...(6.14)

suggested that step *b* of Eq. 6.14 is rate limiting. The relative values of V_{max} for hydration of fumarate, fluorofumarate, and difluorofumarate (104, 410, and 86 mol ml^{-1} min^{-1} mg^{-1}) also seemed to support the carbocation mechanism. However, it has become clear that protons removed from a substrate to a basic group in a protein need not exchange rapidly with solvent.

In fact, the proton removed by fumarate hydratase from malate is held by the enzyme for relatively long periods of time. Its rate of exchange between malate and solvent is slower than the exchange of a bound fumarate ion on the enzyme surface with another substrate molecule from the medium. Thus, the overall rate is determined by the speed of dissociation of products from the enzyme and we cannot yet decide whether removal of a proton precedes or follows loss of OH^-. Two new lines of evidence suggest that proton abstraction comes first.

A careful study of both ^{18}O and ^{2}H isotope effects[60] supports the carbanion intermediate, as does the strong inhibition by anions of 3-nitropropionate and 3-nitro-2-hydroxypropionate. To provide a good electron sink the carboxylate group adjacent to the proton that is removed by fumarate hydratase must either be actually protonated in the enzyme-substrate (ES) complex or paired with and hydrogen bonded to a positively charged group.

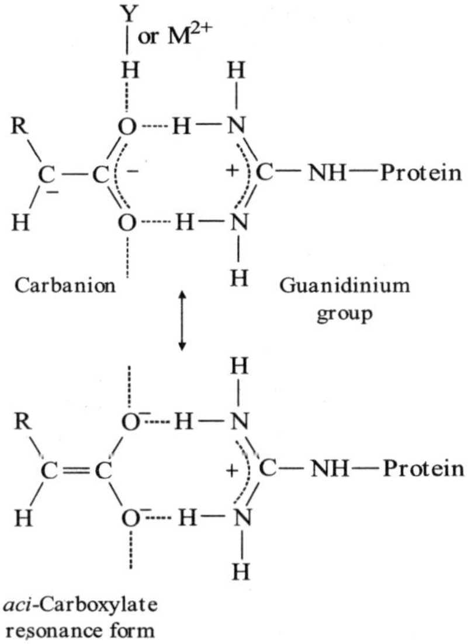

Y or M^{2+}

C—NH—Protein

Carbanion

Guanidinium group

C—NH—Protein

aci-Carboxylate resonance form

The anion formed by removal of the 3–H is analogous to the enolate anion of Eq. 6.6 and has a strong structural similarity to the readily formed anions of organic nitro compounds. The nitronate anions may, perhaps, be regarded as transition state inhibitors.

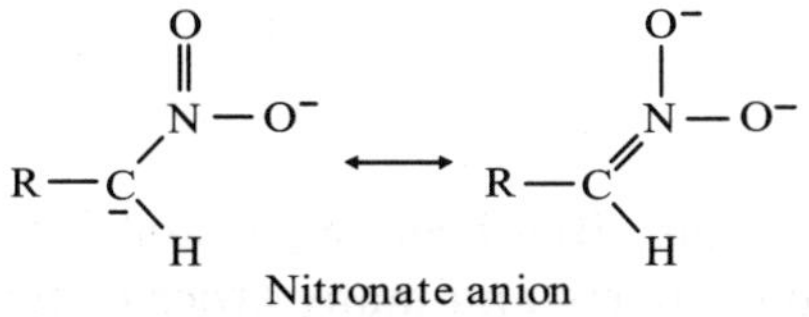

Nitronate anion

There is still a third possible mechanism for the fumarate hydratase reaction. The proton and hydroxyl groups may be added *simultaneously* in a concerted reaction. However, observed kinetic isotope effects are not consistent with this mechanism. In 1997 the structure of fumarase C of *E. coli* was reported. Each active site of the tetrameric enzyme is formed using side chains from three different subunits. The H188 imidazole is hydrogen bonded to an active site water molecule and is backed up by the E331 carboxylate which forms a familiar catalytic pair. However, these results have not clarified the exact mode of substrate binding nor the details of the catalytic mechanism. Structural studies of fumarate hydratase from yeast and the pig are also in progress.

Some Other Fumarate–Forming Reactions

Other enzymes that catalyze elimination reactions that produce fumarate are *aspartate ammonialyase* (aspartase), *argininosuccinate lyase* and *adenylosuccinate lyase*. In every case it is NH_3 or an amine, rather than an OH group, that is eliminated. However, the mechanisms probably resemble that of fumarate hydratase. Sequence analysis indicated that all of these enzymes belong to a single *fumarase-aspartase family*. The three-dimensional structure of aspartate ammonia-lyase resembles that of fumarate hydratase, but the catalytic site lacks the essential H188 of fumarate hydratase. However, the pK_a values deduced from the pH dependence of V_{max} are similar to those for fumarase. *3-Methylaspartate lyase* catalyzes the same kind of reaction to produce ammonia plus *cis*-mesaconate. Its sequence is not related to that of fumarase and it may contain a dehydroalanine residue.

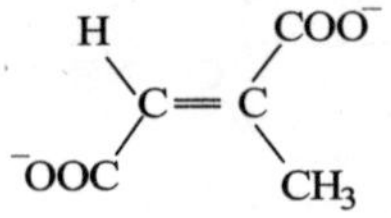

H COO⁻
C=C
⁻OOC CH₃

cis-Mesacondate

Enolase

A key reaction in the metabolism of sugars is the dehydration of 2-phosphoglycerate to form *phosphoenolpyruvate* (PEP), the phospho derivative of the enolic form of pyruvic acid:

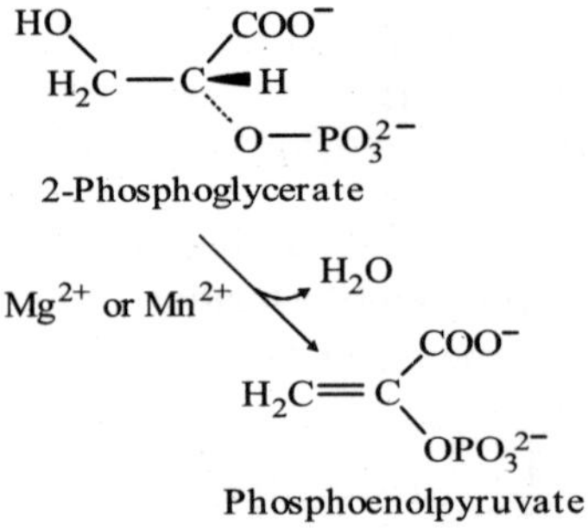

$$HO-H_2C-C(H)(COO^-)-O-PO_3^{2-}$$

2-Phosphoglycerate

$$\xrightarrow[Mg^{2+} \text{ or } Mn^{2+}]{} H_2O$$

$$H_2C=C(COO^-)OPO_3^{2-}$$

Phosphoenolpyruvate ...(6.15)

A carbanionic intermediate has often been suggested for this enzyme. However, despite measurements of kinetic isotope effects and many other experiments it has been difficult to establish a detailed mechanism. Enolase has a complex metal ion requirement, usually met by Mg^{2+} and Mn^{2+}. From NMR studies of the relaxation of water protons, it was concluded that a Mn^{2+} ion coordinates two rapidly exchangeable water molecules in the free enzyme. When substrate binds, one of these water molecules may be immobilized and may participate in an addition reaction that forms phosphoenolpyruvate (reverse of reaction 6.15).

A tightly bound "conformational" metal ion is located in the known three-dimensional structure in such a way that it might function in this manner. A more loosely bound "catalytic" metal ion is also essential.

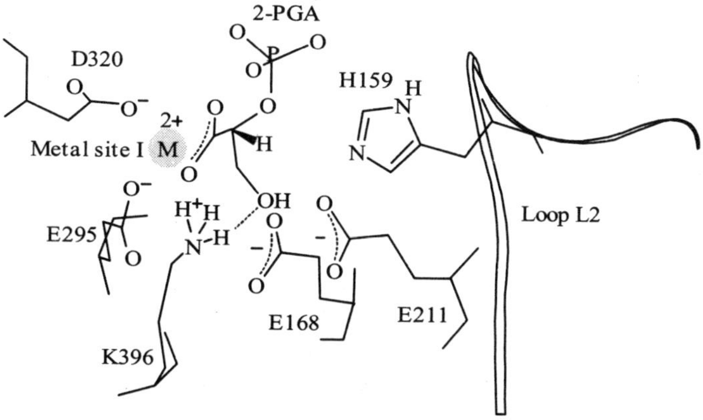

Fig. 6.2. View of the active site of yeast enolase containing a bound molecule of 2-phospho-D-glycerate. The catalytic magnesium ion is at the left but the "conformational" metal is not visible here. The imidazole group of His 159 serves as the catalytic base and the -NH_3^+ of Lys 396 or Lys 345 as the catalytic acid.

Figure 6.2 shows a view of the active site of yeast enolase occupied by a molecule of bound 2-phospho-glycerate. Histidine 159 is probably the catalytic base that removes the α-proton to form an *aci*-anion which is stabilized by interaction with the catalytic Mg^{2+} ion. A protonated lysine 396 amino group may be the catalytic acid. The active site is surrounded by a complex hydrogen-bonded network. As is discussed in Section B, a large number of other enzymes belong to an *enolase superfamily* of enzymes. Among them is *glucarate dehydratase,* which initiates a pathway for catabolism of D-glucarate and galactarate in *E. coli.*

Pectate lyase and related enzymes

A group of polysaccharide lyases cleave the chains of polymers of uronic acids with 1,4 linkages such as pectins, hyaluronan, heparin, and dermatan sulfate Fig. 6.3. These bacterial enzymes also employ an elimination mechanism. The geometry of the β-linked galacturonic acid units of pectin is favorable for *anti* elimination of the 5-H and the O-glycosyl group in the 4 position (Eq. 6.16). However, the corresponding *hyaluronate lyase* (hyaluronidase) acting on glucuronic acid residues causes a *syn* elimination. These results suggest the formation of anionic intermediates which can eliminate a substituent from either the equatorial or axial position of the sugar ring.

Hyaluronate lyase from the pathogenic *Streptococcus pneumonias* apparently utilizes an imidazole group as the catalytic base and a tyrosine side chain as the proton donor in the reaction. *Pectate lyase* C from the plant pathogen *Erwinia,* which causes softrot in many different plants, has a parallel β barrel structure (Fig. 6.3) which is similar to that of the tailspike protein shown in and represents what may be a very large structural family of proteins. The location of the active site is not obvious from the structure but on the basis of the effects of a large number of mutations, it is thought to be near the Ca^{2+}-binding site labeled in Fig. 6.3. Replacement of any of three Asp or Glu and three Lys residues in this region leads to loss of or reduction in catalytic activity.

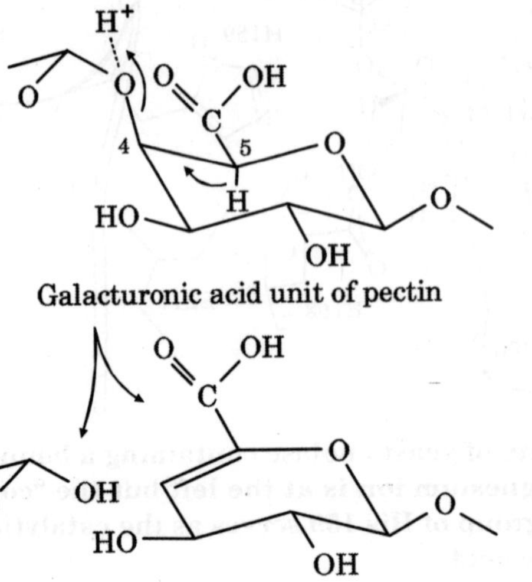

Galacturonic acid unit of pectin

...(6.6)

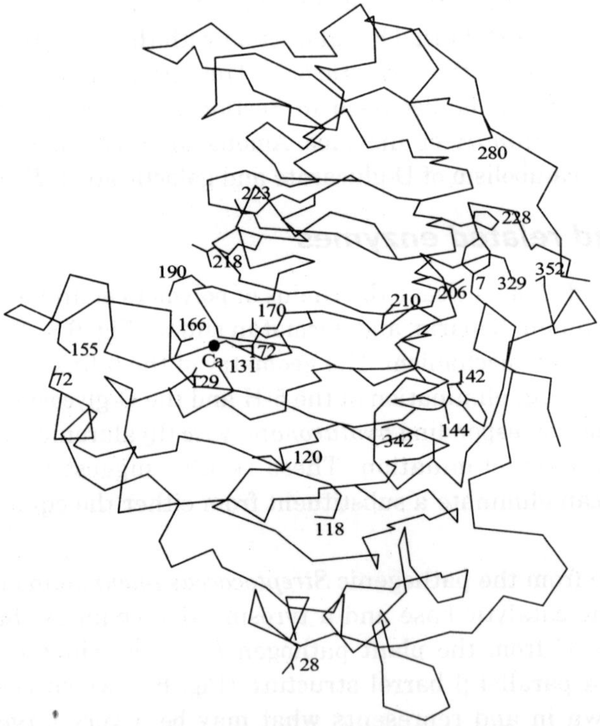

Fig. 6.3. The three-dimensional structure of pectate lyase C showing locations of amino acids substituted by oligonucleotide-directed mutation of the cloned gene. The green dot labeled Ca is the Ca^{2+} binding site.

Aconitase and Related Iron – Sulfur Proteins

Two consecutive reactions of the citric acid cycle (Fig. 6.6), the dehydration of citrate to form *cis*-aconitate and the rehydration in a different way to form isocitrate (Eq. 6.17), are catalyzed by aconitase (aconitate hydratase). Both reactions are completely stereo-specific. In the first (Eq. 6.17, step *a*), the *pro–R* proton from C–4 (stereochemical numbering) of citrate is removed and in step *c* isocitrate is formed. Proton addition is to the *re* face in both cases. As with fumarate hydratase, the enzyme holds the abstracted proton (for up to 7×10^{-5}s) long enough so that a *cis*-aconitate molecule sometimes diffuses from the enzyme and (if excess *cis*-aconitate is present) is replaced by another.

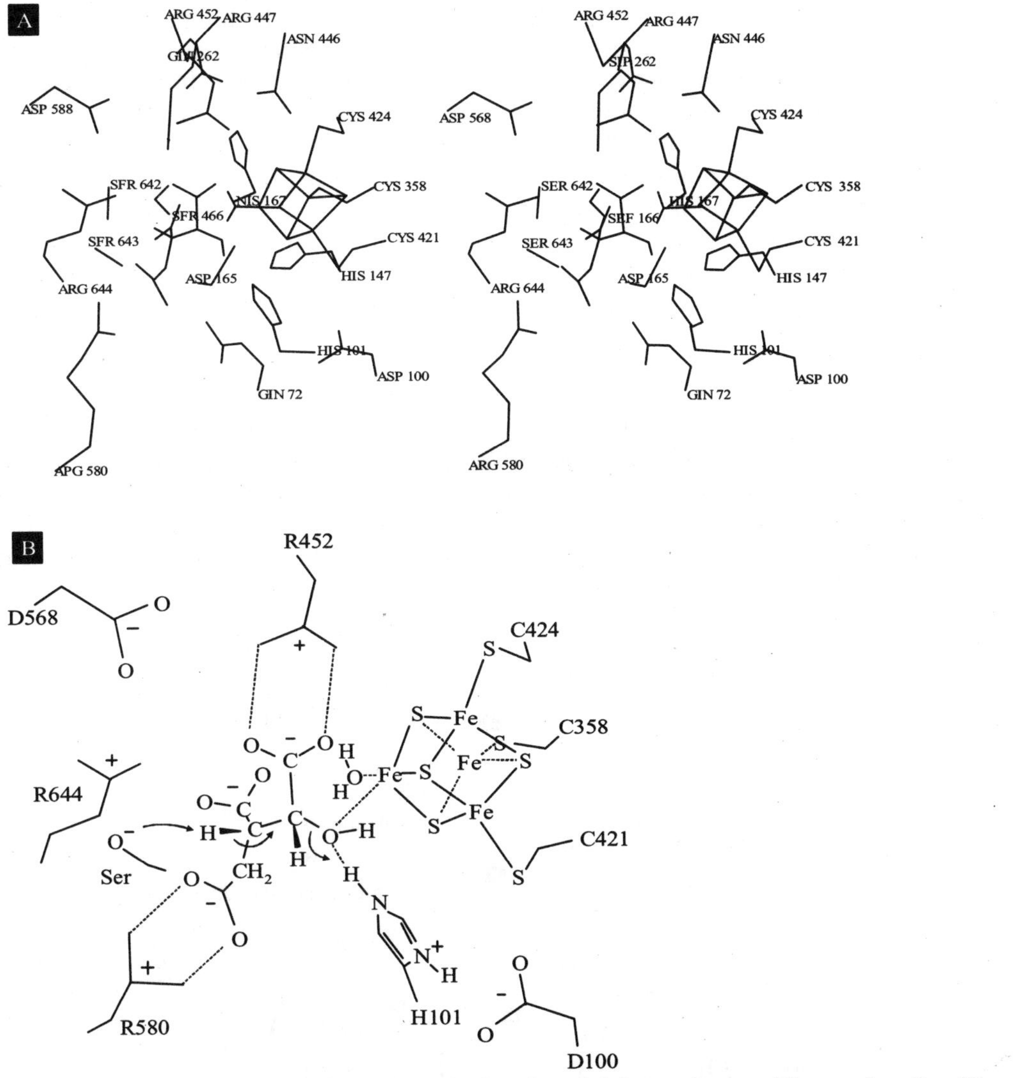

Fig. 6.4. (A) Stereoscopic view of the active site of mitochondrial aconitase with a molecule of L-isocitrate placed by modeling next to the Fe_4S_4 cluster. This cluster still has a water molecule bound to one iron atom as in the free enzyme. Courtesy of C. D. Stout. (B) Interpretive diagram.

The result is that the new *cis*-aconitate molecule sometimes receives the proton (intermolecular proton transfer). The proton removed from citrate is often returned to the molecule in Eq. 6.17, step *b*, but the position of reentry is different from that of removal. Apparently, after the initial proton removal, the *cis*-aconitate that is formed "flips over" so that it can be rehydrated (Eq. 6.17, step *c*) with participation of the same groups involved in dehydration but with formation of the new product. Aconitase contains iron in the form of an Fe_4S_4 iron–sulfur cluster (Fig. 6.4). However, the enzyme is usually isolated in a form that does not show its maximum activity until it has been incubated with ferrous iron (Fe^{2+}).

The inactive form of the enzyme is thought to contain an Fe_3S_4 cluster which is converted back to the Fe_4S_4 cluster by the incubation with Fe^{2+}. From Mössbauer spectroscopy it was deduced that the iron atom that is lost upon inactivation is the binding site for the –OH group of citrate or isocitrate and of an adjacent carboxylate. Since iron can engage in oxidation–reduction processes, aconitase could act by a mechanism different from others discussed here. However, an Fe–OH group could act much as does the Zn–OH of carbonic anhydrase (Eq. 6.4). Any redox chemistry may be involved in control of the enzyme.

X-ray studies have confirmed the proximity of the Fe_4S_4 cluster to bound substrate analogs. The isocitrate shown in Fig. 6.4 was fitted into the active site by modeling. It shows the water molecule bound to the Fe_4S_4 cluster as observed for the free enzyme. Notice that this cluster is held by three cysteine side chains (C358, C421, and C424) rather than the usual four. The fourth iron atom is free to bind to water. When the enzyme acts on isocitrate this water must be displaced by the one generated from the substrate. The Fe probably acts as a polarizing electrophile that assists in the elimination but it is apparently the histidine–aspartate pair (H101–D100) that serves as the catalytic acid in generating the H_2O.

What is the catalytic base? The only candidate seen in the structure is the Ser 642 –OH, dissociated to $-O^-$. Can this be correct? Mutational analysis supports the essential role of this side chain. The peptide NH and guanidinium groups of R644 appear to provide an "oxyanion hole" that stabilizes the negative charge. Mutations also support the role of the His–Asp pair. Another His–Asp pair (H167–D165) is located directly behind the bound substrate and Fe–OH_2 in Fig. 6.4A and is apparently also essential.

R580 ⊕ Citrate

cis-Aconitate

Aconitate molecule flips 180°

cis-Aconitate

2R, 3S-Isocitrate

...(6.17)

Aconitase exists as both mitochondrial and cytosolic isoenzyme forms of similar structure. However, the cytosolic isoenzyme has a second function. In its apoenzyme form, which lacks the iron—sulfur cluster, it acts as the much–studied *iron regulatory factor,* or iron–responsive element binding protein (IRE–BP). This protein binds to a specific stem–loop structure in the messenger RNA for proteins involved in iron transport and storage. Other enzymes in the aconitase family include *isopropylmalate isomerase and homoaconitase* enzymes functioning in the chain elongation pathways to leucine and lysine, both of which are pictured.

There are also iron–sulfur dehydratases, some of which may function by a mechanism similar to that of aconitase. Among these are the two fumarate hydratases, *fumarases A and B,* which are formed in place of fumarase C by cells of *E. coli* growing anaerobically. Also related may be bacterial L–serine and L–threonine dehydratases. These function without the coenzyme pyridoxal phosphate but contain iron–sulfur centers. *A lactyl–CoA dehydratase* and some related iron–sulfur enzymes may act via a mechanism related to that of vitamin B_{12}.

Addition to or Formation of Isolated Double Bonds

Only a few examples are known in which an enzyme induces addition to a double bond that is *not conjugated* with a carbonyl or carboxyl group. Pseudomonads have been observed to catalyze stereospecific hydration of oleic acid to D-10-hydroxystearate. The addition is *anti* and the proton enters from the *re* face.

Conjugative and Decarboxylative Elimination Reactions

Elimination can occur if the electrophilic and nucleophilic groups to be removed are located not on adjacent carbon atoms but rather are separated from each other by a pair of atoms joined by a double bond. Such a conjugative elimination of phosphate is the last step in the biosynthesis of *chorismate* (Eq. 6.18, step *a).* Chorismate is converted to *prephenate* which undergoes a conjugative and decarboxylative elimination (Eq. 6.18, step *c*) with loss of both water and CO_2 to form *phenylpyruvate,* the immediate biosynthetic precursor of phenylalanine. These reactions provide a good example of how *elimination reactions can be used to generate aromatic groups.*

In fact, this is the usual method of synthesis of aromatic rings in nature. Notice the stereochemistry of Eq. 6.18, step *a.* Orbital interaction rules predict that if the elimination is a concerted process it should be *syn.* The observed anti elimination suggests a more complex mechanism involving participation of a nucleophilic group of the enzyme. Elimination usually involves loss of a proton together with a nucleophilic group such as –OH, $-NH_3^+$, phosphate, or pyrophosphate. However, as in Eq. 6.18, step *c,* electrophilic groups such as $-COO^-$ can replace the proton.

Another example is the conversion of *mevalonic acid–5–pyrophosphate to isopentenyl pyrophosphate* (Eq. 6.19): This is a key reaction in the biosynthesis of isoprenoid compounds such as cholesterol and vitamin A. The phosphate ester formed in step a is a probable intermediate and the reaction probably involves a carbo–cationic intermediate generated by the loss of phosphate prior to the decarboxylation.

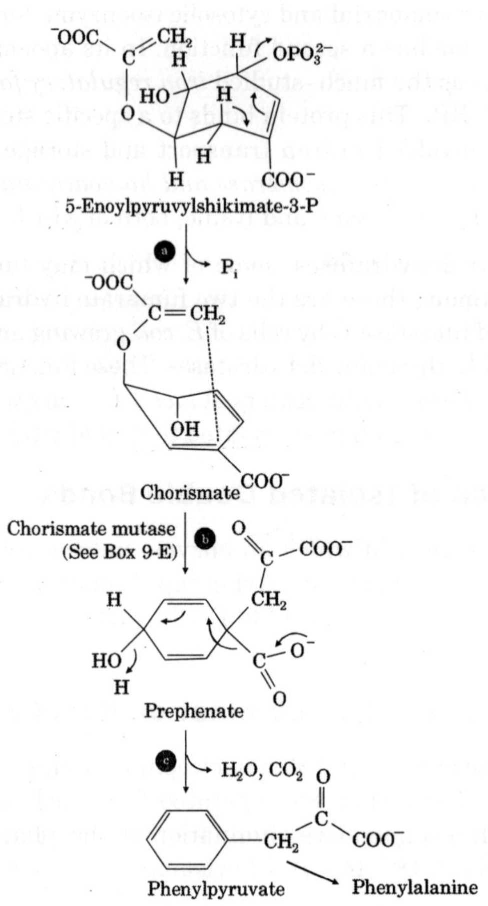

5-Enoylpyruvylshikimate-3-P

a → P_i

Chorismate

Chorismate mutase (See Box 9-E) b

Prephenate

c → H_2O, CO_2

Phenylpyruvate → Phenylalanine

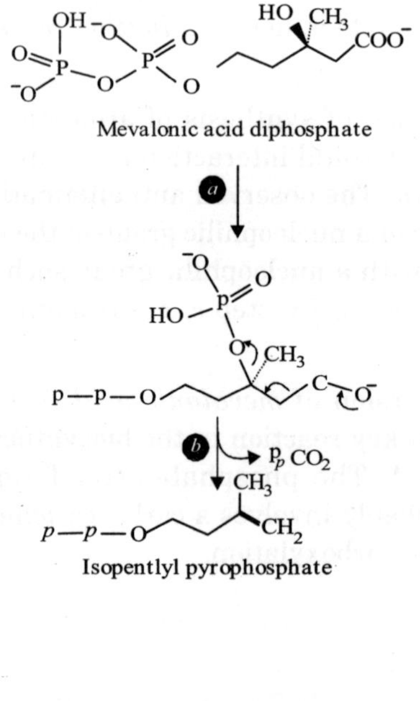

Mevalonic acid diphosphate

a

b → P_i, CO_2

Isopentlyl pyrophosphate

...(6.18)

Isomerization Assisted by Addition

An interesting use is made of addition to a double bond by glutathione–dependent *cis–trans isomerases.* One of them converts *maleate* to fumarate with a turnover number of 300 s^{-1}. Similar enzymes, which participate in bacterial breakdown of aromatic compounds isomerize *maleylacetoacetate* and *maleylpyruvate* to the corresponding fumaryl derivatives (Eq. 6.20). The – SH group of bound glutathione is thought to add to the double bond. Rotation can then occur in the enolic intermediate. Thiocyanate ion catalyzes the isomerization of maleic acid nonenzy–matically, presumably by a similar mechanism.

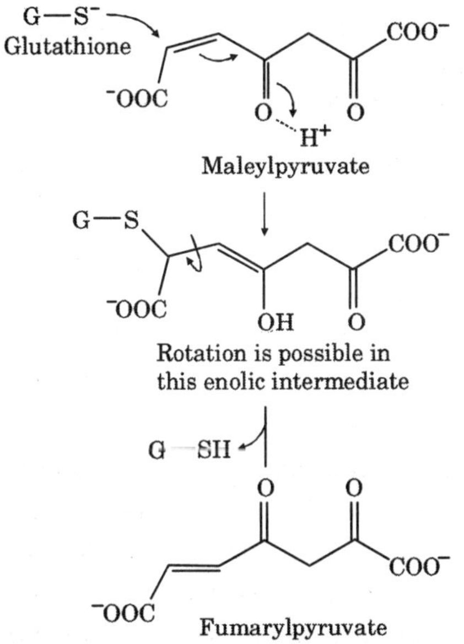

G—S⁻ Glutathione

Maleylpyruvate

Rotation is possible in this enolic intermediate

G—SH

Fumarylpyruvate ...(6.20)

Reversibility or Addition and elimination Reactions

Many addition and elimination reactions, *e.g.*, the hydration of aldehydes and ketones, and reactions catalyzed by lyases such as fumarate hydratase are strictly reversible. However, biosynthetic sequences are often nearly irreversible because of the elimination of inorganic phosphate or pyrophosphate ions. Both of these ions occur in low concentrations within cells so that the reverse reaction does not tend to take place. In decarboxylative eliminations, carbon dioxide is produced and reversal becomes unlikely because of the high stability of CO_2. Further irreversibility is introduced when the major product is an aromatic ring, as in the formation of phenylpyruvate.

Enolic Intermediates in Enzymatic Reactions

Enzymologists have freely proposed enolate anions, enols, and enamines as intermediates for many years. Such intermediates have been demonstrated for some nonenzymatic acid– or base–catalyzed reactions, but how can enzymes form enolates at pH 7 without the use of strong acids or bases? The microscopic pK_a value of an α–hydrogen in *a* ketone or aldehyde is about 17–20.

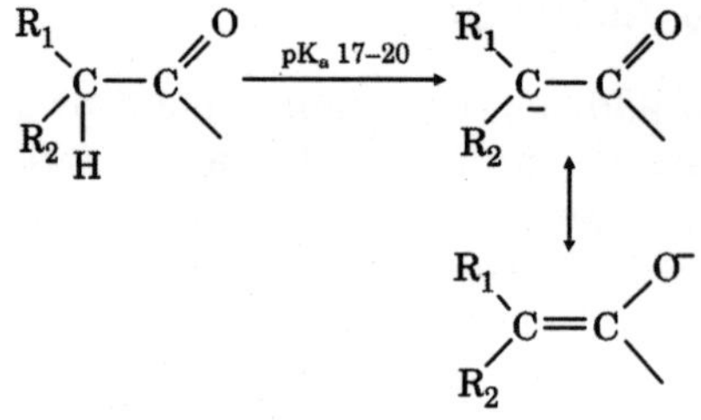

$$\xrightarrow{pK_a\ 17\text{–}20} \qquad (6.21)$$

It will be similar for an acyl-CoA. However, for a carboxylic acid, protonated on oxygen, the pK_a will be much higher (~22–25) and for carboxylate anions even higher (~29–32).

Mandelate Racemase and Related Enzymes

The degradation of mandelic acid by the bacterium *Pseudomonas putida* is initiated by mandelate racemase, another $(\alpha/\beta)_8$-barrel protein. X-ray structures of bound inhibitors together with modeling suggest that the side chain of Lys 264 is the catalytic base that abstracts the α–H from S–mandelate (Fig. 6.5) and that the catalytic pair of His 297 and Asp 270 acts as proton donor, or, in the reverse direction, as catalytic base for deprotonation of R–mandelate. The enzyme is structurally a member of the enolase superfamily and requires Mg^{2+} for activity.

The pH dependence of the reaction velocity k_{cat} reveals two pK_a values in the ES complex; 6.4 and ~10.0. These are the same for S-and *R*-isomers. How can the pK_a values be assigned? Do they each belong in part to His 297 and in part to Lys 166? Do other adjacent groups also share? In the K166R mutant the lower pK_a is raised to 8.0. The D270N mutant has lost all but 0.01% of its catalytic activity. The carboxylate group of the mandelate interacts with side chains of E317, K264, and a Mg^{2+} ion (Fig. 6.5). These may serve both to protonate the carboxylate and also to help stabilize an *aci* anion formed upon dissociation of the α-hydrogen (Eq. 6.22).

Both pK_a values for mandelic acid and its enolic form and the equilibrium constants for enolization in water are known and are given beside the arrows in Eq. 6.22. It is difficult to imagine how a base with a pK_a near neutrality could remove an α-proton with a pK_a of 22.0, which is 14 pK units away from the pK_a of the catalytic base. This corresponds to a thermodynamic barrier of 14 × 5.7 = 80 kJ/ mol, making the reaction impossibly slow. In addition to this thermodynamic barrier the *rates* of dissociation of carbon acids are known to be slow, presumably because of the lack of hydrogen bond formation between the C–H proton and the catalytic base. This *intrinsic* barrier for simple ketones has been estimated as 45 kJ/ mol for a total barrier of ~125 kJ. However, the observed $\Delta G^{\ddagger}$ as estimated from only ~57 kJ/ mol. The enzyme must catalyze the reaction by lowering the very high thermodynamic barrier and perhaps also by lowering the intrinsic barrier.

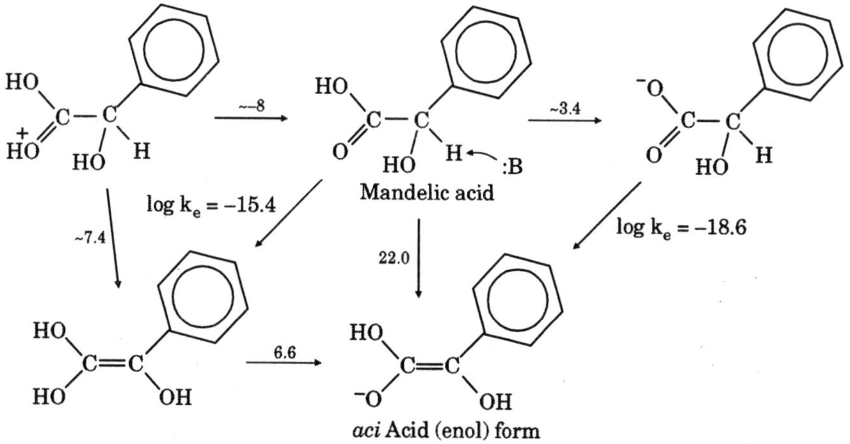

Mandelic acid

aci Acid (enol) form

...(6.22)

Protonation of the carboxylate greatly decreases the microscopic pK_a for loss of the α-hydrogen as a proton to form the enolic *aci* acid (Eq. 6.22). Double protonation, although depending upon a pK_a *of* –8, 1 would reduce the pK_a of the α-hydrogen to ~7.4, making it very easy to enolize and solving the thermodynamic problem. However, what forces could keep the substrate molecule in this unlikely state of protonation? Gerlt and Gassman proposed formation of a strong, short hydrogen bond to a carboxylate oxygen.

Formation of such a bond would lead to an increased positive charge on the alpha carbon, lowering the pK_a and therefore the thermodynamic barrier. It could also lower the intrinsic barrier, for example, by permitting, to some extent, the formation of a C—H···N hydrogen bond in the transition state. Polarization by the Mg^{2+} ion may also be involved. The very strong electrostatic forces within the active site may be sufficient to explain the formation of enolate anions or enols as intermediates in many different enzymes. A number of other racemases and epimerases may function by similar mechanisms. While some amino acid racemases depend upon pyridoxal phosphate several others function without this coenzyme.

These include racemases for aspartate, glutamate, proline, phenylalanine, and diaminopimelate epimerase. Some spiders are able to interconvert D and L forms of amino acid residues in intact polypeptide chains. Another enzyme of the mandelate pathway of degradation of aromatic rings (Fig. 6.8) is the *cis,cis–muconate lactonizing enzyme* which catalyzes the reaction of Eq. 6.23. It has a three-dimensional structure almost identical to that of mandelate racemase but has incorporated an additional feature that allows formation, by addition, of the intermediate *aci*-acid.

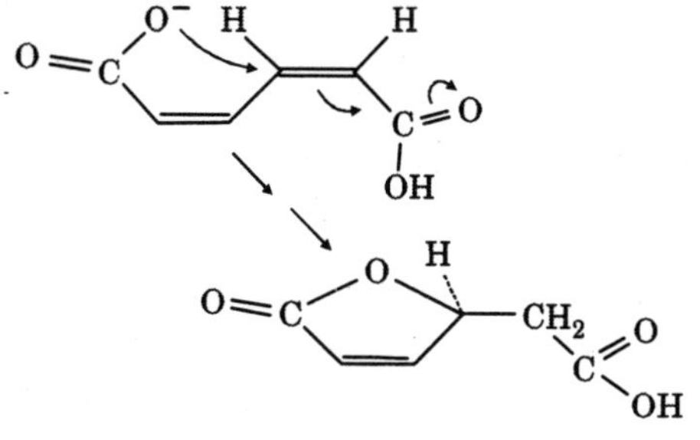

...(6.23)

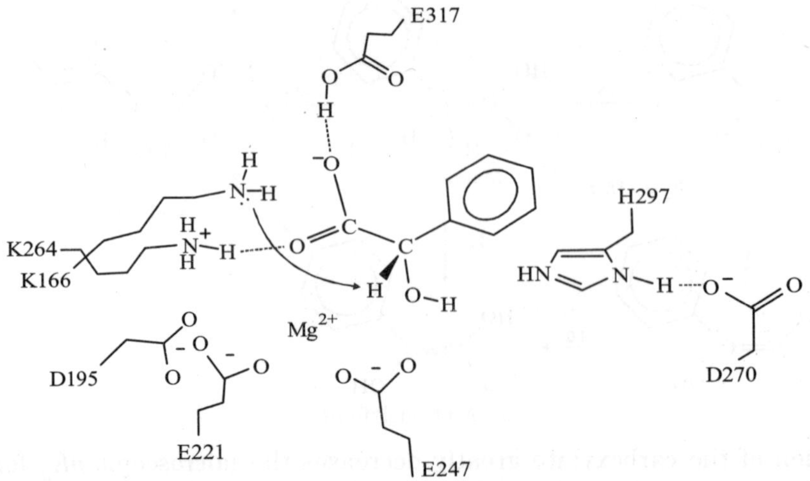

Fig. 6.5. An S-mandelate ion in the active site of mandelate racemase. Only some of the polar groups surrounding the active site are shown. The enzyme has two catalytic acid-base groups. Lysine 166 is thought to deprotonate S–mandelate to form the *aci* anion, while His 297 deprotonates R-mandelate to form the same anion.

Isomerases

The isomerases that catalyze the simplest reactions are *tautomerases* that promote the oxo-enol (keto-enol) transformation. The widely distributed oxaloacetate tautomerase (Eq. 6.24) is especially active in animal tissues. Oxaloacetate exists to a substantial extent in the enolic form: at 38°, ~ 6% enol, 13% oxo, and ~ 81% covalent hydrate. A mammalian *phenylpyruvate tautomerase* has also been investigated.

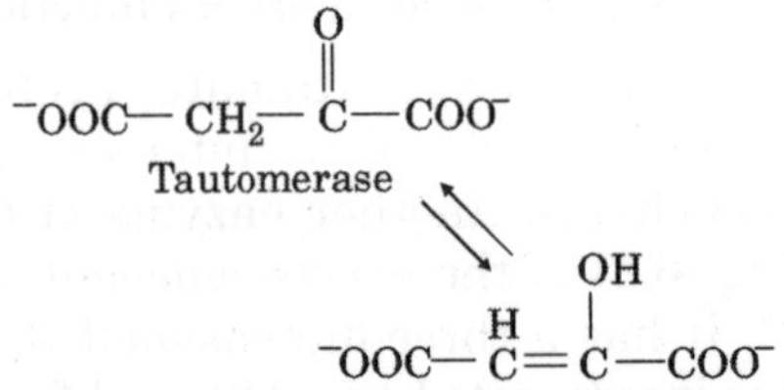

$$^{-}OOC-CH_2-\overset{O}{\overset{\|}{C}}-COO^{-} \underset{\text{Tautomerase}}{\rightleftharpoons} {}^{-}OOC-\overset{H}{C}=\overset{OH}{\overset{|}{C}}-COO^{-} \quad \text{...(6.24)}$$

The oxidation of one functional group of a molecule by an adjacent group in the same molecule is a feature of many metabolic sequences. In most cases an enolic intermediate, formed either from a ketone as in Eq. 6.25 or by a dehydration reaction is postulated.

Aldose–Ketose Interconversions

A metabolically important group of enzymes catalyze the interconversion of aldose sugars with the corresponding 2-ketoses. Several sugar phosphates undergo rapid isomerization. *Glucose 6-phosphate isomerase* (Eq. 6.25) appears to function in all cells with a high efficiency. The 132 kDa dimeric protein from muscle converts glucose 6-phosphate to fructose 6-phosphate with a turnover number of ~10^3 S^{-1}. Hereditary defects in this enzyme cause a variety of problems that range from mild to very severe. *Mannose 6-phosphate isomerase* also forms fructose 6-phosphate, while *ribose 5-phosphate isomerase* interconverts the 5-phosphates of D-ribose and D-xylulose in the pentose phosphate pathways. Other enzymes, most often metalloenzymes, catalyze the isomerization of free sugars.

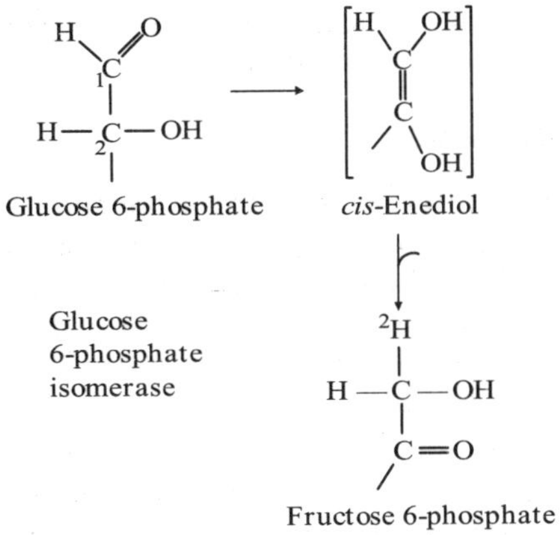

H–C_1=O, H–C_2–OH → [H(OH)C=C(OH)]

Glucose 6-phosphate → *cis*-Enediol

Glucose 6-phosphate isomerase

^{2}H–C(H)–OH, C=O

Fructose 6-phosphate

These enzymes vary widely in secondary and tertiary structure. Mannose-6-phosphate isomerase is a 45 kDa Zn^{2+}-containing monomer. The larger 65 kDa *L-fucose isomerase,* which also acts on D-arabinose, is a hexameric Mn^{2+} dependent enzyme. *L Arabinose isomerase* of *E. coli,* which interconverts arabinose and L-ribulose, is a hexamer of 60-kDa subunits while the *D-xylose isomerase* of *Streptomyces* is a tetramer of 43-kDa subunits. The nonenzymatic counterpart of the isomerization catalyzed by the enzyme is the base-catalyzed *Lobry deBruyn-Alberda van Ekenstein transformation* (Eq. 6.25). In 1895, Emil Ficher proposed an enediol interme-diate for this isomerization.

As would be expected, the enzyme-catalyzed isomerization of glucose-6-phosphate in ^2H_2O is accompanied by incorporation of deuterium into the product fructose 6-phosphate at C–1. In the reverse reaction ^{2}H-containing fructose 6-phosphate was found to react at only 45% of the rate of the ^{1}H-containing compound. Thus, the primary deuterium isotope effect expected for a rate-limiting cleavage of the C–H bond was observed. When fructose 6-phosphate containing both ^{2}H and ^{14}C in the 1 position was isomerized in the presence of a large amount of nonlabeled fructose 6-phosphate, the product glucose 6–phosphate contained not only ^{14}C but also ^{2}H, and the distribution indicated that the ^{2}H had been transferred from the C–1 position into the C–2 position.

It was concluded that in over half the turnovers of the enzyme, ^{2}H removed from C–1 is put back on the same molecule at C–2. This intramolecular transfer of a proton suggests a syn transfer, the proton being removed and put back on the same side of the molecule. The carrier of the proton may be a histidine side chain. This result, together with the known configuration of glucose at C–2, indicates that the intermediate is the *cis*-enediol and that addition of a proton at either C–1 or C–2 of the enediol is to the *re* face. However, mannose-6-phosphate isomerase catalyzes addition to the *si* face.

Catalysis of Ring Opening by Isomerases

Glucose–6–phosphate isomerase catalyzes a second reaction, namely, the opening of the ring of the α–anomer of glucose 6–phosphate (one–half of the mutarotation reaction; Eq. 6.88). Noltmann suggested a concerted acid–base catalysis by two side chain groups as is indicated. An NMR shady showed that the isomerization of the α anomer occurs at least ten times faster than that of the β. Thus, β-glucose 6-phosphate is first converted to α-glucose 6-phosphate before it can be isomerized to fructose 6-phosphate.

Triose Phosphate Isomerase

This dimeric 53-kDa enzyme interconverts the 3-phosphate esters of glyceraldehyde and dihydroxyacetone and is the fastest enzyme participating in glycolysis. Its molecular activity at 25° is ~2800 s^{-1} in the direction shown in Eq. 6.26 and ~250 s^{-1} in the reverse direction (the predominant direction in metabolism) and is thought to operate at the diffusion-controlled limit. Each of the identical subunits consists of a striking $(\alpha/\beta)_8$ barrel (Fig. 6.6) with an active site at the carboxyl ends of the β strands.

Structures of the enzyme containing bound inhibitors such as phosphoglycolo-hydroxamate have also been determined (Fig. 6.6). Triose phosphate isomerase is also one of the most investigated of all enzymes. Not only are its catalytic properties unusual but also there are known defects in the human enzyme and it is also a potential target for antitrypanosomal drugs.

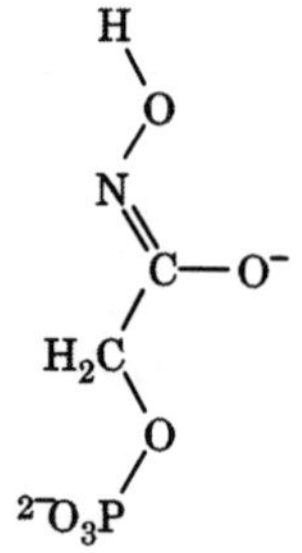

H
O
N
C—O^-
H_2C
O
$^{2-}O_3P$

Phosphoglycolohydroxamate (PGH), a tight-binding inhibitor (K_d = 15μM) that resembles enediolate 1 of Eq. 6.26

Although its high catalytic activity might favor intramolecular transfer of the proton removed by triose phosphate isomerase, little such transfer has been observed. This suggested that a relatively weak base such as a carboxylate group might serve as the proton acceptor at the C–3 position of dihydroxyacetone phosphate. Covalent labeling experiments and later X-ray studies have implicated Glu 165, whose carboxylate group is thought to remove the *pro–R* hydrogen atom from the hydroxymethyl group of dihydroxyacetone phosphate as indicated in Eq. 6.26. When this residue is replaced by Asp, most activity is lost. Kinetic studies suggest that this carboxyl group has a pK_a of ~3.9 in the free enzyme. However, k_{cat} is independent of pH up to pH 10.

The pK_a of the phosphate also affects the binding (K_m), with only the phosphate dianion being a substrate. Study of the infrared spectrum of dihydroxyacetone bound to the enzyme revealed a shift of the carbonyl bands at 1733 cm^{-1} by about 20 cm^{-1} to ~ 1713 cm^{-1}. This might indicate a polarization and stretching of the carbonyl group by a positively charged histidine or lysine side chain of the protein. The His 95 side chain is appropriately placed (Fig. 9.6) to function in this way and its replacement by glutamine decreases catalytic activity by a factor of 400. The $-NH_3^+$ group of lysine 12, another essential residue, is apparently needed for substrate binding. After product is released, an isomerization within the enzyme is usually required to prepare it for acceptance of a new substrate. This may involve movement of protons between side chain groups or a conformational change or both. In the case of triose phosphate isomerase, a very rapid isomerization requiring about ~ 1 ms has been detected.

To investigate further the function of His 95, Lodi and Knowles recorded the ^{13}C and ^{15}N NMR spectra of the three histidine rings both in unligated enzyme and in the

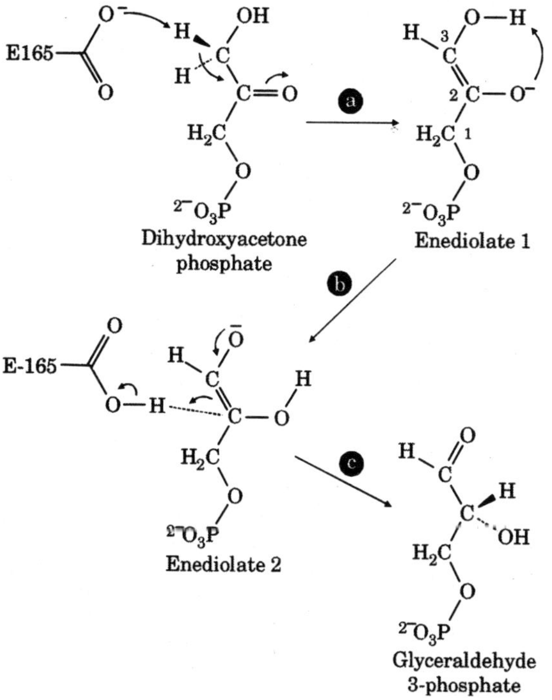

E165 Dihydroxyacetone phosphate (a) Enediolate 1 (b) E-165 Enediolate 2 (c) Glyceraldehyde 3-phosphate ...(3.26)

phosphoglycolohydroxamate (PGH) complex. The results were a surprise. The His 95 resonances did not change at all when the pH was changed from below 5 to 9.9, and the chemical shift values (shown above) indicate clearly that the ring is *unprotonated* in both free enzyme and in the complex. The key chemical shift values are shown on the diagram to

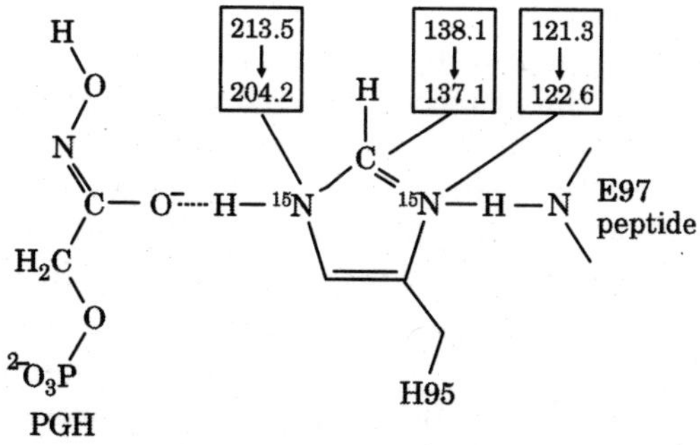

213.5 → 204.2; 138.1 → 137.1; 121.3 → 122.6; E97 peptide; PGH; H95

the left; the arrows indicate the changes upon formation of the PGH complex. These results suggest that H95 is never protonated and that if it acts as the catalytic acid it does so by dissociating to an imidazolate ion (Eq. 6.27).

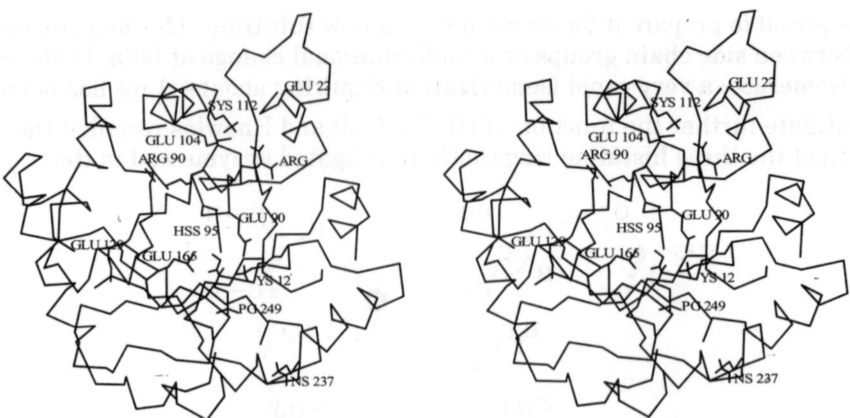

Fig. 6.6. Stereoscopic view into the active site of triose phosphate isomerase showing side chains of some charged residues; PGH, a molecule of bound phosphoglycolohydroxamate, an analog of the substrate enolate. The peptide backbone, as an alpha-carbon plot, is shown in light lines. The $(\alpha/\beta)_8$-barrel structure is often called a TIM barrel because of its discovery in this enzyme.

Bound enolate intermediate

Imidazolate

Enediol

(6.27)

The imidazole would donate a proton to the enolate ion and then remove a proton from the resulting enediol. Although the proposed chemistry is unusual, it is argued that the high pK_a of the neutral imidazole acting as a proton donor would be matched with the high pK_a of the enol, permitting rapid reaction. Nevertheless, two other possibilities exist. Theoretical calculations support the idea that the proton transfer between the two oxygens in the enediolate may occur *without catalysis by a proton donor* as indicated in step *b* of Eq. 6.26.

Another possibility is that a transient protonation of H95 by the adjacent E97 carboxyl group occurs and allows the histidine to participate in the proton transfer. Study of kinetic isotope effects has indicated coupled motion of protons and proton tunneling. Another detail should be mentioned.

The active site of triose phosphate isomerase is formed by a series of loops connecting the a helices and β strands of the barrel. One of those loops, consisting of residues 167–176, folds over the active site after the substrate is bound to form a hinged lid that helps to hold the substrate in the correct orientation for reaction. When the lid, which can be seen in Fig. 6.6, closes, the peptide NH of G171 forms a hydrogen bond to a phosphate oxygen atom of the substrate. This is only one of many known enzymes with deeply buried active sites that close in some similar fashion before a rapid reaction occurs.

$$\underset{\text{Enediolate 2}}{H{-}C(O^-){=}C(OH){-}H_2C{-}O{-}PO_3^{2-}} \xrightarrow{-P_i} H{-}C(=O){-}C(OH){=}CH_2 \longrightarrow \underset{\text{Methylglyoxal}}{H{-}C(=O){-}C(=O){-}CH_3} \quad ...(6.28)$$

Although enzymes tend to be extremely specific they are not always completely able to avoid side reactions. Triose phosphate isomerase releases small amounts of *methylglyoxal* (Eq. 6.28), presumably as a result of elimination of phosphate from enediolate 2 of Eq. 6.26. Deletion of four residues from the hinged lid produced an enzyme in which this side reaction was increased 5.5-fold.

Xylose Isomerase and the Hydride Shift Mechanism

This bacterial enzyme, which isomerizes D-xylose to D-xylulose, has an $(\alpha/\beta)_8$ barrel similar to that of triose phosphate isomerase. It is of industrial importance because it also catalyzes isomerization of D-glucose to the sweeter sugar D-fructose, a reaction used in preparation of hig-fructose syrup. For this purpose the isomerase is often immobilized on an insoluble matrix such as diethylaminoethyl cellulose. This enzyme, as well as other isomerases that act on free sugars, requires a metal ion such as Mg^{2+}, Co^{2+}, or Mn^{2+}. The three-dimensional structures, solved in several different laboratories, show that there are two metal ions (ordinarily Mg^{2+}) about 0.5 nm apart and held by an array of glutamate and aspartate side chains.

One glutamate carboxylate forms short ionic bonds to both metal ions and an essential histidine is also coordinated with metal ion 2. The three–dimensional structure is superficially similar to that of triose phosphate isomerase but there are major differences in properties. Xylose isomerase requires metal ions, acts on unphosphorylated non–ionic substrates, and does not catalyze detectable exchange of protons with the solvent. Furthermore, *X-ray structures do not show the presence of any catalytic base that could initiate formation of an intermediate enolate ion.* These facts suggested an alternative isomerization mechanism, one involving a hydride ion shift and well known from studies of the Cannizarro reaction. Because the non-ionic substrates and inhibitors bind weakly it has been difficult to obtain a clear picture of events in the active site. The substrate is the α-anomer of D-xylopyranose and the enzyme catalyzes the opening of the sugar ring (step *a,* Eq. 6.29). The details of this process are not clear but acid-base catalysis as in is probable.

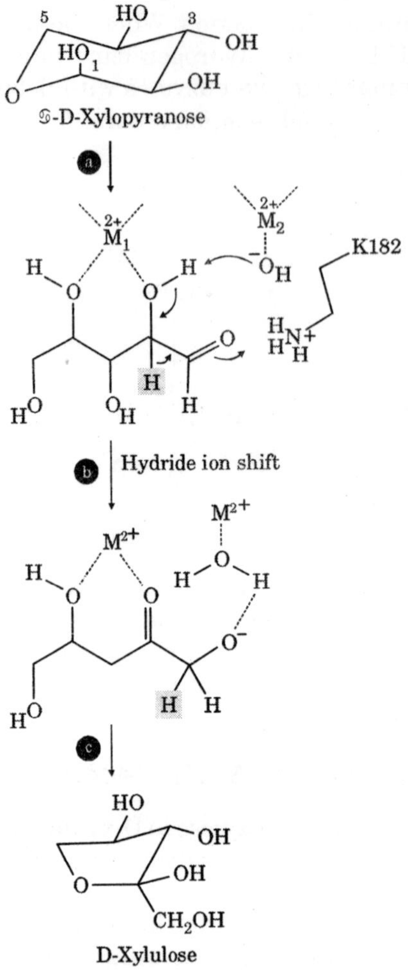

β-D-Xylopyranose

Hydride ion shift

D-Xylulose

...(6.29)

There is a consensus that the open form of the sugar is bound to one of the two metal ions as shown in Eq. 6.29 and that a bound hydroxide ion on the second metal provides the catalytic base to remove a proton from the –OH at position 2. The isomerization occurs with the shaded hydrogen atom being shifted as H^- to the carbon atom of the carbonyl group. During the reaction metal ion 2 moves apart from metal ion 1 by ~0.09 run, another fact that has made analysis of X-ray data difficult. Protonation of the alkoxide ion at C–2 is followed by ring closure (step *c*, Eq. 6–29).

3–Oxosteroid Isomerases

Cholesterol serves in the animal body as a precursor of all of the steroid hormones, including the 3-oxosteroids progesterone and testosterone. The 3-hydroxyl group of cholesterol is first oxidized to an oxo group. This is followed by an essentially irreversible migration of the double bond in the 5, 6-position into conjugation with the carbonyl group (Eq. 6.30) catalyzed by a 3-oxo-Δ^5-steroid isomerase. The small 125-residue enzyme from the bacterium *Pseudomonas testosteroni* has been studied by a great variety of methods. The subunits associate as dimers or higher oligomers. The enzyme has a high content of nonpolar amino acid residues and is soluble in high concentrations of ethanol. This property is compatible with the location of the eukaryotic enzyme in the endoplasmic reticulum.

The substrate binding site is a hydrophobic cavity. Oxosteroid isomerase has a remarkably high molecular activity (~0.75×10^5 s^{-1}). Substrates containing ^2H in the 4 position react only one–fourth as rapidly as normal substrates. The large isotope effect suggested that cleavage of the C–H bond to form an enzyme-stabilized enolate anion is rate limiting (Eq. 6.30, step *a*). The proton in the axial position at C–4 is removed preferentially but without complete stereo-specificty. The abstracted proton must be carried by a group on the enzyme and returned to the 6 position of the substrate, again in an axial orientation (Eq. 6.30, step *b*), a suprafacial transfer. Studies of the binding of spin-labeled substrate analogs, molecular modeling, and directed mutation indicate that Asp 38 is this proton carrier.

Little exchange of the proton with solvent is observed, presumably because of the extreme rapidity of the isomerase action. However, competitive inhibitors such as nortestosterone, whose double-bond arrangement is that of the product in Eq. 6.30, undergo exchange of one of the hydrogens at C–4 with the medium. The ultraviolet absorption band of the inhibitor at 248 nm is shifted to 258 nm upon combination with the enzyme, presumably as a result of formation of the enolate anion. Tyrosine 14 has been identified as the catalytic acid, able to stabilize the enolate anion as depicted in Eq. 6.30. A high-resolution NMR structure and X-ray crystal structures have revealed an adjacent Asp 99 side chain that may assist.

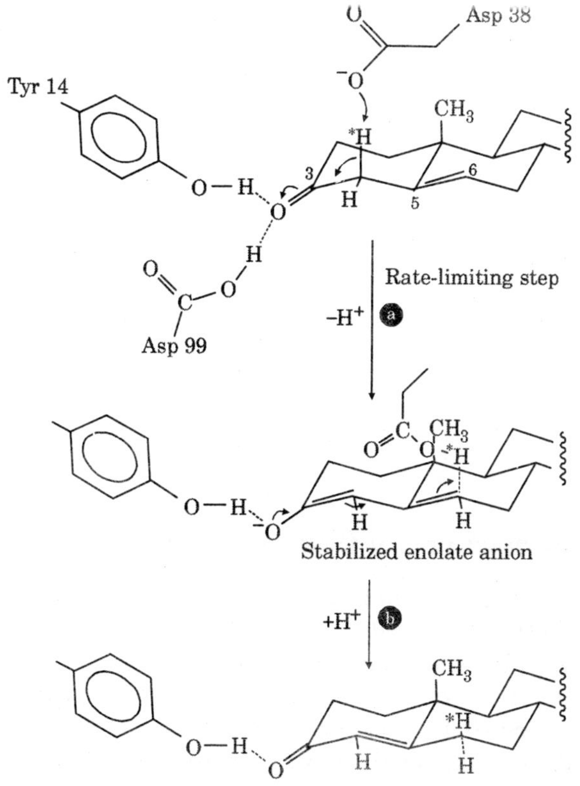

Asp 38
Tyr 14
CH_3
*H
3
6
O–H
O
H
5
H
O
C
O
Asp 99
Rate-limiting step
$-H^+$ a
CH_3
C
O
O
*H
O–H
^-O
H
H
Stabilized enolate anion
$+H^+$ b
CH_3
*H
O–H
O
H
H

...(6.30)

Enzymatic isomerization of *cis*-aconitate to *trans*–aconitate apparently also involves proton abstraction, with resonance in the anion extending into the carboxylic acid group. Its mechanism may be directly related to that of the oxosteroid isomerase. However, there are other 1, 3-proton shifts in which neither a carbonyl nor a carboxyl group is present in the substrate.

4-Oxalocrotonate Tautomerase

This bacterial enzyme, which functions in the degradation of toluene, is actually an isomerase. It catalyzes rapid interconversion of an unconjugated unsaturated α-oxoacid such as 4-oxalocrotonate with an intermediate enol (which may leave the enzyme) and the isomeric conjugated oxoacid (Eq. 6.31). A related 5-carboxymethyl-2-hydroxymuconate isomerase catalyzes the same reaction when R = $-CH_2COO^-$ in Eq. 6.31. It is an unusually small enzyme consisting of only 62 amino acid residues.

The pH dependence reveals pK_a values of 6.2 and 9.0 in the free enzyme and 7.7 and 8.5 in the ES complex. The pK_a of 6.2 has been associated with the amino-terminal proline 1. - Oxalocrotonate tautomerase is one of a small group of enzymes that have been synthesized nonenzymatically with both L. amino acids and as a mirror image constructed with D amino acids.

The enol 2-Hydroxymuconate

4-Oxalocrotonate 2-Oxo-3-hexenedioate ...(6.31)

Internal Oxidation–Reduction by Dehydration of Dihydroxyacids

When a carboxylic acid contains hydroxyl groups in both the α and β positions, dehydration leads to formation of an enol that can tautomerize to 2-oxo-3-deoxy derivatives of the original acid (Eq. 6.32). Thus, phosphogluconate dehydratase yields 2-oxo-3-deoxyphosphogluconate as the product. When the reaction is carried out in ^2H_2O the ^2H is incorporated with a random configuration at C–3, indicating that the enzyme catalyzes only the dehydration and that the tautomerization of the enol to the ketone is nonen-zymatic.

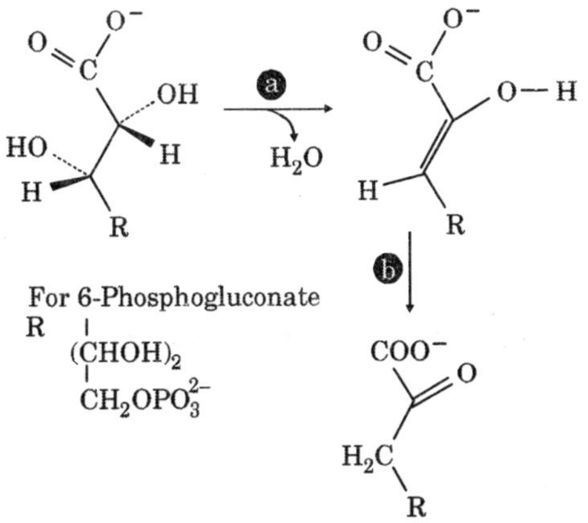

a: H_2O released; b

For 6-Phosphogluconate R = $(CHOH)_2CH_2OPO_3^{2-}$

...(6.32)

The reaction of Eq. 6.32 initiates a unique pathway of sugar breakdown, the Entner-Douderoff pathway, in certain organisms. The 6-phosphoglu-conate is formed by oxidation of the aldehyde group of glucose 6-phosphate. This pattern of oxidation of a sugar to an aldonic acid followed by dehydration according to Eq. 6.32 occurs frequently in metabolism. A related reaction is the dehydration of 2-phosphogly-cerate by enolase (Eq. 6.15). In this case the product is phosphoenolpyruvate, a stabilized form of the enolic intermediate of Eq. 6.32 (when R = H).

Formation and Metabolism of Methylglyoxal (Pyruvaldehyde)

The rather toxic methylglyoxal is formed in many organisms and within human tissues. It arises in part as a side reaction of triose phosphate isomerase (Eq. 6.28) and also from oxidation of acetone or aminoacetone, a metabolite of threonine. In addition, yeast and some bacteria, including *E. coli,* have a *methylglyoxal synthase* that converts dihydroxyacetone to methylglyoxal, apparently using a mechanism similar to that of triose phosphate isomerase. It presumably forms eiiediolate 2 of Eq. 6.26, which eliminates inorganic phosphate to yield methylglyoxal as in Eq. 6.28. Methylglyoxal is converted to D-lactate by the two-enzyme *glyoxalase* system (Eq. 6.33).

The combined action of methylglyoxal synthase and the glyoxalases provides a bypass to the usual glycolysis pathway. Although it does not provide energy to the cell, it releases inorganic phosphate from sugar phosphates that may accumulate under conditions of low phosphate because the free P_i concentration is too low to support the gly-ceraldehyde 3-phosphate dehydrogenase reaction. Glyoxalase acts not only on methylglyoxal but also on other α-oxo-aldehydes. It is thought to be an important enzyme system that protects cells against these potentially dangerous metabolites. Glyoxalase consists of a pair of enzymes, *glyoxalase I* and *glyoxalase II,* which catalyze the reactions of Eq. 6.33. Each subunit of the 183 residue human glyoxalase I contains one tightly bound Zn^{2+} ion. However, the *E. coli* enzyme is inactive with Zn^{2+} and maximally active with Ni^{2+}.

The enzyme requires glutathione as a cofactor, and in step *a* of the reaction (Eq. 6.33) the glutathione adds to form a thiohemiacetal which is then isomerized in step *b*. During this step retention of the abstracted proton is so complete that it was earlier thought to function by an intramolecular hydride ion shift as in xylose isomerase. More recent evidence favors an enolate anion intermediate. However, the three–dimensional structure is not related to that of other isomerases and the exact mechanism remains uncertain. Glyoxylase II, which catalyzes step *c* of Eq. 6.33, is an esterase.

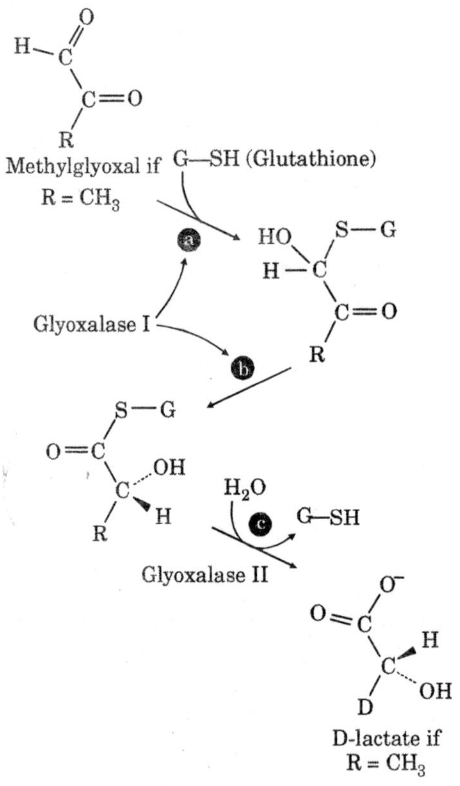

Methylglyoxal if R = CH_3

G—SH (Glutathione)

Glyoxalase I (a), (b)

H_2O (c) G–SH

Glyoxalase II

D-lactate if R = CH_3

...(6.33)

Beta Cleavage and Condensation

In the preceding sections of this chapter we examined displacement and addition reactions involving nucleophilic centers on O, N, or S. Bonds from carbon to these atoms can usually be broken easily by acidic or basic catalysis. The breaking and making of C–C bonds does not occur as readily and the "carbon skeletons" of organic molecules often stick together tenaciously. Yet living cells must both form and destroy the many complex, branched carbon compounds found within them. A major mechanistic problem in cleavage or formation of carbon-carbon bonds is the creation of a nucleophilic center on a carbon atom.

The problem is most often solved by using the *activating influence of a carbonyl group to generate a resonance–stabilized enolate anion.* Just as the presence of a carbonyl group facilitates cleavage of an adjacent C–H bond, so it can also assist the cleavage of a C–C bond. The best known reactions of this type are the *aldol cleavage* and the *decarboxylation of β–oxo acids*. The latter has been referred to as *β* decarboxylation and its reverse as β carboxylation. In this book these terms have been extended to include other reactions by which bonds between the α and β carbon atoms of a carbonyl compound are broken or formed, and these will be referred to as β *cleavage and β condensation.*

Cleavage of C–H bond facilitated by a carbonyl group

β Cleavage with release of CO_2

The β *condensation* reactions consist of displacement or addition reactions in which an enzyme-bound enolate anion acts as the nucleophile. We can group these condensation reactions into three categories as indicated by reaction types 5A, 5B, and 5C.

Displacement on a Carbonyl Group

A β-oxo acid, with proper catalysis, is susceptible to hydrolysis by attack of water on the carbonyl group. An example is the reaction catalyzed by *oxaloacetate acetylhydrolase* which has been isolated from *Asper gillus niger* (Eq. 6.34). A related cleavage is catalyzed by ribulose bisphosphate carboxylase (see Eq. 6.48).

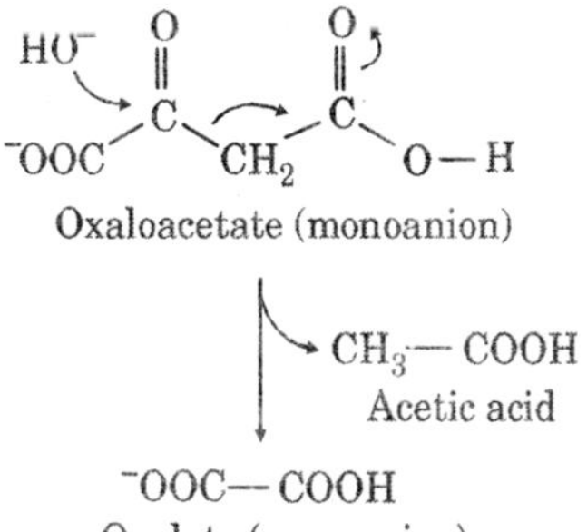

Oxaloacetate (monoanion)

CH_3—COOH
Acetic acid

$^-$OOC—COOH
Oxalate (monoanion)

...(6.34)

The *thiolases* are lyases that cleave β-oxoacyl derivatives of CoA by displacement with a thiol group of another CoA molecule (Eq. 6.35). This is the chain cleavage step in the β oxidation sequence by which fatty acid chains are degraded. Biosynthetic thiolases catalyze the condensation of two molecules of acetyl-CoA to form acetoacetyl-CoA, a precursor to cholesterol and related compounds and to poly-β-hydroxybutyrate. Because acetyl-CoA is a thioester, the reaction is usually described as a *Claisen condensation.*

A related reaction, involving decarboxylation of malonyl-CoA, is a step in fatty acid synthesis. Since the thiolases are inhibited by –SH reagents it has been suggested that a thiol group in the enzyme reacts initially with the β-carbonyl group as in Eq. 6.35 to give an enzyme-bound S-acyl intermediate. The acyl group is then transferred to CoA in a second step.

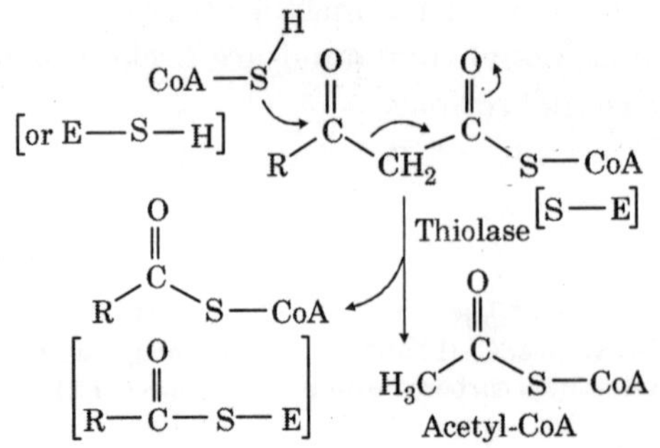

CoA—S—H [or E—S—H]

R—C(=O)—CH$_2$—C(=O)—S—CoA [S—E]

Thiolase

R—C(=O)—S—CoA [R—C(=O)—S—E]

H$_3$C—C(=O)—S—CoA
Acetyl-CoA

...(6.35)

A very similar reaction is catalyzed by 3-hydroxy-3-methylglutaryl-CoA lyase (HMG–CoA lyase), which functions in the formation of acetoacetate in the human body and also in the catabolism of leucine and in the synthesis of *3-hydroxy-3-methylglutaryl-CoA,* the presursor of cholesterol.

Addition of an Enolate Anion to a Carbonyl Group or an Imine; Aldolases

The *aldol condensation* (Eq. 6.36), which is also illustrated in is one of the more common reactions by which C–C bonds are formed and, in the reverse reaction, cleaved in metabolism. Aldolases are classified into two major types. The type II aldolases are metal-ion dependent, the metal ion stabilizing the intermediate enolate ion (Eq. 6.36, steps *a* and *b*). Type I aldolases, which include the most studied mammalian enzymes, have a more complex mechanism involving intermediate Schiff base forms (Eq. 6.36, steps *a′*, *b′*, *c′*, *d′*).

The best known members of this group are the *fructose bisphosphate aldolases* (often referred to simply as aldolases), which cleave fructose-l, 6-P$_2$ during glycolysis. These enzymes have been found in all plant and animal tissues examined and are absent only from a few specialized bacteria. Three closely related isoenzymes are found in vertebrates. The much studied rabbit muscle aldolase A is a 158-kDa protein tetramer of identical peptide chains. Aldolase B, which is lacking in hereditary fructose intolerance, predominates in liver and isoenzyme C in brain. Treatment with sodium borohydride of the enzyme-substrate complex of aldolase A and dihydroxyacetone phosphate leads to formation of a covalent linkage between the protein and substrate.

This and other evidence suggested a Schiff base intermediate (Eq. 6.36). When ^{14}C-containing substrate was used, the borohydride reduction labeled a lysine side chain in the active site. The radioactive label was followed through the sequence determination and was found on Lys 229 in the chain of 363 amino acids. The enzyme is another $(\alpha/\beta)_g$-barrel protein and the side chain of Lys 229 projects into the interior of the barrel which opens at the C-terminal ends of the strands. The conjugate base form of another lysine, Lys 146, may represent the basic group B in Eq. 6.36.

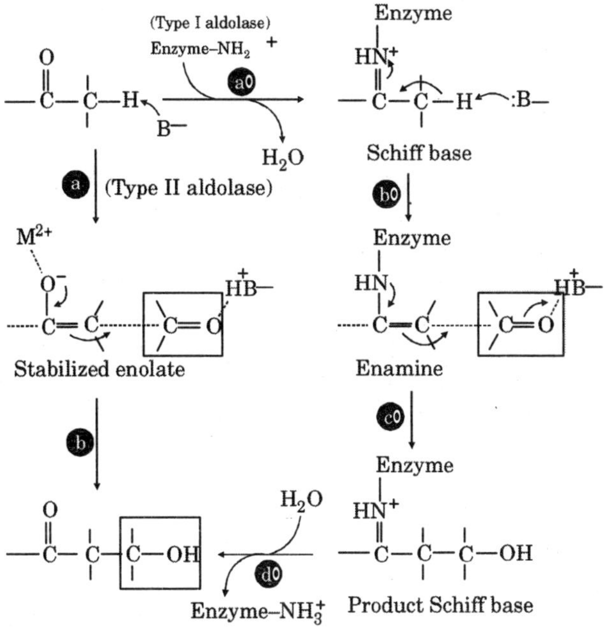

(Type I aldolase)
Enzyme–NH_2 +
a0
H_2O
Enzyme
HN^+
Schiff base
a (Type II aldolase)
b0
M^{2+}
Stabilized enolate
Enamine
b
c0
Enzyme
Product Schiff base
d0
Enzyme–NH_3^+

...(6.36)

step *b*. Another possibility is that the adjacent phosphate group of the substrate acts as the acid-base catalyst for this step. Aldolase A has been altered by mutations into a monomeric form that retains high catalytic activity, something that has not often been accomplished for oligomeric enzymes.

The *type II aldolases* are not inactivated by sodium borohydride in the presence of substrate. A probable function of the essential metal ion is to polarize the carbonyl group as indicated in Eq. 6.36. In both yeast and *E. coli* aldolase the Zn^{2+} is held by 3 imidazole groups. An arginine side chain is a conserved residue involved in substrate binding in several class II aldolases. Some blue-green algae contain both types of aldolase, as do the flagellates *Euglena* and *Chlamydomonas.*

The catabolism of L-fucose by *E. coli* requires cleavage of L-fuculose-1-phosphate to form dihydroxy acetone phosphate and D–lactaldehyde by a class II aldolase.

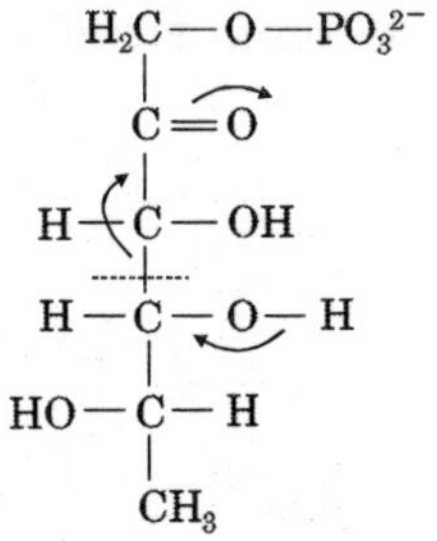

L-Fuculose-1-phosphate

The mechanism of chain cleavage proposed on the basis of the structure and modeling is illustrated in Fig. 6.7.

Special aldolases of both classes cleave and form C–C bonds throughout metabolism. Several of them act on 2-oxo-3-deoxy substrates forming pyruvate as one product (Eq. 6.37). The aldehyde product varies. In the Entner-Doudoroff pathway of carbohydrate metabolism 3-deoxy-2-oxo-6-phospho-gluconate (KDPG) is cleaved to pyruvate and 3-phos-phoglyceraldehyde. The same products arise from the corresponding phosphogalactonate derivative.

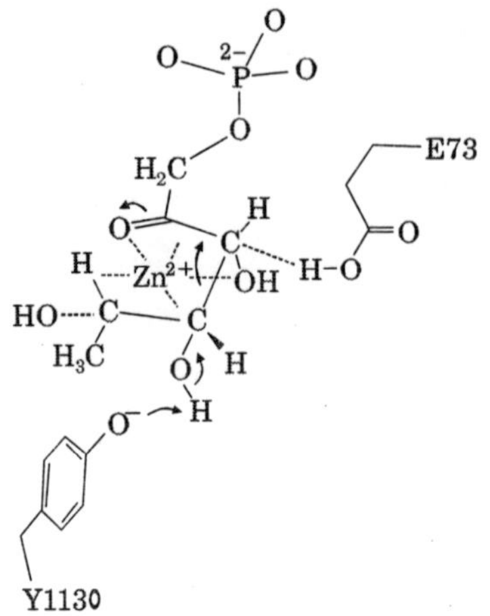

Fig. 6.7. Interaction of the bound zinc ion of L-fuculose-1-phosphate aldolase and catalytic side chains with the substrate in the acitve site of the enzyme as revealed by X-ray crystallography and modeling.

The subunits of the trimeric KDPG aldolase have an $(\alpha/\beta)_8$-barrel structure similar to that of eukaryotic fructose 1, 6-bisphosphate aldolase. The 8-carbon sugar acid "KDO" of bacterial cell walls is cleaved by another aldolase. The catabolism of hydroxy-proline leads to 4-hydroxy-2-oxoglutarate, which is cleaved to pyruvate and glyoxylate. In the catabolism of deoxynucleotides, another aldolase converts 2-deoxyribose 5-phosphate to acetaldehyde and gly-ceraldehyde 3-phosphate. Closely related to aldolases is *transaldolase,* an important enzyme in the pentose phosphate pathways of sugar metabolism and in photosynthesis.

The mechanism of the transaldolase reaction is similar to that used by fructose-l, 6-bisphosphate aldolase with a lysine side chain forming a Schiff base and catalytic aspartate and glutamate side chains.

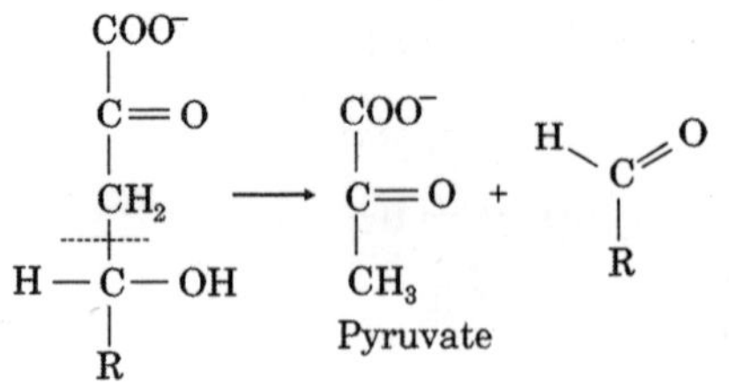

$$COO^- - C{=}O - CH_2 - CH(OH) - R \longrightarrow \underset{\text{Pyruvate}}{COO^- - C{=}O - CH_3} + H - C({=}O) - R \qquad \text{...(6.37)}$$

Table 6.1. Products Arising from Reactions of Acetyl-CoA with a Second Substrate with Catalysis by a Polycarboxylate Synthase

Ketone substrate		*Product*		*Further metabolites*
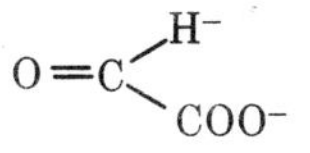 $O{=}C(H^-)(COO^-)$ Glyoxylate	→	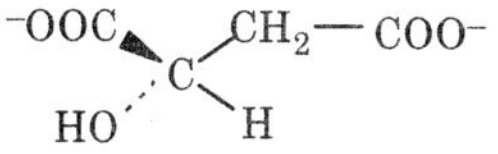 ^-OOC, $CH_2—COO^-$, HO, H on C L-Malate	→	Carbohydrates, etc,. via glyoxylate pathway
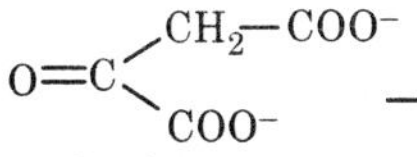 $O{=}C(CH_2—COO^-)(COO^-)$ Oxaloacetate	*Si*-Citrate synthase →	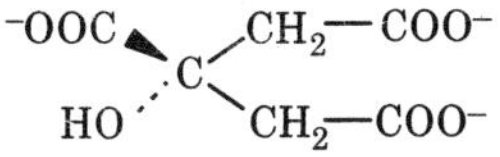 ^-OOC, $CH_2—COO^-$, HO, $CH_2—COO^-$ on C Citrate	◄······	2-Carbon unit from acetyl-CoA occupies *pro-S* position
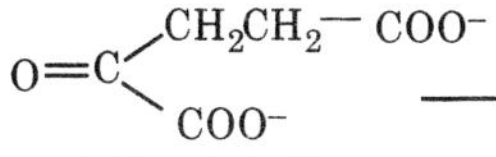 $O{=}C(CH_2CH_2—COO^-)(COO^-)$ 2-Oxoglutrate	→	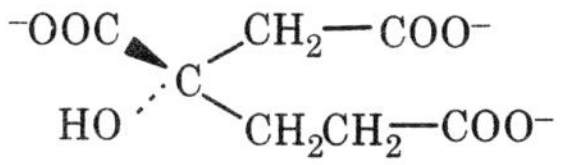 ^-OOC, $CH_2—COO^-$, HO, $CH_2CH_2—COO^-$ on C Homocitrate	→	Lysine via α-aminoadipic acid
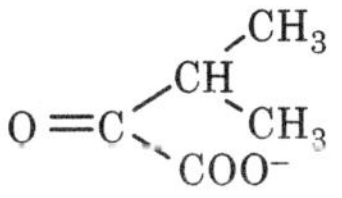 $O{=}C(CH(CH_3)CH_3)(COO^-)$ 2-Oxoisovalerate	→	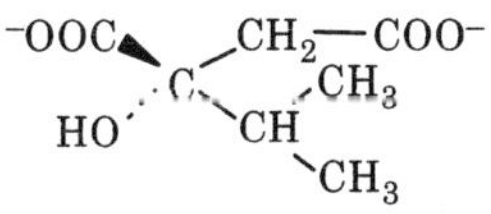 ^-OOC, $CH_2—COO^-$, HO, $CH(CH_3)CH_3$ on C α-Isopropylmalate	→	Leucine
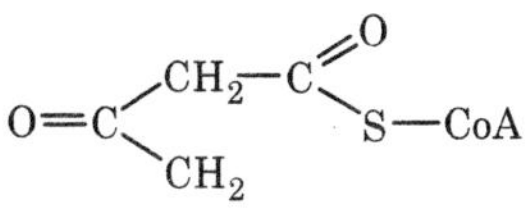 $O{=}C(CH_2—C(=O)—S—CoA)(CH_3)$ Acetoacetyl-CoA	→	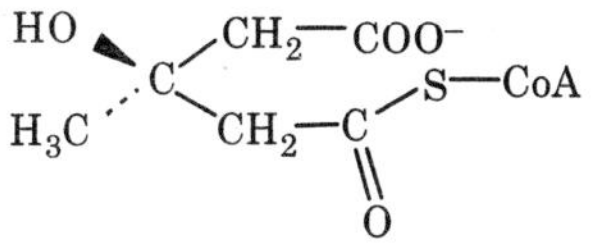 HO, $CH_2—COO^-$, H_3C, $CH_2—C(=O)—S—CoA$ on C S-3-Hydroxy-3methylglutaryl-CoA (HMG-GoA)	↗ ↘	Free acetoacetate Isoprenoid compounds

Polycarboxylic acid synthases. Several enzymes, including **citrate synthase,** the key enzyme which catalyzes the first step of the citric acid cycle, promote condensations of acetyl-CoA with ketones (Eq. 6.38). An α-oxo acid is most often the second substrate, and a thioester intermediate (Eq. 6.38) undergoes hydrolysis to release coenzyme A. Because the substrate acetyl-CoA is a thioester, the reaction is often described as a Claisen condensation. The same enzyme that catalyzes the condensation of acetyl-CoA with a ketone also catalyzes the second step, the hydrolysis of the CoA thioester. These polycarboxylic acid synthases are important in biosynthesis. They carry out the initial steps in a *general chain elongation process.* While one function of the thioester group in acetyl-CoA is to activate the methyl hydrogens toward the aldol condensation, the subsequent hydrolysis of the thioester linkage provides for overall irreversibility and "drives" the synthetic reaction. The stereochemistry of the reaction is also illustrated in Eq. 6.38. These enzymes may be classified by designating the face of the carbonyl group to which the enolate anion adds.

The *si* face is up in Eq. 6.38. The common citrate synthase of animal tissues and that of *E. coli* condense with the si face and are designated (*si*)-citrate synthases. A few anaerobic bacteria use citrate (*re*)-synthase having the opposite stereochemistry. Many citrate synthases are ~100-kDa dimers but some are hexamers and are allosterically inhibited by NADH. The second substrates and products of several related reactions of acetyl-CoA are summarized in Table 6.1.

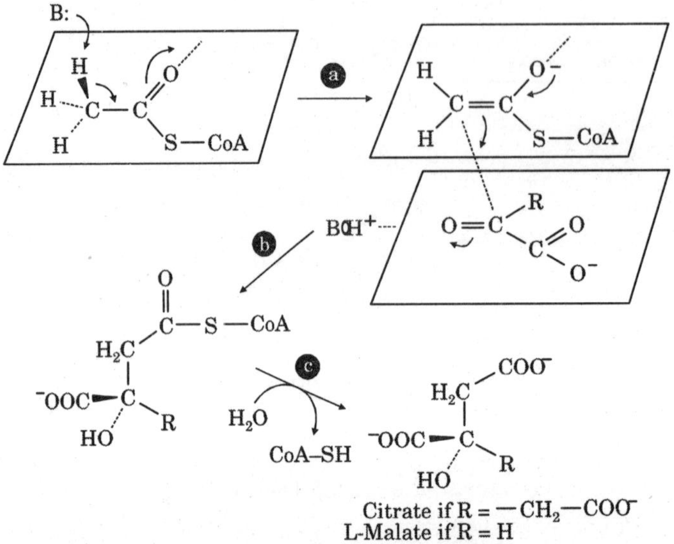

Citrate if R = $-CH_2-COO^-$
L-Malate if R = H

...(6.38)

The *si*–citrate synthase of pigs is a dimer of 437-residue chains, each of which is orga-nized into a large rigid domain and a smaller flexible domain. The active site lies between the domains. When oxaloacetate binds into the cleft the smaller domain undergoes a complex motion that closes the enzyme tightly around the substrate (Fig. 6.8). The bound oxaloacetate is shown in Fig. 6.9. It is surrounded by a large number of polar side chains, including several from histidine and arginine residues. Of these, Arg 401 and Arg 421 of the second subunit bind the substrates two carboxylate groups. In this tight complex the imiclazole of His 320 is in the correct position to protonate the carbonyl oxygen of the oxaloacetate.

The domain movement has also brought the groups that bind the acetyl-CoA into their proper position creating, by "induced fit," the acetyl-CoA binding site. This conformational change also accounts for the observation of an ordered kinetic mechanism with oxaloacetate binding before acetyl-CoA. The imidazole of His 274 is correctly oriented to abstract a proton from the methyl group of acetyl-CoA to generate the intermediate enolate anion. When oxaloacetate alone is bound into the active site the carbonyl stretching frequency, observed by infrared spectroscopy, is shifted from 1718 cm^{-1} for free oxaloacetate to 1697 cm^{-1}. This decrease of 21 cm^{-1} suggests a strong polarization of the C = O bond by its interaction with His 320 in the ground state. This interaction is seen also in the ^{13}C NMR spectrum, the carbonyl resonance being shifted downfield by 6.8 ppm upon binding to the protein. The enolate anion mechanism is also supported by kinetic isotope effects.

The methyl protons of dithia-acetyl-CoA are much more reactive than those of acetyl-CoA (pK_a ~12.5 vs –20). When citrate acts on this acetyl-CoA analog together with oxaloacetate, the expected enolate anion forms rapidly as is indicated by the appearance of a 306 nm absorption band. However, it condenses very slowly with the oxaloacetate. The binding equilibria, kinetics, and X-ray structures of complexes for a variety of other analogs of acetyl-CoA have also been studied. It appears that Asp 375 and His 274 work together to generate

$$H3C-\overset{\overset{S}{\|}}{C}-S-CoA$$

Dithia-acetyl-CoA

the enolate anion (Fig. 6.9). NMR measurements indicate formation of an unusually short hydrogen bond, but its significance is uncertain.

Malate synthase (Table 6.1) operates with a very similar mechanism but has an entirely different amino acid sequence and protein fold. Other Claisen condensations are involved in synthesis of fatty acids and polyketides and in formation of 3-hydroxy-3-methylglutaryl-CoA, the precursor to the polyprenyl family of compounds. In these cases the acetyl group of acetyl-CoA is transferred by a simple displace-ment mechanism onto an –SH group at the active site of the synthase to form an acetyl-enzyme. The acetyl-enzyme is the actual reactant in step *b* of where this reaction, as well as that of HMG-CoA lyase, is illustrated.

Citrate Cleaving Enzymes

In eukaryotic organisms the synthesis of citrate takes place within the mitochondria, but under some circumstances citrate is exported into the cytoplasm. There it is cleaved by *ATP–citrate lyase*. To ensure that the reaction goes to completion, cleavage is coupled to the hydrolysis of ATP to ADP and inorganic phosphate (Eq. 6.39). The value of G′ given here is extremely dependent upon the concentration of Mg^{2+} as a result of strong chelation of Mg^{2+}

Citrate

CoA ↘ ↙ ATP

↘ ADP + P_i

Acetyl-CoA ↙ ↘ Oxaloacetate ...(6.30)

$\Delta G0(pH\ 7)^* = +0.9\ kJ\ mol^{-1}$

by citrate. The reaction sequence is complex but can be understood in terms of an initial ATP-dependent synthesis of citryl-CoA using a mechanism similar to that. There is evidence for both phosphoenzyme and citryl enzyme intermediates (Eq. 6.40). Native ATP-citrate lyase is a tetramer of 110-kDa subunits. It usually contains some phosphoserine and phosphothreonine residues but they apparently have little effect on activity. Phosphorylation is catalyzed by cAMP-dependent and by insulin-dependent protein kinases. A related reaction is the ATP-dependent cleavage of malate to acetyl-CoA and glyoxylate. It requires two enzymes, malyl-CoA being an intermediate.

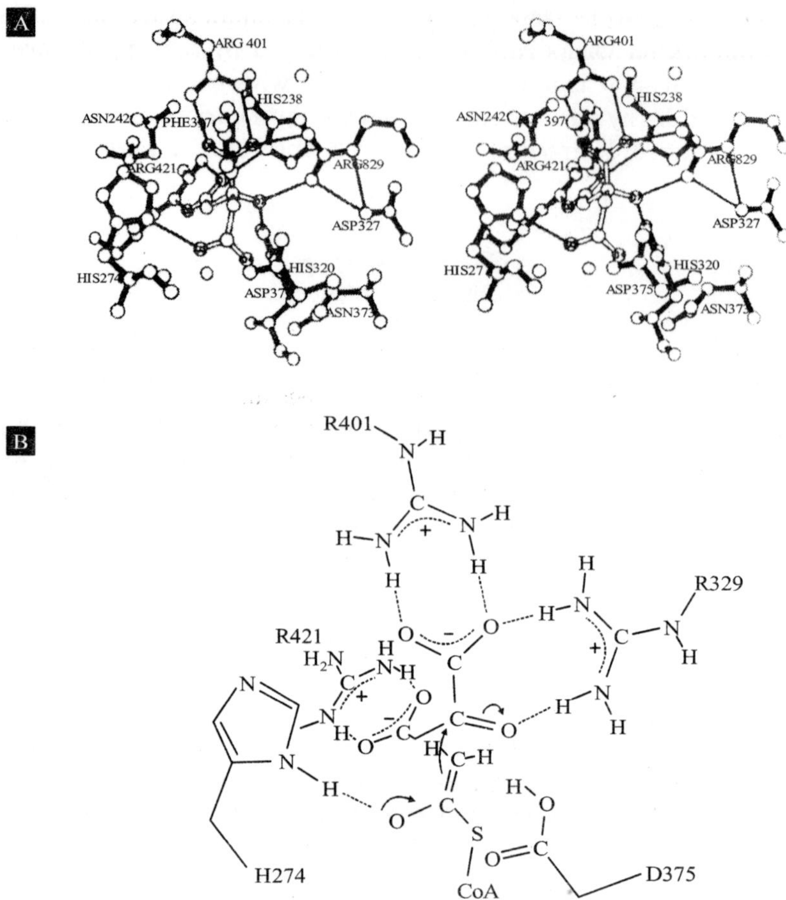

Fig. 6.8. Active site of pig citrate synthase. (A) Stereoscopic view with a molecule of citrate in the active site. (B) Interpretive view of the enolate anion of acetyl-CoA and oxaloacetate bound in the active site.

A substrate-induced citrate lyase found in bacteria such as *E. coli* and *Klebsiella* promotes the anaerobic dissimilation of citrate splitting it to oxaloacetate and acetate. The large ~585-kDa protein from *Klebsiella* has the composition $(\alpha\beta\gamma)_6$, where the α, β, and γ subunits have masses of ~55–, 30–, and 10-kDa, respectively. The subunit carries an unusual covalently bound derivative of coenzyme A . The 10-kDa y subunit serves as an *acyl-carrier protein,* the –SH of its prosthetic group being acetylated by a separate ATP-dependent ligase. The resulting acetyl-enzyme undergoes an acyl exchange (analogous to a CoA-transferase reaction; to form a citryl enzyme before the aldol cleavage takes place (Eq. 6.40).

$$\text{Acetyl-E} \xrightarrow[\text{acetate}]{\text{citrate}} \text{citryl-E} \xrightarrow[\text{oxaloacetate}]{} \text{acetyl-E} \qquad ...(6.40)$$

The first unique enzyme of the important glyoxylate pathway, *isocitrate lyase,* cleaves isocitrate to succinate and glyoxylate (Eq. 6.41). The carboxylate group that acts as electron acceptor would presumably be protonated by the enzyme.

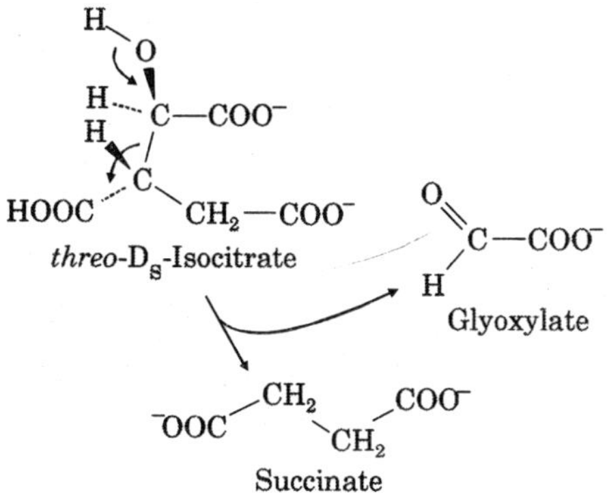

threo-D_s-Isocitrate

Glyoxylate

Succinate

$\Delta G0(pH\ 7) = 8.7\ kJ\ mol^{-1}$ ...(6.41)

Chiral Acetates and Their Use in Stereochemical Studies

Consider the series of enzyme-catalyzed reactions shown in Eq. 6.42. Fumarate is hydrated in ^3H-containing water to malate which is oxidized to oxaloacetate. Hydrolysis of the latter with oxaloacetate acetyl-hydrolase (Eq. 6.34) in ^2H_2O gives oxalate and chiral (R) acetate. The identical product can be obtained by condensing oxaloacetate with acetyl-CoA using citrate (*re*)–synthetase. The resulting citrate is cleaved in ^2H_2O using a citrate lyase having the *si* specificity. Acetate of the opposite chirality can be formed enzy-matically beginning with [2, 3–^3H] fumarate hydrated by fumarate hydratase in ordinary water. Chiral acetates have also been prepared nonenzymatically, and their configuration has been established unequivocally.

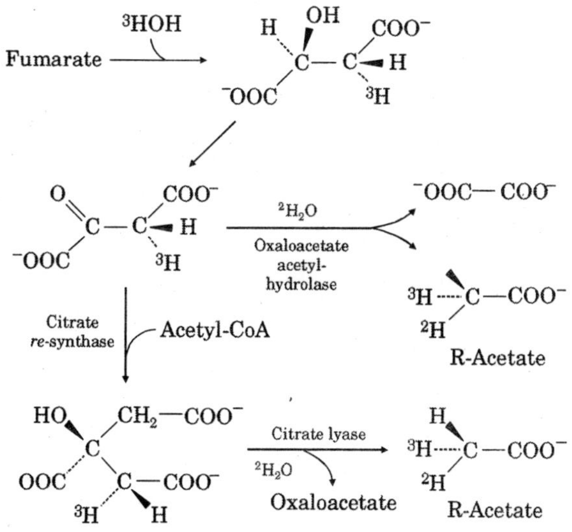

Fumarate $\xrightarrow{^3HOH}$

^2H_2O Oxaloacetate acetyl-hydrolase

$^-OOC—COO^-$

R-Acetate

Citrate *re*-synthase — Acetyl-CoA

Citrate lyase ^2H_2O Oxaloacetate

R-Acetate ...(6.42)

During the action of both oxaloacetate acetyl-hydrolase and citrate lyase (Eq. 6.42) inversion of configuration occurs about the carbon atom that carries the negative charge in the departing enolate anion. That inversion also occurs during catalysis by citrate (*re*)-synthase and other related enzymes has been demonstrated through the use of chiral acetates. The findings with malate synthase are illustrated in Eq. 6.38. Presumably, a basic group B of the enzyme removes a proton to form the planar enolate anion. The second substrate glyoxylate

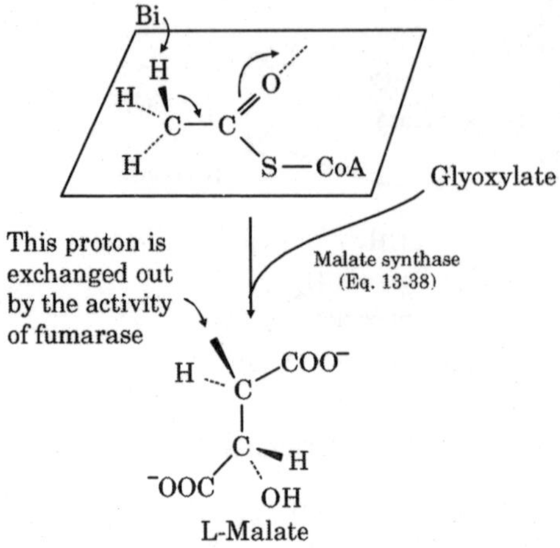

Bi

S—CoA   Glyoxylate

This proton is exchanged out by the activity of fumarase

Malate synthase (Eq. 13-38)

L-Malate ...(6.43)

approaches from the other side of the molecule and condenses as is shown. Since any one of the three protons in either R or S chiral acetyl-CoA might have been abstracted by base B, several possible combinations of isotopes are possible in the L-malate formed. One of the results of the experiment using chiral (R) acetyl-CoA is illustrated in Eq. 6.43. The reader can easily tabulate the results of removal of the ^2H or ^3H. However, notice that if the base – B: removes ^2H (D) or ^3H (T) the reaction will be much slower because of the kinetic isotope effects which are expected to be $^{H}k/^{D}k \approx 7$ and $^{H}k/^{T}k = 16$.

A second important fact is that the *pro- R* hydrogen at C–3 in malate is specifically exchanged out into water by the action of fumarate hydratase. From the distribution of tritium in the malate and fumarate formed using the two chiral acetates, the inversion by malate synthase was established. See Kyte for a detailed discussion.

Addition of an Enolate Ion to Carbon Dioxide and Decarboxylation

The addition of an enolate anion to CO_2 to form a β-oxoacid represents one of the commonest means of incorporation of CO_2 into organic compounds. The reverse reaction of *decarboxylation* is a major mechanism of biochemical formation of CO_2. The equilibrium constants usually favor decarboxylation but the cleavage of ATP can be coupled to drive carboxylation when it is needed, *e.g.*, in photosynthesis.

Detarboxylation of β-Oxoacids

Beta–oxoacids such as oxaloacetic acid and acetoacetic acid are unstable, their decarboxylation being catalyzed by amines, metal ions, and other substances. Catalysis by amines depends upon Schiff base formation, while metal ions form chelates in which the metal assists in electron withdrawal to form an enolate anion.

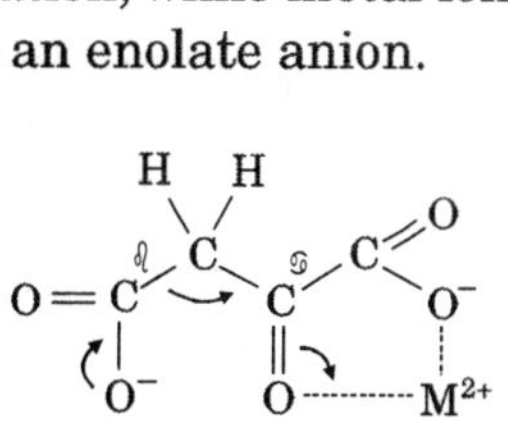

Can we apply any of this information from non-enzymatic catalysis to decarboxylating enzymes? Some decarboxylases do form Schiff bases with their substrates, and some are dependent on metal ions. The acetone-forming fermentation of *Clostridium acetobutylicum* requires large amounts of acetoacetate decarboxylase (Eq. 6.44).

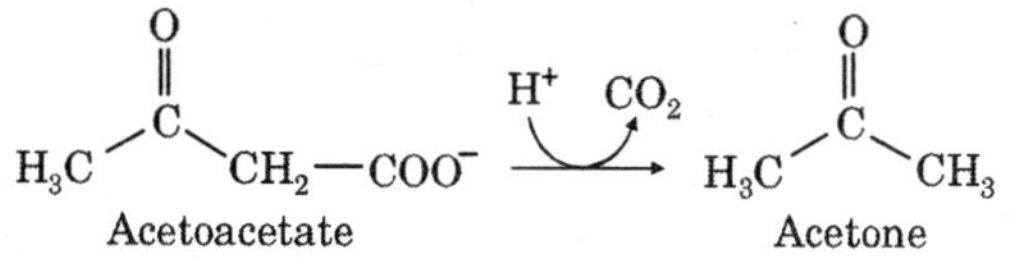

$$H_3C{-}C({=}O){-}CH_2{-}COO^- \xrightarrow{H^+ \quad CO_2} H_3C{-}C({=}O){-}CH_3 \quad ...(6.44)$$

Acetoacetate → Acetone

The enzyme is inactivated by borohydride in the presence of substrate, and acid hydrolysis of the inactivated enzyme yielded ε-N-isopropyllysine. Decarboxylation occurs from a Schiff base by a mechanism analogous to that of the aldol cleavage shown in Eq. 6.36. Mechanistically related is 4-oxalocrotonate decarboxylase.

Linked Oxidation and Decarboxylation

Metabolic pathways often make use of oxidation of a β-hydroxy acid to a β-oxoacid followed by decarboxylation in the active site of the same enzyme. An example is conversion of L-malate to pyruvate (Eq. 6.45). The Mg^{2+} or Mn^{2+}-dependent decarboxylating malic dehydrogenase that catalyzes the reaction is usually called *the malic enzyme.* It is found in most organisms. While a concerted decarboxylation and dehydrogenation may sometimes occur, the enzymes of this group appear usually to operate with bound oxoacid intermediates as in Eq. 6.45.

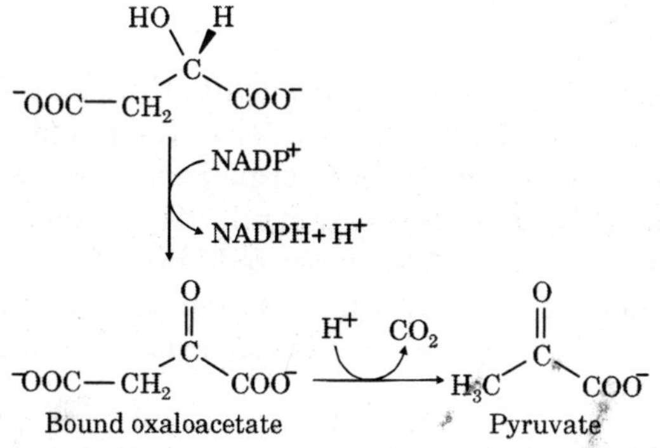

$$^-OOC{-}CH_2{-}CH(OH){-}COO^- \xrightarrow{NADP^+ \rightarrow NADPH + H^+} {}^-OOC{-}CH_2{-}C({=}O){-}COO^- \xrightarrow{H^+ \quad CO_2} H_3C{-}C({=}O){-}COO^- \quad ...(6.45)$$

Bound oxaloacetate → Pyruvate

Other reactions of this type are the oxidation of isocitrate to 2-oxoglutarate in the citric acid cycle oxidation of 6–phosphogluconate to ribulose 5-phosphate and corresponding reactions of isopropylmalate dehydrogenase and tartrate dehydrogenase. Crystallographic studies of isocitrate dehydrogenase using both photolabile "caged" isocitrate and slow mutant forms with polychromatic Laue crystallography have demonstrated the rapid formation of the anticipated intermediate *oxalosuccinate.*

Phosphoenolpyruvate, a Key Metabolic Intermediate

A compound of central importance in metabolism is the phosphate ester of the enol form of pyruvate, commonly known simply as phosphoenol-pyruvate (PEP). It is formed in the glycolysis pathway by dehydration of 2-phosphoglycerate (Eq. 6.15) or by decarboxylation of oxaloacetate. Serving as a preformed enol from which a reactive enolate anion can be released for condensation reactions, PEP is utilized in metabolism in many ways.

$$^{-}OOC-C(-O-PO_3^{2-})=CH_2$$

Phosphoenolpyruvate (PEP)

In animals and in many bacteria, PEP is formed by decarboxylation of oxaloacetate. In this reaction, which is catalyzed by *PEP carboxykinase* (PEPCK), a molecule of GTP, ATP, or inosine triphosphate captures and phosphorylates the enolate anion generated by the decarboxylation (Eq. 6.46). The stereochemistry is such that CO_2 departs from the *si* face of the forming enol. The phospho group is transferred from GTP with inversion at the phosphorus atom. The enzyme requires a divalent metal ion, preferably Mn^{2+}. In fact, kinetic studies of the GTP-dependent avian mitochondrial enzyme indicate two metalbinding sites, one on the polyphosphate group of the bound GTP and one on carboxylate side chains of the protein.

The three-dimensional structure of the ATP-dependent *E. coli* enzyme reveals a nucleotide binding site similar to the ATP site of adenylate kinase. A definite binding site for CO_2 is also present in the enzyme. PEPCK is also activated by low concentrations of Fe^{2+} and this activation depends upon a protein that has been identified as glutathione peroxidase. By destroying H_2O_2 the latter may allow the Fe^{2+} to prevent oxidation of an SH group on PEP carboxykinase.

Synthesis of PEPCK is stimulated by glucagon, evidently through a direct action of cAMP on transcription of the structural gene. Transcription is also stimulated by glucocorticoids and thyroid hormone and is inhibited by insulin. In some organisms, such as the parasitic *Ascaris suum,* PEPCK functions principally as a means of synthesis of oxaloacetate by reaction of PEP with CO_2 and GDP. These organisms lack pyruvate kinase, which allows for buildup of the high PEP concentration needed to drive this reaction. In a similar way PEP can be converted to oxaloacetate by *PEP carboxy-transphosphorylase,* an enzyme found only in propionic acid bacteria and in *Entamoeba.* The

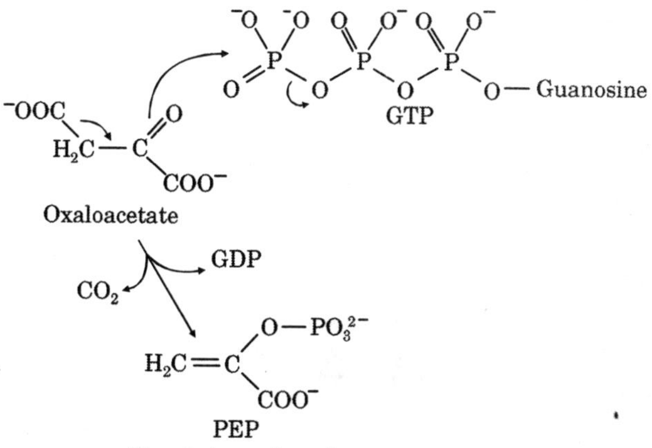

re Face is toward reader ...(6.46)

reaction (Eq. 6.47) is accompanied by synthesis of inorganic pyrophosphate which may be cleaved to "pull" the reaction in the indicated direction.

$$PEP + CO_2 + P_i \rightarrow PP_i + \text{oxaloacetate} \quad ...(6.47)$$

Oxaloacetate is also decarboxylate without phosphorylation of the enolate anion formed but with release of free pyruvate. Both pyruvate kinase and PEPCK can act as oxaloacetate decarboxylases.

In the important reactions discussed in the following sections enolate ions are intermediates in carbon-carbon bond formation. Other examples are given in which C–C bonds are formed by action of PEP as a carbon nucleophile *e.g.,* in aldol–like condensations. Yet another unusual type of reaction, involving formation of an enol ether linkage is illustrated.

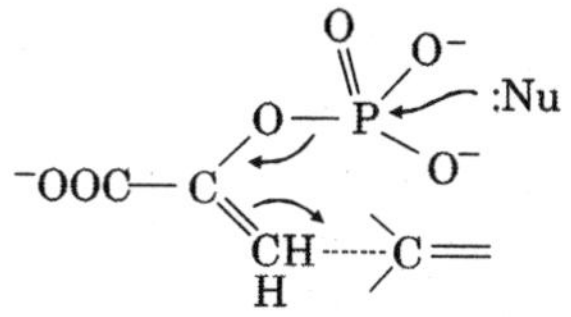

Ribulose Bisphosphate Carboxylase

The major route of incorporation of CO_2 into organic compounds is via photosynthesis. When ^{14}C-labeled CO_2 enters chloroplasts of green plants, the first organic ^{14}C-containing compound detected is 3-phosphoglycerate. Two molecules of this compound are formed through the action of *ribulose-*1, 5*-bisphosphate carboxylase* (abbreviated *rubisco*), an enzyme present in chloroplasts and making up 16% of the protein of spinach leaves. This enzyme, whose structure is illustrated is thought to be the most abundant protein on earth. Because O_2 competes with CO_2 as a substrate, the enzyme also catalyzes an "oxygenase" reaction. It is therefore often called *ribulose-*1, 5*-bisphosphate carboxylase/oxygenase.* The rubisco from most plants is a 500– to 560-kDa L_8S_8 oligomer as shown in Fig. 6.10. The large subunit is encoded in the chloroplast DNA. However, a family of nuclear genes encode the small subunits, which are synthesized as larger precursors, with N-terminal extensions being removed to give the mature subunits. Two types of small subunits occupy different positions in the quasi-symmetric spinach rubisco.

In *Euglena* the small subunit is synthesized as a polyprotein precursor containing eight copies of the subunit. Rubisco from the hydrogen-oxidizing bacterium *Alcaligenes eutrophus* has a similar quaternary structure, but the enzyme from *Rhodospirillum rubrum* is a simple dimer. In dinoflagellates the rubisco gene is present in nuclear DNA rather than in the chloroplasts. The carboxylation reaction catalyzed by rubisco differs from others that we have considered in that the carboxylated product is split by the same enzyme (Eq. 6.48). The mechanism shown in Eq. 6.48 was suggested by Bassham *et al.* Ribulose bisphosphate, for which the enzyme is absolutely specific, is first converted to its enolic form, a 2, 3-enediol. Loss of a proton from the 3-OH group forms the endiolate anion needed for the carboxylation in step *b*. The product of this step is a β-oxoacid which undergoes enzyme-catalyzed hydrolytic cleavage (Eq. 6.48, step *c;* see also Eq. 6.34).

Support for this mechanism came from the observation that 2-carboxyarabinitol bisphosphate (Fig. 6.11 B) is a potent inhibitor, possibly a transition state analog. This inhibitor, bound into the active site of the enzyme from spinach, is seen in both Figs. 6.10B and 6.11A. An expanded version of Eq. 6.48 is given in Fig. 6.12, which is based upon modeling together with a variety of X-ray structures including that shown in Fig. 6.11.

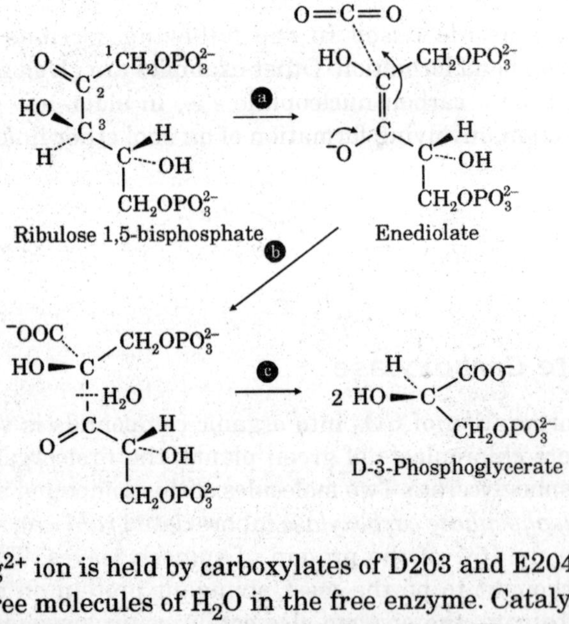

Ribulose 1,5-bisphosphate

Enediolate

D-3-Phosphoglycerate ...(6.48)

An essential Mg^{2+} ion is held by carboxylates of D203 and E204 and by modified K201. It also coordinates three molecules of H_2O in the free enzyme. Catalytic roles for the H294 and H327 imidazole groups are still being elucidated. Lysine 175 protonates C2 of the aci anion generated by C–C bond cleavage (step *e*, Fig. 6.12). Like many enzymes, rubisco exists in two major conformational states: open and closed. Chemical studies also support the indicated mechanism.

For example, the β-oxoacid intermediate formed in step *b* of Eq. 6.48 or Fig. 6.12 has been identified as a product released from the enzyme by acid denaturation during steady-state turnover. Isotopic exchange with ^3H in the solvent and measurement of ^{13}C isotope effects have provided additional verification of the mechanism. The catalytic activity of the enzyme is determined by ionizable groups with pK_a values of 7.1 and 8.3 in the ES complex. The apparent value of K_m for total CO_2 ($CO_2 + HCO_3^-$) is high, 11.30 mM, but for the true substrate CO_2 it is only 0.45 mM. In intact chloroplasts the affinity for substrate is distinctly higher, with the K_m for total CO_2 dropping to –0.6 mM.

The difference appears to result largely from a regulatory reaction of CO_2 in which the side chain amino group of lysine 201 of the large subunit forms a carbamate (Eq. 6.49). Although carbamylation is spontanous, it is enhanced by an ATP-dependent process catalyzed by *rubisco activase.* The carbamylation converts the side chain of Lys 201 into a negatively charged group that binds to an essential divalent metal, usually Mg^{2+}, in the active center as is shown in Fig. 6.11, A and B. The nature of the reaction is uncertain. One possibility is that the activase is a chaperonin. It appears to assist the enzyme in removing inhibitory sugars that arise by side reactions in the active site. Rubisco is also regulated by the level of a natural inhibitor which has been identified as 2-carboxyarabinitol 1-phosphate. This is the same as the inhibitor shown in Fig. 6.11 but with one less phosphate group and consequent weaker binding.

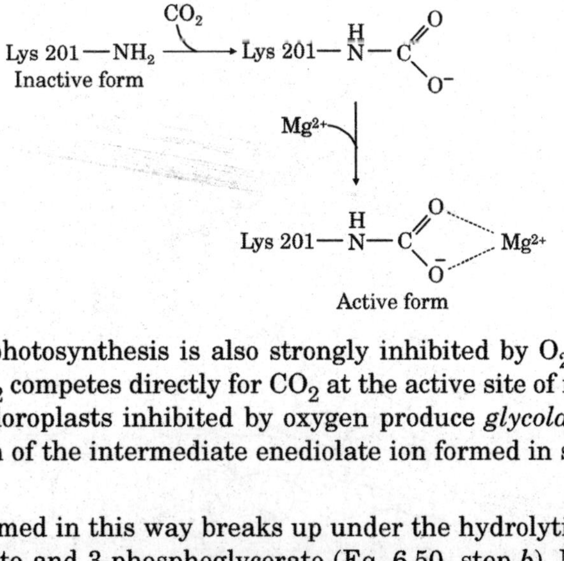

$$\underset{\text{Inactive form}}{\text{Lys 201—NH}_2} \xrightarrow{CO_2} \text{Lys 201—NH—C}(=O)O^- \xrightarrow{Mg^{2+}} \underset{\text{Active form}}{\text{Lys 201—NH—C}(=O)O^- \cdots Mg^{2+}} \qquad ...(6.49)$$

In most plants photosynthesis is also strongly inhibited by O_2. This observation led to the discovery that O_2 competes directly for CO_2 at the active site of rubisco in a process called *photorespiration* Chloroplasts inhibited by oxygen produce *glycolate* in large amounts as a result of the reaction of the intermediate enediolate ion formed in step *b of* Eq. 6.48 with O_2 (Eq. 6.50, step *a).*

The peroxide formed in this way breaks up under the hydrolytic action of the enzyme to form phosphoglycolate and 3-phosphoglycerate (Eq. 6.50, step *b*). Molecular oxygen usually reacts rapidly with only those organic substrates, such as dihydroflavins, that are able to form stable free radicals. However, the endiolate anion of Eq. 6.50 may be able to donate a single electron to O_2 to form a superoxide-organic radical pair prior to formation of the peroxide. Similar oxygenase side reactions have been observed for a variety of other enzymes that utilize carbanion mechanisms. The reaction of rubisco with O_2 is of both theoretical and practical interest, the latter because of its significance in lowering the yield in photosynthesis.

Phosphoglycolate 3-Phosphoglycerate ...(6.50)

The simpler dimeric rubisco from *Rhodospirillum* is very inefficient in carboxylation and catalyzes much more oxygenation than do rubiscos of higher plants. Mutant enzymes that have impaired carboxylase and enhanced oxygenase are also known.

The small subunits of rubisco may help suppress undesirable side reactions. For example, the following deoxypentodiulose phosphate can be formed by β elimination from the second intermediate of Eq. 6.48.

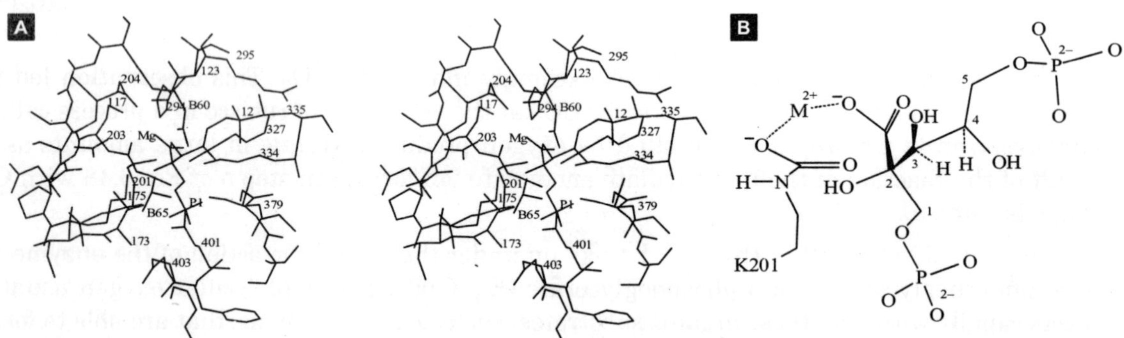

Fig. 6.9. (A) Overview of the active site of spinach rubisco showing bound 2-carboxy-D-arabinitol 1, 5-bisphosphate and Mg^{2+} and residues within hydrogen-bonding distance of these ligands. The hydroxyl groups at C2 and C3 of the inhibitor are in cis conformation. (B) Structure of the inhibitor 2-carboxy-D-arabinitol 1, 5-bisphosphate. A part of the carbamylated lysine 201 and the essential metal ion are also shown.

Some side products of the normal oxgenase reaction arise by elimination of a peroxide ion from the first intermediate of Eq. 6.50. The resulting dicarbonyl bisphosphate can rearrange to give a carboxytetritol bisphosphate (Eq. 6.51).

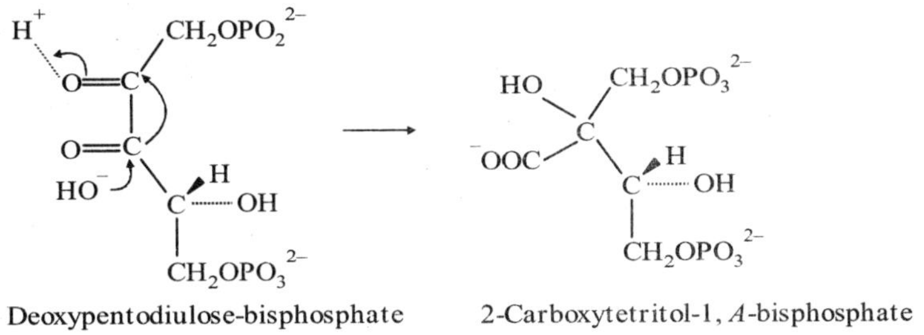

$$\text{Deoxypentodiulose-bisphosphate} \longrightarrow \text{2-Carboxytetritol-1,4-bisphosphate} \qquad ...(6.51)$$

Deoxypentodiulose-bisphosphate 2-Carboxytetritol-1, 4-bisphosphate

Carbon Dioxide or Bicarbonate Ion

An important question in the consideration of carboxylation and decarboxylation reactions is whether the reactant or the product is CO_2 or HCO_3^-. An approach to answering the question was suggested by Krebs and Roughton, who pointed out that the attainment of equilibrium between free CO_2 and HCO_3^- (Eq. 6.52) may require several seconds. If an enzyme produces

$$CO_2 + H_2O \xrightarrow{a} H_2CO_3 \xrightarrow{b} H^+ + HCO_3^- \qquad ...(6.52)$$

CO_2 as a product and progress of the reaction is followed manometrically, the pressure will rise higher than the equilibrium value as the CO_2 is evolved. Later when substrate is exhausted and the CO_2 equilibrates with bicarbonate, the pressure will fall again. The addition of carbonic anhydrase, which catalyzes Eq. 6.52, step *a,* abolishes the "overshoot." If bicarbonate is the primary product of a decarboxylation, there is a lag in the appearance of free CO_2.

A second approach is to use [^{18}O]bicarbonate and to follow the incorporation of ^{18}O into a carboxylated substrate. If CO_2 is the primary substrate only two labeled oxygen atoms enter the compound, whereas if HCO_3^- is the reactant three are incorporated. A third technique is measurement of the rate of incorporation of CO_2 or bicarbonate in the carboxylated product. Over a short interval of time, *e.g.*, 1 min, different kinetics will be observed for the incorporation of CO_2 and of bicarbonate. Using these methods, it was established that the product formed in Eq. 6.46 and the reactant in Eq. 6.17 is CO_2. However, the carboxylation enzymes considered in the next section use bicarbonate as the substrate.

Incorporation of Bicarbonate into Carboxyl I Groups

An important enzyme with a biosynthetic function in many bacteria and in all higher plants is *PEP carboxylase,* which catalyzes the reaction of Eq. 6.53. This enzyme, in effect, accomplishes the reverse of Eq. 6.46 by converting the three-carbon PEP, by reaction with bicarbonate, into the four-carbon oxaloacetate. The latter is needed for "priming" of the citric acid cycle and for biosynthesis of such amino acids as aspartate and glutamate. That the enzyme functions in this way is indicated by the fact that mutants of *Salmonella* defective in the enzyme do not grow unless oxaloacetate or some other intermediate in the citric acid cycle is

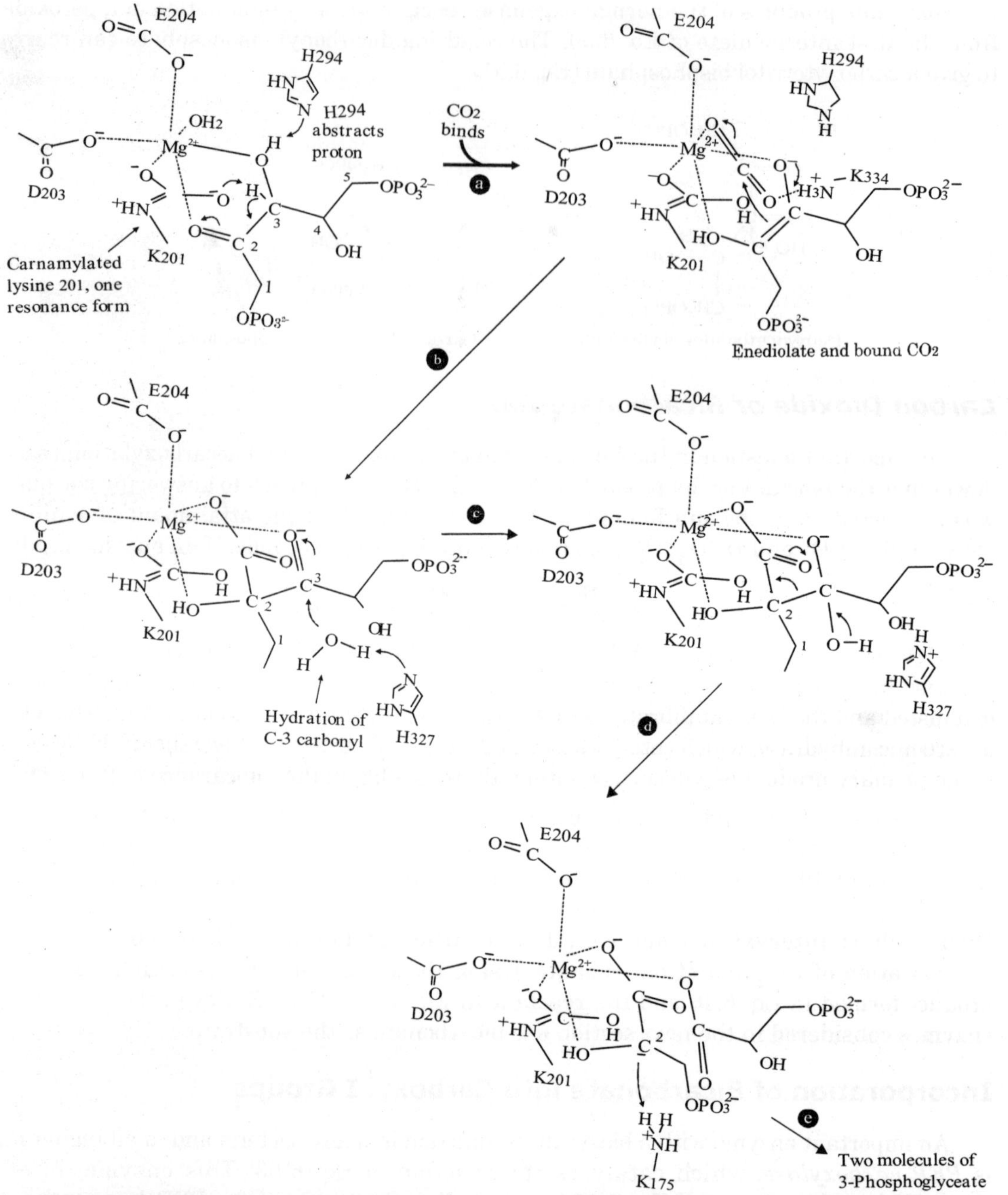

Fig. 6.10. Proposed mechanism of action of ribulose bisphosphate carboxylase (rubisco). This is an abbreviated version of the mechanism as presented by Taylor and Andersson. The binding of ribulose 1, 5-bisphosphate occurs after carbamylation of lysine 201 and binding of a magnesium ion. Formation of an enediolate intermediate in step *a* is probably catalyzed by the carbamate group as indicated. Removal of a proton from the 3-OH, perhaps by H294, and addition of a proton to form an –OH group at C2 are also necessary. The CO_2 may bind to the Mg^{2+} and be polarized by interaction with other side chains prior to reaction. Carboxylation occurs by addition of the enediolate to CO_2 in step *b*. The hydration of the resulting 3–oxo group (step *c*) is necessary for cleavage of the C–C bond in step *d*. The participation of the newly formed carboxylate as an *ad* anion coordinated to Mg^{2+} is presumably involved. Protonation, with stereochemical inversion, is thought to involve K175, as shown. The two product molecules dissociate in step *e*.

added to the medium. The enzyme from *S. typhimurium* is a 400-kDa tetramer with complex regulatory properties. The corresponding enzyme from spinach has 12 subunits and 12 bound Mn^{2+} ions. The enzyme also has a special function in the C_4 plants, in which it is a component of a carbon dioxide concentrating system.

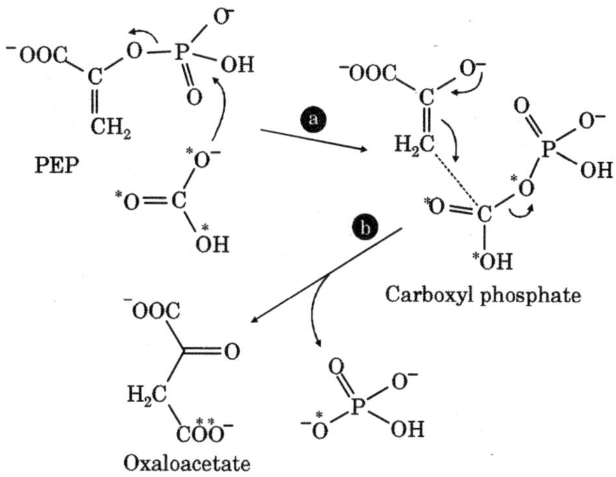

PEP

Carboxyl phosphate

Oxaloacetate

...(6.53)

When [^{18}O]bicarbonate is a substrate, two labeled oxygen atoms enter the oxaloacetate, while the third appears in P_i. A concerted, cyclic mechanism could explain these results. However, study of kinetic isotope effects, use of a substrate with a chiral thiophospho group, and additional ^{18}O exchange studies have ruled out this possibility. A transient carboxyl phosphate (Eq. 6.53) is evidently an intermediate. The incorporation of the ^{18}O from bicarbonate into phosphate is indicated by the asterisks. The carboxyl group enters on the *si* face of PEP. However, there is another possibility. The carboxyl phosphate, while in the active site adjacent to the enolate anion, may eliminate phosphate, the enolate ion adding to the resulting CO_2 to form the final product. According to this mechanism the group transfer potential of the phosopho group in PEP is utilized to concentrate the CO_2 and to localize it next to the enolate anion generated by the same process.

PEP carboxylase is lacking from animal tissues and fungi. In these creatures PEP is converted to pyruvate, which is then carboxylated to oxaloacetate with coupled cleavage of ATP by the action of *pyruvate carboxylase*, an enzyme that not only utilizes bicarbonate ion but also contains *biotin*. However, there are mechanistic similarities between its action and that of PEP carboxylase.

PEP *Mutase and the Synthesis of Phosphonates*

The lipids of some organisms, such as *Tetrahymena,* contain aminoethylphosphonate, a compound with a C–P bond. There are also many other naturally occurring phosphoric compounds and huge quantities of synthetic phosphonates, present in detergents, herbicides, and insecticides, are metabolized by bacteria. Here we will consider only one step in the biosynthesis of phosphonates, the conversion of PEP into phosphonopyruvate (Eq. 6.54), a reaction catalyzed by *PEP mutase.* The phospho group is moved from the oxygen atom of PEP to the methylene carbon atom. When attached to carbon the phospho group is designated

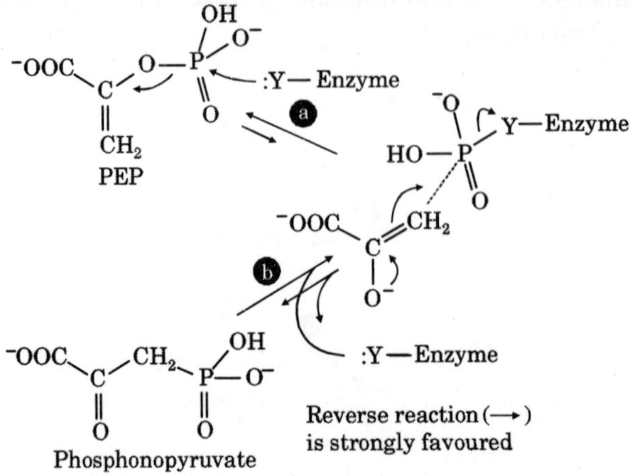

PEP

:Y— Enzyme

Y—Enzyme

:Y—Enzyme

Phosphonopyruvate

Reverse reaction (⟶) is strongly favoured

...(6.54)

phosphono. The reaction occurs with retention of the configuration around phosphorus and a phosphoenzyme intermediate seems to be involved as shown in Eq. 6.54. The equilibrium in this reaction strongly favors PEP. One further product formed from phosphonopyruvate is phosphonoethyl-amine, a component of phosphonolipids. It is easy to imagine a synthetic route involving transamination followed by decarboxylation. Another mutase, similar to PEP mutase, shifts the carboxyphospho group $^-OOC–PO_2$— exactly as in Eq. 6.54 as part of the biosynthetic pathway of a natural herbicide formed by *Streptomyces hygroscopicus.* Soil bacteria contain specialized enzymes that catalyze the hydrolytic cleavage of P–C bonds.

Some Isomerization and Rearrangement Reactions

A few metabolic reactions do not fit into any of the categories discussed so far and apparently do not depend upon a coenzyme. Some of these involve transfer of alkyl groups or of hydrogen atoms from one carbon to another. The hydrogen atoms move by direct transfer without exchange with the medium. All of the reactions could involve carbocations but there is often more than one mechanistic possibility.

A simple 1, 3-proton shift is shown in as reaction type 6A. An example is the isomerization of oleic acid to £rans-Δ^{10}-octadecenoic acid (Eq. 6.55) catalyzed by a soluble enzyme from a pseudomonad.

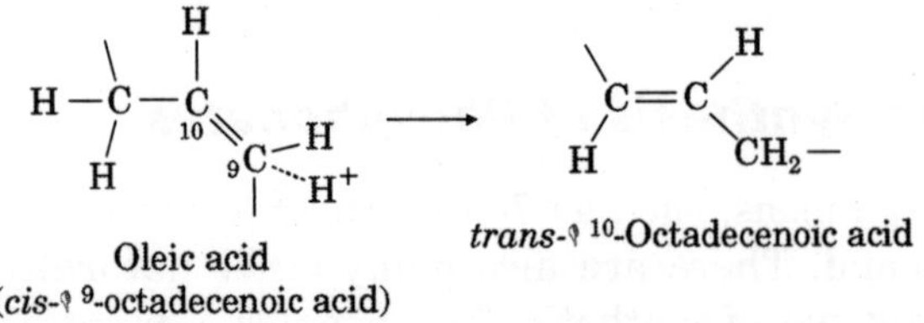

Oleic acid (*cis*-Δ^9-octadecenoic acid)

trans-Δ^{10}-Octadecenoic acid

...(6.55)

A second example is isomerization of *isopentenyl diphosphate to dimethylallyl diphosphate* (Eq. 6.56). The stereochemistry has been investigated using the ^3H–labeled compound shown in Eq. 6.56. The *pro-R* proton is lost from C–2 and a proton is added to the *re* face at C–4. When the reaction was carried out in ^2H_2O a chiral methyl group was produced as shown. A concerted proton addition and abstraction is also possible, the observed *trans* stereochemistry being expected for such a mechanism. However, the fact that the enzyme is strongly inhibited by the cation of 2-(diethylamino) ethyl pyrophosphate, an analog of a probable carbocation intermediate. Cysteine and glutamate side chains are essential.

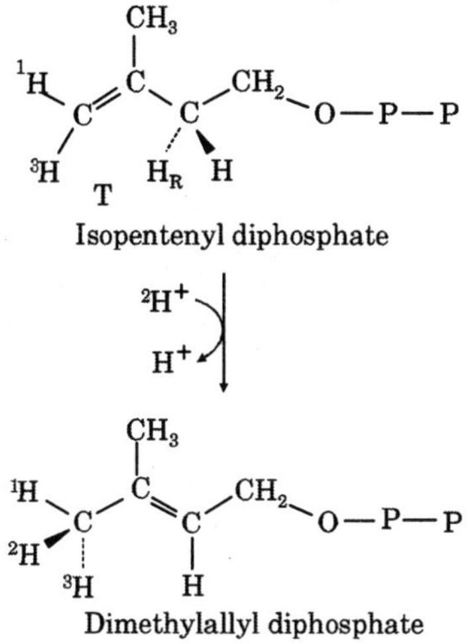

Isopentenyl diphosphate

$^2H^+$ H^+

Dimethylallyl diphosphate ...(6.56)

Reaction type 6B of is allylic rearrangement with simultaneous condensation with another molecule. The reaction, which is catalyzed by *prenyl-transferases,* occurs during the polymerization of polyprenyl compounds. Experimental evidence favors a carbocation mechanism for all of these reactions.

Reaction type 6C occurs during the biosynthesis of leucine and valine. The rearrangement is often compared with the nonenzymatic acid-catalyzed pinacol-pinacolone rearrangement in which a similar shift of an alkyl group takes place (Eq. 6.57). The enzyme-catalyzed rearrangement presumably gives the structure drawn in brackets. The same enzyme always catalyzes reduction with NADH to the diol, the Mg^{2+}-dependent enzyme being called *acetohydroxy acid isomeroreductase.* Rearrangement has never been observed without the accompanying reduction. More complex rearrangements that occur during biosynthesis of sterols are described.

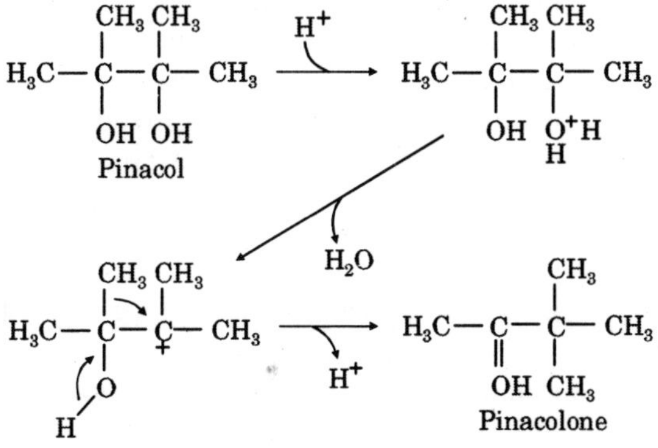

Pinacol

H^+ H_2O H^+

Pinacolone ...(6.57)

ZINC

The average human ingests 10–15 mg of zinc a day. Although it is poorly absorbed, the concentrations of zinc in tissues are relatively high and the metal plays an essential role in a multitude of enzymes. The total zinc content of a 70–kg person is 1.4–2.3 g. A typical tissue concentration of Zn^{2+} is 0.3–0.5 mM; an unusually high content of ~15 mM is found in the prostate gland. Zinc ion is much more tightly bound to most organic ligands than is Mg^{2+}. It has a filled 3*d* shell and tends to form four ionic bonds with a tetrahedral geometry, often with nitrogen or sulfur-containing ligands. Unlike Mg^{2+}, which interacts rapidly and reversibly with enzymes, Zn^{2+} tends to be tightly bound within over 300 *metalloenzyntes.*

A common feature is the surrounding of the Zn^{2+} at the active center by three imidazole groups, the fourth coordination position being free for interaction with substrate. The second nitrogen of the imidazole ring in many instances is hydrogen bonded to a main chain carbonyl group of the peptide, a feature that is also shared by histidines in other metalloproteins. The most important chemical function of Zn^{2+} in enzymes is probably that of a Lewis acid providing a concentrated center of positive charge at a nucleophilic site on the substrate. This role for Zn^{2+} is discussed for carboxypeptidases and thermolysin, alkaline phosphatase RNA polymerases, DNA polymerases, carbonic anhydrase (Fig. 6.1) class II aldolases (Fig. 6.7), some alcohol dehydrogenases, and superoxide dismutases.

Zinc ions in enzymes can often be replaced by Mn^{2+}, Co^{2+}, and other ions with substantial retention of catalytic activit. In addition to its function in catalysis, zinc often plays an important structural role, *e.g.*, in the *zinc finger* transcriptional regulators. Zinc ions bind to insulin and stabilize its hexameric structure. Six Zn^{2+} ions are present in the hexagonal tail plate of the T-even bacteriophage and appear to be essential for invasion of bacteria. In carnivores, the *tapetum,* the reflecting layer behind the retina of the eye of many animals, contains crystals of the Zn^{2+}-cysteine complex.

Since zinc ions have no color their presence has often been overlooked. Zinc ions will doubtless be found in many more places within cells. Zinc is usually the major component of the bound metals in the *metallothioneins.* These small 6.6-kDa proteins which contain ~33% cysteine and bind as many as six ions of Cd^{2+}, Hg^{2+}, Cu^{2+}, or Zn^{2+} per molecule are present in all animal tissues as well as in plants and some bacteria. From a nutritional viewpoint, Cu^{2+} competes with zinc ion, as does the very toxic Cd^{2+}.

The latter accumulates in the cortex of the kidney. Dietary cadmium in concentrations less than those found in human kidneys shortens the lives of rats and mice. However, some marine diatoms contain *a* cadmium-dependent carbonic anhydrase. Although zinc deficiency was once regarded as unlikely in humans, it is now recognized as occurring under a variety of circumstances and is well-known in domestic animals. Consumption of excessive amounts of protein as well as alcoholism, malabsorption, sickle cell anemia, and chronic kidney disease can all be accompanied by zinc deficiency.

EPSP SYNTHASE AND THE HERBICIDE GLYPHOSATE

The reversible reaction of phosphoenolpyruvate (PEP) with shikimate 3-phosphate is a step in the synthesis of the aromatic amino acids. The chemical mechanism indicated was proposed by Leaven and Sprinson in 1964. Step *a* is unusual because it involves protonation on the methylene carbon of PEP and addition of a nucleophile at C–2, the opposite of the addition in the enolase reaction (Eq. 6.15, reverse). It is likely that formation of a cationic intermediate

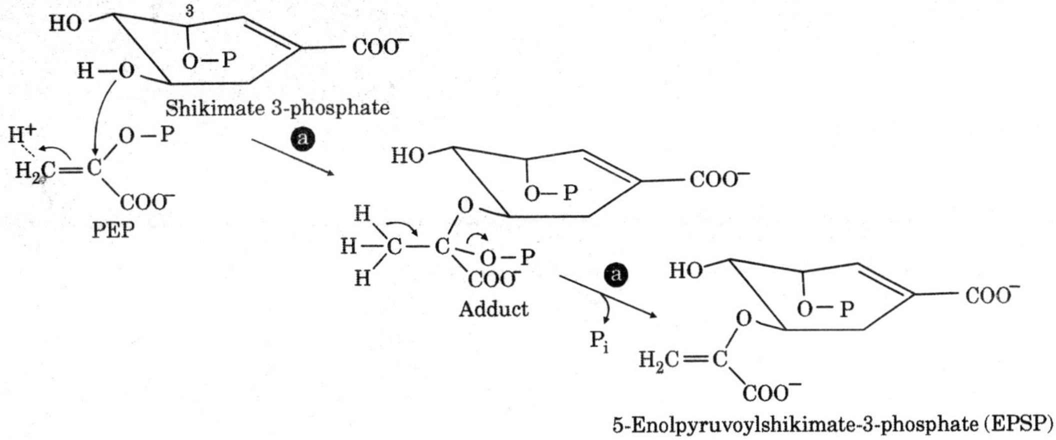

Shikimate 3-phosphate

PEP

Adduct

5-Enolpyruvoylshikimate-3-phosphate (EPSP)

precedes that of the adduct shown. However, the structure of the adduct has been confirmed by isolation from the active site and by synthesis and it has been observed in the active site by NMR spectroscopy. The three-dimensional structure of EPSP synthase is known. It is inhibited by the commercial herbicide *glyphosate* [N-(phosphonomethyl) glycine] whose structure is somewhat similar to that of the proposed carbocation that arises from PEP. The inhibitor (Z)-3-fluorophosphoenolpyruvate is a pseudosubstrate that reacts in step *a* to give a stable adduct unable to go through step *b* to form a product. Glyphosate was for many years viewed as a transition–state analog but more recently has been shown to be a tight-binding noncompetitive inhibitor.

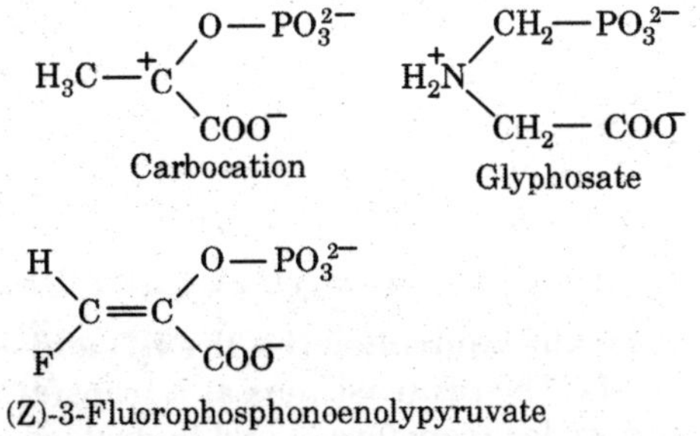

Carbocation

Glyphosate

(Z)-3-Fluorophosphonoenolypyruvate

A related mechanism is utilized in the biosyn-thesis of UDP-muramic acid. There is an enolpyruvoyl adduct analogous to that of EPSP synthase; a proposed enolpyruvoyl-enzyme adduct with Cys 115 is not on the major path. However, this enzyme is not inhibited by glyphosate.

Biosynthesis of Peptides

The basis for the classification of proteases as enzymes is their biological function as hydrolases. As catalysts, the proteases must also catalyse the reverse reaction, *i.e.*, the formation of a peptide bond. For some time this was treated as a curiosity rather than a property that could be used to gain more knowledge of the catalytic function of proteases, or for peptide synthesis *in vitro*. Recently, their potential use for the latter has led to a marked increase in studies on the synthetic function of protease.

Proteases catalyse the synthesis of peptides in an equilibrium–controlled or a kinetically controlled process. The different processes are shown schematically in Fig. 7.1. Their time dependence and dependence on enzyme properties, temperature and pH are given in Fig. 7.2.

In the kinetically controlled process (I–III), an activated substrate (peptide or amino acid) is used and the "enzyme" catalyses the transfer of the acyl group to a nucleophile (amino acid or peptide) in reaction I. In this reaction, the protease acts as a transferase. Competing with

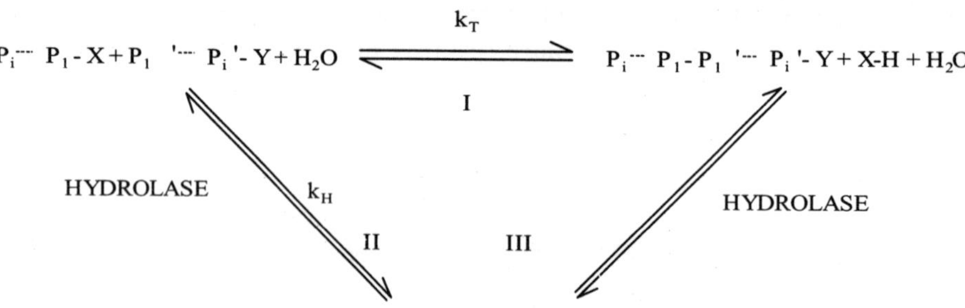

Fig. 7.1. Protease–catalysed kinetically controlled (I + II + III) and equilibrium–controlled (III) synthesis of peptides P_i ... P_1 – P_1 ...P_i. The protease acts as a transferase in (I) and as a hydrolase in (II, III): k_r, k_H, apparent second–order transferase and hydrolase rate constant of the enzyme used; P_i ... P_1, amino acid residues; –X, group used to activate the carboxyl end of peptide P_1 – P_2; –Y, group used to protect the carboxyl end of peptide (amino acid) P_i' ... P_1'–.

this reaction is the hydrolysis of the activated substrate (reaction II). The product peptide is also a substrate for the protease, and can be hydrolysed in reaction III. The product yield depends on the apparent ratio of the transferase to hydrolase rate constant $(k_T/k_H)_{app}$ and the rate with which the product peptide is hydrolysed. When the ratio $(k_T/k_H)_{app}$ is large, the product $-P_1-P_1'-$ can be formed in high yields even when the thermodynamically favourable products are $-P_1OH-$ and $-P_1'$..., *i.e.* the eguilibrium favours hydrplysis This formation of non-equilibrium concentration of peptides (and other condensation products) requires activated substrates.

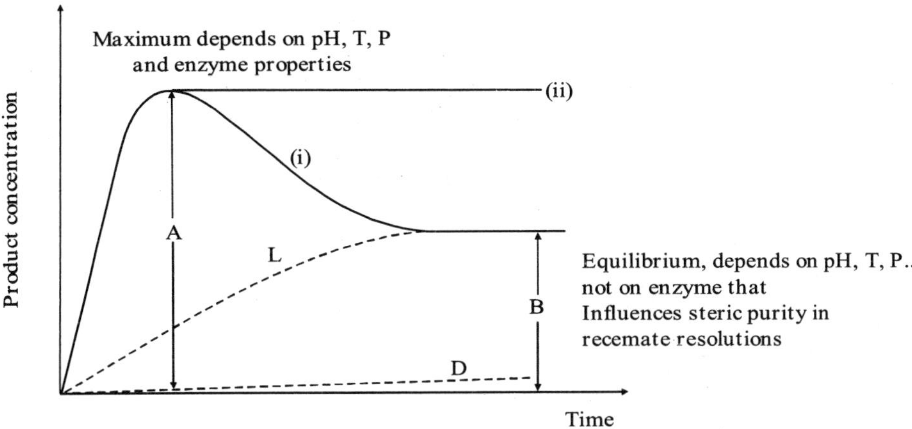

Fig. 7.2. Time dependence in kinetically (solid line) and equilibrium (dashed line) controlled peptide synthesis. For the kinetically controlled synthesis two cases are considered: *(i)* synthesis rate of the same order of magnitude as the rate of hydrolysis of product peptide, and *(ii)* synthesis rate >> hydrolysis rate. For the equilibrium–controlled synthesis the time dependence for the condensation of a racemic mixture to a L–amino acid (peptide) using a L–specific protease is given.

Contrary to the equilibrium–controlled process, the product yield depends on the properties of the enzyme used (k_T/k_{app}) and the substrate specificity. The enzymes used to catalyse the synthesis of peptides (in vivo transferases and ligases with a transferase function as amino acid–tRNA synthetase) may also catalyse the hydrolysis reactions (II or HI in Fig. 7.1). Examples here are hydrolytic steps in proof-reading mechanisms–amino acid activation or DNA–synthesis.

The ratio of the apparent transferase to hydrolase rate constant for these transferases can be estimated to be > 10^6 from the few experimental data published here. Such high ratios are necessary to minimize the energy waste involved in the acyl or peptidyl transfer to H_2O (≈ 55 M) at the low concentrations (< 10^{-3} M) of peptides in vivo. Only proteases that form covalent acyl–enzyme intermediates can be used as catalysts in kinetically controlled peptide synthesis. They have values of $(k_T/k_H)_{app}$ the range 10^2–10^4. In an equilibrium–controlled process reactrion (reaction II or III in Fig. 7.1) the protease is used only to increase the rate at which the equilibrium is established. The enzyme connot influence this equilibrium.

The stereospecificity of a protease gives information on the enantiomer (L or D) that is preferentially converted in the enzyme–catalysed reaction. The stereo–specificity of an enzyme may differ in the $S_1$and S_2'–binding subsites. Recently. it has been shown that enzymes are not strictly stereospecific. To describe this, a quantitative measure for the stereospecificity–the stereoselectivity or enantio–selectivity—has been introduced. It is a ratio of the rate constants involved in the enzyme–catalysed reaction for the L– to the D–enantiomer. As a ratio of rate constants, it should depend on temperature. This has also recently been shown to apply for proteases. Proteases can therefore also be used for equilibrium or kinetically controlled synthesis of peptides with both L– and D–amino acids and racemate resolutions.

In this case, the steric purity of the roducts depends onthe stereoselectivity of the protease used. Since the first edition of this book the use of proteases in peptide synthesis has been extensively covered in books and reviews. This reflects the continued interest in evaluating the potential of proteases as catalysts for peptide synthesis as an alternative or complement to chemical peptide synthesis.

Different proteases are now used for the synthesis of peptide antibiotics and hormones, neuropeptides, peptides used as sweeteners such as aspartame, and modifications of recombinant proteins such as in the production of human insulin. To evaluate the further potential of proteases as catalysts in peptide synthesis, the quantitative analysis of the yield–controlling factors, their P_1 and P_1' specificity, and stereoselectivity of their $! and S{ binding subsites, not covered in existing reviews, will be stressed injhis chapter.

PROPERTIES INFLUENCING THE PRODUCT YIELD

Kineticall controlled Synthesis

Mechanism

Studies on the kinetically controlled synthesis of condensation products such as oligo-saccharides, peptides and β–lactam antibiotics using hydrolases as catalysts have presented kinetic evidence that the nucleophile NH must be bound to the acyl– (or glycosyl–) enzyme before it can deacylate this inter-mediate. The reactions determining the yield of the condensation product from the acyl–enzyme are shown Fig. 7.3. This mechanism also applies for other condensation products. Therefore, a more general designation will be used here for the activated substrate (AB) and the nucleophile (NH) than that in Fig. 7.1.

Product yield

The peptide formed in the kinetically controlled synthesis is a substrate for the hydrolase. The hydrolysis of this product influences the maximum product yield when the hydrolysis rate is of the same order of magnitude or larger than the rate of the synthesis of the condensation product AN. Thus, two cases must be considered.

(*i*) The hydrolysis rate of the condensation product is not negligible when compared with the synthesis rate. In this case the product concentration is maximal when its rate of hydrolysis equals the synthesis rate

(*ii*) The hydrolysis rate of AN is negligible compared with the synthesis rate.

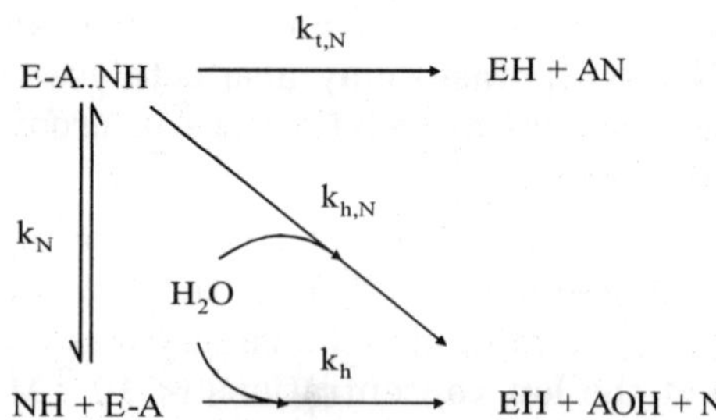

Fig. 7.3. Mechanism of kinetically controlled synthesis of *a* peptide AN from the acyl–enzyme E–A, involving the binding of the nucleophile NH (amino acid or peptide). It can only deacylate the acyl–enzyme when it has been bound to it. The formation of the peptide competes with the deacylation of the acyl–enzyme by water (hydrolysis).

In both cases, the apparent ratio of the rate of condensation (AN) to hydrolysis product (AOH) formation can be derived from Fig.3. When the rate of equilibration of E–A, NH and E–A–NH is much faster than the other reaction rates the following relation is derived.

$$(k_T/k_H)_{app} = k_{t,N}/(k_h K_N + k_{h.N}[NH]) \quad ...(7.1)$$

The quantity $(k_T/k_H)_{app}$ can easily be determined from the ratio of the initial rates for the formation of the transferase, v_{AN}, and hydrolase, v_{AOH} product :

$$(v_{AN}/v_{AOH}) = (k_T/k_H)_{app}([NH]/H_2O]) \qquad ...(7.2)$$

In other studies, a partition constant p has been used as a measure for the apparent hydrolase to transferase activity of the proteinase. The relation between p and$(k_T/k_H)_{app}$ is

$$(k_T/k_H)_{app} = [H_2O]/p \qquad ...(7.3)$$

In Case i (Table 7.1) the condensation product concentration is maximal when its rate of hydrolysis equals the rate of synthesis of AN. Expressions for the maximum yield of condensation product are given in Table 7.1. In contrast to the maximal product yield in equilibrium–controlled peptide synthesis, the maximum product yield in kinetically controlled peptide synthesis thus depends on $(k_T/k_H)_{app}$ and the specificity constants, i.e. the properties of the enzyme.

In Case ii (Table 7.1) the hydrolysis of AN can be neglected. When $(NH)_0 >> [AN]_{max}$ the nucleophile content and $(k_T/k_H)_{app}$ is practically constant during the synthesis. For this condition, the expression for $[AN]_{max}$ given in Table 7.1. can be derived. In contrast to Case i,

Table 7.1. Expressions for the maximum yield of the condensation product AN in the knetically-controlled synthesis. The constants aredefined in Fig. 7.1. and Equation 7.2. subscript 'O' denotes initial concentratnos.

	Maximum concentration of Product peptide $[AN]_{max}$	*Has been shown to apply for the kinetically controlled synthesis of*
Case I Rate of product hydrolysis ~ rate of product synthesis	$\left(\dfrac{\left(\frac{k_T}{k_H}\right)_{app}[NH]}{[H_2O]+\left(\frac{k_T}{k_H}\right)_{app}[NH]}\right)\dfrac{\left(\frac{k_{cat}}{k_m}\right)_{AB,NH}}{\left(\frac{k_{cat}}{k_m}\right)_{AN,NH}}[AB]$	b-Lactam antibiotics Oligosaccharides
Case II Rate of product hydrolysis << rate of product synthesis	$\dfrac{\left(\frac{k_T}{k_H}\right)_{app}[AB]_0+[H_2O]\ \text{in}\left(1-\frac{[AN]_{max}}{[NH]_0}\right)}{\left(\frac{k_T}{k_H}\right)_{app}}$ for $[NH]_0>>[AB]_0$ $\dfrac{\left(\frac{k_T}{k_H}\right)_{app}[AB]_0+[NH]_0}{\left(\frac{k_T}{k_H}\right)_{app}[NH]_0+[H_2O]}$	Peptides with proteases whose whose esterase activity >> amidase activity

the maximum yields do not depend on theratio of the substrate and condensation product specificity constants. In both cases, the maximum does not depend on the enzyme content. The latter only influences the time required to reach the maximum.

Whether a specific synthesis can be described by Case (*i*) or Case ii in Table 7.1 cannot generally be stated. It has to be determined from the time course off [At] in the kinetically controlled synthesis. Generally, Case (*ii*) applies for peptide synthesis using activated esters and proteases with large ratios of the ester to amide hydrolysis rate constants.

The relations for $[AN]_{max}$ given in Table 7.1 can be used for rational selection of the optimal enzyme for the kinetically controlled synthesis of a given peptide bond based on experimental values on $(k_T/k_H)_{app}$ and specificity constants; or optimization of the maximal yield for a given peptide bond with one enzyme.

Steric purity

The steric purity is determined by the stereospecificity of the enzyme. This qualitative property gives information on which of the D– or L–enantiomers that are preferred by the enzyme. The stereoselectivity or enantioselectivity, E, has been introduced as a quantitative measure for the stereospecificity of enzymes. It is defined as the ratio of rate constants for the enzyme catalysed reactions with the L– or o–enantiomer, respectively. For the kinetically controlled synthesis shown in Figure 7.1 the stereoselectivity E_s is defined as

$$E_S = \frac{\left(\dfrac{k_T}{k_H}\right)_{app.L}}{\left(\dfrac{k_T}{k_H}\right)_{app.D}} \quad ...(7.4)$$

From Figure *1* follows that it is a measure for the stereoselectivity in the S_1'–binding subsite. For the hydrolysis of a substrate the stereoselectivity E_H is defined as

$$E_H = \frac{\left(\dfrac{k_{cat}}{k_m}\right)_L}{\left(\dfrac{k_{cat}}{k_m}\right)_D} \quad ...(7.5)$$

Depending on whether the chiral centre is located in the P_1 or P_i' part of the substrate it is a measure for the stereoselectivity in the S_1 or S_1'–binding subsite. For a given activated substrate and nucleophile E_S and E_H are properties of the enzyme. Kinetically controlled synthesis can *be* used for the kinetic resolution of a racemic nucleophile or activated substrate. From Table 7.1 and Equations 7.4 and 7.5 it follows that the steric purity of the product are influenced by both E_S and E_H.

Equilibrium–controlled peptide synthesis

Yield

The maximum concentration (yield) of a peptide AN formed in equilibrium–controlled synthesis is where K_{3pp} is the apparent equilibrium (association) constant. It can easily be determined from measurements of the total concentrations given in The enzyme used to catalyse the formation of AN cannot influence $[AN]_{max\cdot e'}$ it only influences the rate with which

$$[AN]_{max,e'} = K_{app} \frac{[AOH][NH]}{[H_2O]} \quad ...(7.6)$$

the equilibrium is established. From Equation 6 it follows that the maximal peptide yield can be increased by the law of mass action, or by increasing K_{app}. The dependence on pH, temperature, ionic strength and solvent composition of K_{app} is given by the following relation:

$$k_{app} = K_{th}\frac{f_{UA}f_{UB}}{f_{UP}f_{H_2O}}\frac{\left(\frac{\Sigma[\text{peptide}]}{[\text{uncharged peptide}]}\right)}{\left(\frac{\Sigma[\text{acid}]}{[\text{uncharged acid}]}\right)\left(\frac{\Sigma[\text{base}]}{[\text{uncharged base}]}\right)} \quad ...(7.7)$$

where the sum Σ denotes the total concentration, i.e. the sum of all charged and uncharged forms of the product (reactants); K_{th} is the thermodynamic association constant for the formation of an uncharged peptide UP from the uncharged reactants UA (AOH) and UB (NH) containing the carboxyl and amino group from which the peptide bond is formed; and f is the activity coefficient. The quantity Kth cannot be determined directly, but is independent on solvent properties. The latter influences only the last two factors in Equation 7.7. They can be used for a quantitative analysis of the influence on pH and other system properties on K_{app}.

Steric Purity

The following relation can be derived for K_{app} in equilibrium controlled peptide synthesis (16)

$$K_{app} = \frac{\left(\frac{k_{cat}}{K_m}\right)_{synth}}{\left(\frac{k_{cat}}{K_m}\right)_{hydr}} \quad ...(7.8)$$

where the subscripts synth and hydr denote the synthesis and hydrolysis of the peptide, respectively. From Equation 8 the following relation can be derived for the enantioselectivity in equilibrium–controlled peptide synthesis where the subscripts synth and hydrdenote the

$$E_{S,e} = \frac{\left(\frac{k_{cat}}{K_m}\right)_{synth,L}}{\left(\frac{k_{cat}}{K_m}\right)_{synth,\,D}} = \frac{K_{app,L}\left(\frac{k_{cat}}{K_m}\right)_{hydr,L}}{K_{app.D}\left(\frac{k_{cat}}{K_m}\right)_{hydr,D}} = \frac{K_{app.L}}{K_{app,D}}E_H \quad ...(7.9)$$

where the subscripts L and D denote the synthesis and hydrolysis of peptides with only L–enantiomer or with one D–enantiomer, respectively. From Equation 7.9, it follows that the steric purity, in contrast to the yield in equilibrium–controlled peptide synthesis, depends on E_H, i.e. the properties of the enzyme. No data are available for the ratio ($K_{app\cdot D}/K_{app,L}$).

SELECTING THE OPTIMAL PROTEASE

Purity of the protease

The properties of the enzyme influence only the yield in kinetically controlled synthesis. The purity of proteases depends on the procedure used for the isolation of the enzyme from a homogenate or a medium (extracellular proteases) and the activation of zymogens. This can give rise to a heterogeneous preparation of enzymes with the same function but with different specificity constants and values of $(k_T/k_H)_{app}$.

Different bovine trypsin preparations have been found to vary with respect to relative amounts of α– and β–trypsin, and porcine trypsin preparations contain several active trypsin forms that can be separated by IEF or affinity chromatography. Recently, bovine α–chymotrypsin preparations with anionic chymotrypsins that were not present in older preparations have appeared. This may be due to changes in the methods used to isolate the enzyme. Penicillin amidase preparations from different sources have also been shown to be heterogeneous. This may result in a poor reproducibility in product yields when using different enzyme preparations for the synthesis of the same product.

A convenient method to analyse the heterogeneity of different enzyme preparations is IEF with substrate overlay techniques to detect active enzyme forms. In cases where the enzyme preparation is heterogeneous, it can be purified by affinity chroma–tography, provided that suitable affinity adsorbents are available, or by high resolution ion–exchange chromatography. Only homogeneous enzyme preparations should be used in fundamental studies.

P_1 and P_1' Specificity

For the synthesis of a peptide bond P_1–P_1'– it is necessary to use a protease with a high P_1'– or P_1–specificity for the amino acid that should be added to a given P_1 or P_1' amino acid. Data on the P_1–P_1' specificities of proteases, based on extensive studies on hydrolysis of different substrates, have been compiled. The P_1' specificity in peptide synthesis is given by the ratio $(k_T/k_H)_{app}$. For P_1' as a nucleophile in the kinetically controlled synthesis of P_1–P_1', increases with (k_{cat}/K_m) for substrates P_1–P_1', where P_1 is constant and P_1' varies. Thus, hydrolysis data could be used to determine the P_1'–specificity in the synthesis reactipn. The substrates used in the studies on hydrolase P_1'–specificity have been mainly esters (–0–X in P_1', where X = CH_3, C_2H_5) or amides (with NH_2 or chromogenic amines as *p*-nitroanilides in P_1').

Consequently, they are of limited use for the determination of the specificities of amino acids or peptides in the P_1' (..P′) positions. However, data on limited proteolysis of peptides or proteins (P_1–P_1' specificity), $(k_T/k_H)_{app}$ and maximal yields in kinetically controlled synthesis of different peptides can be used as a measure of the P_1'–specificity. Data for the P_1, P_1'–P_1' and P_1' specificities of different proteases that are commercially available and some proteases from extremely thermophilic organisms for which such data are available are given in Fig. 7.4 and 7.5. In Fig. 7.4 the data for P_1 and P_1'-specificity, derived from separate studies on hydrolysis and synthesis, correlate well with the P_1P_1'–specificity, derived from limited proteolysis of proteins and peptide hydrolysis. Fig. 4 can thus be used for a qualitative selection of the proteases that may be used for the synthesis of a specific peptide bond.

From these, the optimal protease for kinetically controlled peptide synthesis can be selected. It is the one with the largest $(k_T/k_H)_{app}$ value. The few existing data for proteases from extremely thermophilic organisms show that their p_1'-specificities are lower than for the best proteases from mesophilic organisms. From the data in Fig. 7.4 and 7.5 it follows that the proteases that can be used for kinetically controlled synthesis have broader P_1' than P_1 specificities. For the endopeptidases, only P_1' amino acids with protected α–carboxylate groups can be used. These enzymes have a negative charge near the Si binding site that repels α–carboxylate groups and limits their use as exopeptidases.

Unprotected Pi amino acids can be used in kinetically controlled synthesis catalysed by the proteases serine type carboxypeptidase and papain, and the amidase penicillin

Table. 7.2. List of the proteases whose P_1 and P_1'-specificities are given in Fig. 7.4 commercially available enzymes, proteases from extremely thermophilic organisms, an amidase for which such specificities have been determined, are included

Proteases forming covalent acyl–enzyme Intermediates that can be used to catalyse both klnetlcally and equilibrium controlled peptlde synthesis (serin*–, cystelne–proteases)

1. EC 3.4.16.1 Serine type carboxypeptidase (former carboxypeptidase Y)
2. EC 3.4.21.1 Chymotrypsin
3. EC 3.4.21.4 Trypsin
4. EC 3.4.21.5 Thrombin
5. EC 3.4.21.6 Coagulation Factor Xa
6. EC 3,4.21.19 Glutamyl endopeptidase (former V 8 protease)
7. EC 3.4.21.26 Prolyl oligopeptidase
8. EC 3.4.21.36 Pancreatic elastase
9. EC 3.4.21.50 Lysil endopeptidase (former *Achromobacter* protease)
10. EC 3.4.21.62 Subtilisin
11. EC 3.4.21.64 Endopeptidase K (former Proteinase K)
12. EC 3.4.21.66 Thermitase
13. EC 3.4.21.68 t–Plasminogen activator
14. EC 3.4.21.73 u–Plasminogen activator (former Urokinase)
15. EC 3.4.21.93 Proprotein convertase
16. EC 3.4.22.2 Papain
17. EC 3.4.22.8 Clostripain
18. Non–classified serine protease from extremely thermophilic organisms
    (a) Themus Rt41A
    (b) Jhemococcus stetteri
    (c) Streptomyces pactum

Proteases without covalent acyl–enzyme Intermediates that can only be used for equilibrium controlled synthesis (aspartlc and metallo–proteases):

19. EC 3.4.17.1 Carboxypeptidase A
20. EC 3.4.17.2 Carboxypeptidase B
21. EC 3.4.23.1 Pepsin
22. EC 3.4.23.4 Chymosin
23. EC 3.4'.23.15 Renin
24. EC 3.4.24.3 Microbial collagenase
25. EC 3.4.24.11 Neprilysin (former Enkephalinase)
26. EC 3.4.24.27 Thermolysin

Amidase forming covalent acyl-enzyme Intermediates that can be used to catalyse both klnetlcairy and equilibrium–controlled peptlde synthesis

27. EC 3.5.1.11 Penicillin amidase

P_1'	*Ala*	*Arg*	*Asn*	*Asp*	*Cys*	*Gln*	*Glu*	*Gly*	*His*	*Ile*
P_1										
Ala	21		16			2	16	16	16	
Arg	3							3, 4, 10, 11		3, 5, 21
Asn	10, 16		2	10		10				
Asp	10		10, 16				21	10		
Cys										
Gln	10, 21	21	21	10			10		10,11, 21	2
Glu	2, 16, 21		16, 21			16, 21	10			16, 10
Gly	16, 21	21	10, 21	21			10, 21	10		16, 21
His	2, 16 21							2		2
Ile			2							
Leu						2, 10 21	10, 16	10, 21	21	2, 21, 26
Lys	2, 3,	10	16	3				2, 3, 16	3	3, 16
Met	2	25	2	10			10	2		2
Phe	2, 25, 26	10		21			2	2, 26		21
Pro	7									
Ser	2, 10, 16		10	10, 21			21	10	10, 11, 16	21
Thr	2, 10, 21		2				10, 21	2		
Trp	2								2, 10, 21	2, 10
Try			2, 10, 21						1, 10, 21	
Val	21	2	10, 21						21	
P_1'-spec. from synthesis	1, 2, 8, 18, 17, 27	2, 3, 8, 16, 17, 18	1	3, 27		1	3, 16, 27	1, 2, 16, 17. 24	1, 16, 17, 27	1, 16 26

Fig. 7.4. P_1–P_1', p_1 and P_1'-specificity for some proteases, The data for P_1 specificity (last column) and P_1' specificity [last row_1, $(k_T/k_H)_{app}$ > 100 from fig. 7.5 were obtained from studies on substrate hydrolysis and protease-catalysed peptide synthesis, respectively. The data for P_1– P_1'-specificity (first 20 rows, columns) were obtained from studies onlimited proteolysis and hydrolysis of peptides. The blank spaces indicate lack of information. The proteases are given by the number in Table. 7.2.

Leu	*Lys*	*Met*	*Phe*	*Pro*	*Set*	*Thr*	*Trp*	*Try*	*Val*	P_1*spac, hydrolysis*
16, 21	10		25		10		10		16, 21	9, 11, 21, 26
21					3	5			3, 10, 13	3, .4, 12, 14
16	2		10, 16, 21		2, 10	2, 10, 21	21		2	
		16	21	21						6
21	2		10, 21		2				2, 21	
21	16, 21		21					21	21	6
26	10, 21		21, 26		10, 16	10			16, 21	16, 26
10, 16, 21	2	21	2						2, 21	16
21	21		16							8, 22
23, 26	2, 21				2, 16, 21	16		10, 11	10, 11, 21	2, 8, 10, 21, 23, 26
3, 21	2, 3, 14, 16, 20			3	3		16	1, 3		3, 9, 16, 17
	2, 10		21		2				2	
2, 26		22	1, 10 21, 26	2	2, 3, 21	10, 16		1, 10, 21, 26	10, 21	2, 21, 25, 26
7	1								10	7, 26
	10		21		21	10, 21			21	
21, 26	2			1	2	21		16	2, 10, 21	
3, 10					26		21		2, 10, 21	2, 10
10, 21	10		16		2, 10, 21	1, 2, 10, 11		2	21	2, 10, 16, 21
21	2					21		21	2, 21	8
1, 2, 3, 8, 16, 17, 26, 27	1, 3, 17, 18	1, 3, 8, 17, 18	1, 2, 3, 16, 17, 18, 26, 27	17	2, 17, 27	1, 3, 16	1, 17	17	1, 2, 3, 8, 16 17, 26	

proteases serine type carboxypeptidase and papain, and the amidase penicillin amidasce. The last enzyme can also be classified as a protease. For the proteases, the yields are lower with unprotected than with protected amino acids. This docs not apply for penicillin amidase that can be used to protect and deprotect amino groups of amino acids by adding or removing a phenylacetyl group. The cysteine proteases have a preference for apolar P_1' amino acids and have, at similar nucleophile concentrations, $(k_T/k_H)_{app}$ values that are an order of magnitude

larger than for the best serine proteases. Compared with the cysteine proteases, α–chymotrypsin and trypsin have a preference for basic P_1' amino acids.

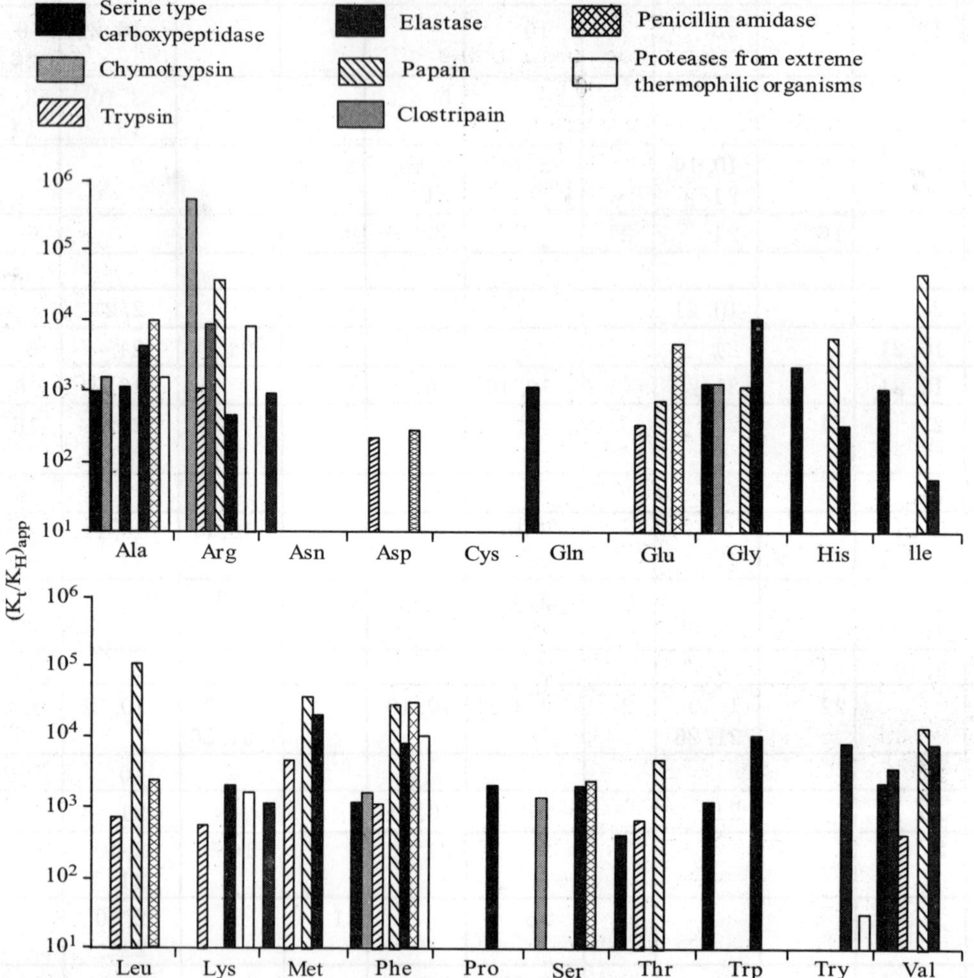

Fig. 7.5. P_L'-specificity of serine and cysteine proteases, proteases from extreme thermophilic organisms and the amidase penicillin amidase expressed as $(k_T/k_H)_{app}$ for different L–amino acid amides or amino acids (penicillin amidase) used as nucleophiles, Conditions were as follows. Serine type carboxypeptidase Y (EC 3.4.16.1), P_i = Ala. 25°C, pH 9.5 (pH–Stat, 10% MeOH) (26); bovine α–chymotrypsin (EC 3.4,21,1) (27), pH 9.0, 25°C, P_1 – Phe (pH–stat, 0.2 M NaCl) (20) or Tyr bicarbonate buffer. *I 0.2* M) (28,29); bovine trypsin (EC 3.4.21.4), P_1 ≈ Arg, 25°C, pH 8.0 (50 mM Tris–HCl. 5 mw $CaCl_2$, 0.5 M NaCl) (30); pancreas elastase (EC 3.4.21.36), P_1 = *Leu,* 25°C, pH 9.0 (pH–stat, total ionic strength 0.2 M), (31); papain (EC 3.4.22.8), P_1 = Ala, 25°C, pH 9.0 (pH–stat, 0.2 M KCl, 5 mw DTT) (32): clostripain (EC 3.4.22.8). P_1 = Arg, 25°C_1 pH 7.3 (0.2 m Tris–HCl). (33); penicillin amidase (EC 3.5.1.11), P_1 = phenylacetyl and β–mandelyl (P_1' – His), 25°C. pH 9.0 (bicarbonate buffer, I = 0.2 M) (28.29). Proteases from extreme thermophilic organisms: *Thermus* Rt41A P_1 Ala, P_1' = Tyr, 40°C, pH 10 (50% DMF) (34); *Thermococcus stetteri,* P_1' = Arg, P_1' = Phe, 55°C, pH 9 (0.2 M carbonate buffer) (35); *St. paeturn,* P_1 = Arg, P_1' Ala, Arg. Lys, ancf PMe, 40°C, pH 9.0 (0.2 M carbonate buffer) (36).

The exopeptidase serine–type carboxypeptidase (former carboxypeptidase Y) has the broadest P_1' specificity. For this enzyme the $(k_T/k_H)_{app}$ values are generally lower than for the other enzymes. Of the enzymes discussed here, only trypsin has been used for the synthesis of peptides with Asp or Glu in the P_1' position. Only carboxypeptidase Y could transfer acyl or peptidyl groups to cysteine. For proline in P_1', only clostripain has been shown to be a suitable catalyst. For endopeptidase K, $(k_T/k_H)_{app}$ was found to be < 100 for Asp–, Arg–, Ser, Ala–, and Phe–NH_2. For both equilibrium and kinetically controlled synthesis, enzymes with a high P_1P_1' specificity for a desired peptide should be used as catalysts. For the kinetically controlled synthesis, the selection of the optimal enzyme depends on whether the synthesis can be described by Case i or ii conditions.

Direct kinetic evidence that the kinetically controlled synthesis using these enzymes and an ester as the activated substrate can be described as a Case ii system (product hydrolysis rate << product synthesis rate) has been given for chymotrypsin trypsin, papain and carboxypeptidase Y. The peptide yield should then increase with $(k_t/k_H)_{app}$. For experimental studies where $(k_T/k_H)_{app}$ and maximal yields have been determined under the same conditions, the observed yield equals the values calculated from Table 7.1. within the experimental error. For kinetically controlled synthesis where the product hydrolysis rate is approximately equal to product synthesis rate (Case i), more enzyme properties than for Case ii influence the maximal product yield.

To obtain maximal yield of a desired product as a function of enzyme, substrate activation and nucleophile content it is therefore, contrary to Case ii, not sufficient to use the activated substrate and enzyme with the highest value of $(k_T/k_H)_{app}$. This is demonstrated by the following experimental observations. In the semisynthesis of the peptide antibiotic ampicillin and cephalexin, higher yields were obtained with D–phenylglycine–O–Me than with D–phenyl–glycine–O–Et or –NH_2, even when the latter are more specific substrates and have larger $(k_T/k_H)_{app}$ values for the enzyme used (*Escheridiia coli* penicillin amidase). The selection of a suitable activated substrate should, in this case, also consider that the yield can be increased when the product can be continuously removed from the reaction mixture by precipitation, adsorption to product–selective adsorbents or extracted by suitable solvents. Such a process integration should always be considered in protease catalysed peptide synthesis on a larger scale. In Fig. 7.6. the maximal observed stereoselectivities the S_1 and S_1'–binding subsites of some proteases and penicillin amidase are given. For most proteases, quantitative data on their stereoselectivity are missing. Observed stereoselectivities are much larger in the S_1 subsite than the S_1'–subsite. Proteases can therefore be used for the synthesis of L–D peptides.

The available data show that the proteases from extremely thermophilic organisms have stereoselectivities of magnitudes similar to most of the proteases from mesophilic organisms. Papain and bovine α–chymotrypsin, where most data are available, have, however, much higher stereoselectivities. For penicillin amidase with a D–specific S_1 subsite and an L–specific S_1'–subsite, the stereoselectivities in both subsites allows its use for racemate resolution. This has been applied on an industrial scale.

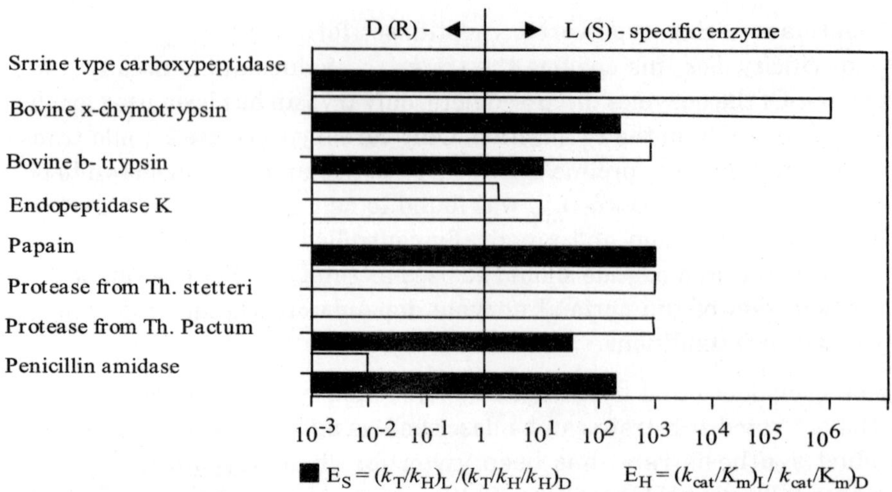

Fig. 7.6. Maximal observed stereoselectivities in the S_1–(E_H) and S_1'–(E_s) binding subsites for the proteases serine type carboxydase, bovine chymotrypsin, trypsin andunpublished data for E_H), endopeptidase K, proteases from the extremely thermophilic organisms Thermococcus. stetteri and Thermococcus pactum and penicillin amidase. For E > 1 theenzyme is L-specific, for E < 1 it is D-specific.

When proteases are used for racemate resolution of the acyl donor P_1 or the nucleophile P_1', a high steric purity of the product requires that the stereoselectivity (E_H, E_s or $E_{S,e}$) is < 0.01 for D–specific or > 100 for L–specific enzymes, respectively.

STERIC PURITY

Once the enzyme has been selected for the kinetically controlled synthesis, $(k_T/k_H)_{app}$ can be further optimized by changes in the group used to protect the P_1'–carboxyl end, pH, temperature, solvent composition, ionic strength or the use of unconventional systems (reactions in ice or suspensions). For the equilibrium–controlled synthesis, k_{app} can be optimized by the use of P_1-amino–blocked and P_1'–carboxyl–protected amino acids, and changes in pH, ionic strength, solvent composition and temperature.

Protection and Activation of the P_1-Carboxyl Group

For the proteases where this has been studied $(k_T/k_H)_{app}$ and the product yield in kinetically controlled syntheses has been shown to be much larger for the ammo acid amide than for the corresponding ester. This may be due to the better binding of $-NH_2$ than –OX (=P'_2) to the S'_2 binding site. This applies also for serine type carboxypeptidase that can be used for synthesis of peptides with unprotected P_1 amino acids. The latter applies also for the amidase penicillin amidase that can be used for the production of N–phenylacetyl protected amino acids that are used as P_1 groups in peptide synthesis. As most proteases that can be used for kinetically controlled synthesis are better esterases than amidases, the P_1-carboxyl group should be esterified for optimal yields.

The synthesis then follows scheme (ii) in Fig. 7.2 where the product peptide is hardly hydrolys. In kinetically controlled peptide synthesis with peptides P_1'–P_n' as nucleophiles, $(k_T/k_H)_{app}$ and the product yield is also influenced by the interactions between P_1' (i = 2,...n) and the S_i'–binding subsites on the used protease. With constant P_i' both decreases and increases have been observed for different P'_2 amino acids. The focus in this review is on the P_1 and P_1' specificity, therefore the P_i' specificity ($i \geq 2$) will not be analysed further. Data on this have, however, been presented in only a few reports. Generally, it has been found that the stereoselectivities E_H and E_s of proteases increase with the specificity constant (k_{cat}/K_m) and the nucleophile specificity $(k_T/k_H)_{app}$, respectively. From this, it follows that for kinetically controlled peptide synthesis with racemic activated aminoacid or P_1'–X, the steric: purity of the product increases with E_H and E_S.

Prproteases that are more specific for L– than D–amino acids at the S_1' binding site can be used for the kinetically controlled synthesis of L–D peptides. The relative importance of a productive binding, when it can deacylate the acyl–enzyme, depends on the relative magnitude of the S_1'-P_1' and S_2'–P_2' interaction. This determines the stereospecificity. When they are of equal magnitude, as for amides, productive binding of D–nucleophiles to i.–specific enzymes is less likely. When the S'_1–P_1' interaction is weaker than the S_1'–P_1' interaction, as for esters, the probability of productive binding of D–nucleophiles and the L–D peptide yield should be higher than for the amides. This has been observed in peptide synthesis with α–chymotrypsin.

pH

Equilibrium–Controlled Synthesis

The pH dependence is given by the concentration ratios in the last factor in Equation. The concentration ratios can be derived from the following equilibria for each of the reactants and condensation products that are acids and bases, where the K values are microscopic dissociation constants:

$$
\begin{array}{ccc}
H_2NR_1COOH & \underset{}{\overset{K_A}{\rightleftharpoons}} & H_2NR_1COO \\
K_D \updownarrow & & \updownarrow K_B \\
H_3^+NR_1COOH & \underset{}{\overset{K_C}{\rightleftharpoons}} & H_3^+NR_1COO^-
\end{array}
\qquad \text{...(7.10)}
$$

With amino–(P_1) and carboxyl–(P_1') protected amino acids the last factor in Equation's 7.5. is

$$
\frac{1}{\left(1+\frac{K_A}{[H^+]}\right)\left(1+\frac{[H^+]}{K_D}\right)} \qquad \text{...(7.11)}
$$

where the K_A and K_D values for the P_1-arboxyl and P_1'–amino group can be found in ref. The pH–dependence of K_{app} for this and other cases has been calculated and compared with experimental data. From these results it is evident that unprotected amino acids (especially P_1) should not be used for peptide synthesis. K_{app} is reduced, owing to charge repulsion, compared with the case where protected amino acids are used. With unprotected amino acids other peptides (P_1-P_1, P_1'-P_1', P_1'-P_1) can be formed in addition to the desired product, P_1-P_1'. This also applies for the kinetically controlled synthesis.

An exception here is the synthesis of ampicillin and cephalexin with unprotected D–phenylglycine using penicillin amidase from *E. coli* as a catalyst. With D–, but not L–, phenylglycine, side–reactions in which undesired dipeptides were formed have not been observed as this enzyme has a D–specific S_1 binding subsite. The calculated pH–dependence in Fig. 7.7 agrees well with experimentally determined K_{app} values as a function of pH. For benzylpenicillin, the experimental data at low pH values (< 5) may be perturbed by the reduced activity of the enzyme at low pH and the chemical instability of β-lactam ring.

For peptide synthesis, the experimental data show that the maximum yield can be increased when the difference between the pK values for the amino group and carboxyl group involved in the condensation reaction are reduced.. This can be achieved by the addition of organic solvents. In some cases, enzyme–dependent equilibrium yields in peptide synthesis have been observed. Here, as for penicillin amidase at low pH, this probability results from enzyme instability at high or low pH and/or non–establishment of the equilibrium at pH values far from the pH optimum of the enzyme used, since the latter cannot influence equilibrium yield.

Kinetically Controlled Synthesis

A pH optimum above the pK value for the P_1' amino group (7–8 for amino acid amides or esters; 9–10 for amino acids is generally observed. Below the pH optimum this is the due to the

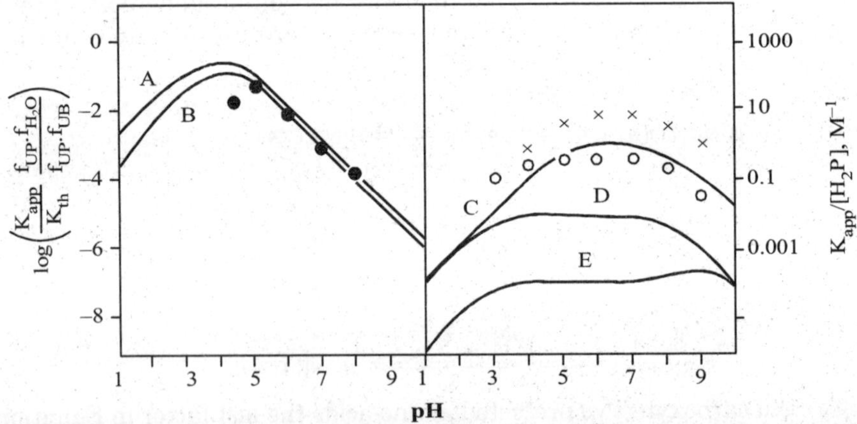

Fig. 7.7. The pH dependence of the association constants for the formation of peptide antibiotics (A, B) and peptides (C, D, E). The last (pH-dependent) factor In Equation 7.7 was calculated as a function of pH using Equations 7.10 and 7.11 and compared with experimental data: pK, values for reactant acid; pK′, values for the reactant base; pK″, values for the product peptide. (A) Ampicillin formation using the values pK_A = 4.11; pK_D' = 6.9, pK_e' = 9.0; pK_A' = 2.6; pK_D', = 4.6, pK_e' = 4.6; pK_A'' = 2.6; pK_D'' = 6.9; pK''_B = 9.0. (B) Benzylpenicillin formation using the values pK_A = 4.2; pK_A'' = 2.6; pK_A' = 2.6; pK_D' = 4.6; pK_B' = 4.6. Experimental data from ref. (filled circles, right-hand scale). Dipeptide synthesis from amino– and carboxyl–protected amino acids. (C) For pK_1 = 3.0; pK_2 = 8.0. (D) For pK_1 = 5.0; pK_2 = 8.0. Experimental data for the synthesis of Cbz-Tty-Gly-NH: 0, in H_20 (pK_1 = 3.60; pK_2 = 8.20); x, in 60% (v/v) trietrtylene glycol (pK_1 = 4.53; pK_2 = 8.17) from ref. (right-hand scale). (E) Dipeptide synthesis from unprotected amino acids for pKi = pKi = pKJ = 3.0; pK_2 = pK_2 = 9.0; pK_2 = 8.0.

decrease in $(k_T/k_H)_{app}$ caused by the decrease in the concentration of the uncharged P_1' amino group that is required for the nucleo–philic deacylation. Above the pH optimum, the decrease is probably due to the increase in the concentration of OH^- as a nucleophile, although this has not been studied in detail. This will increase k_H and reduce $(k_T/k_H)_{app}$.

In cases where Tris has been used as a buffer component, it has been shown that it can deacylate acyl–enzymes. This will reduce the yield of the desired peptide. Consequently, here, as in other enzyme kinetic studies, Tris must not be used as a buffering cornponent. Subtilisin and penicillin amidase (except from *Alcaligenes faecalis),* lack S–S bridges. This reduces their pH stability above pH 9.0 and limits their use for the kinetically–controlled synthesis of peptides in this pH range.

Temperature

Equilibrium–Controlled Synthesis

The temperature dependence of K_{th} and the dissociation constants in the last factor of Equation 7.7 determine the change in K_{app} with temperature. The formation of a peptide rond from uncharged precursors is endothermic. Thus, K_{th} should increase with temperature. For peptide synthesis from protected amino acids, the temperature dependence of the last factor of Equation 7.7 is given by Equation 7.11. This ratio increases with temperature. Thus, in this case, K_{app} and the yield should increase with temperaure. Generally, however, the effect of temperature on K_{app} cannot be predicted but must be studied empirically. Unfortunately, in most studies on protease-catalysed peptide synthesis, the influence of the temperature has not been investigated, although this is of practical and theoretical importance.

Kinetically Controlled Synthesis

Only a few studies have been published on the influence of the temperature. For proteases from mesophilic and extreme thermophilic enzymes both the yield and $(k_t/k_H)_{app}$ have been found to decrease with temperature. This has led to revived interest in studying enzyme-catalysed reactions in ice, where an increase in $(k_T/k_H)_{app}$ and the yields have been found for kinetically controlled peptide synthesis in frozen aqueous systems.

Steric Purity

As a ratio of rate constants the stereoselectivities in the S_1 (E_H) and S_1'–(E_s) binding subsites of enzymes should depend on temperature. This was recently verified by experimental data. The steric purity of the products in equilib-rium and kinetically controlled peptide synthesis should therefore be temperature dependent. In the few studies on the temperature dependence of the stereoselectivity of proteases it has been observed that E_H generally decreases with temperature, since it is mainly determined by the activation energy of k_{cat}. For E_S an increase with temperature has been observed for chymotrypsin. The opposite has been observed for the amidase penicillin amidase in the kinetically controlled synthesis of phenyl acetyl–amino acids. This indicates that the binding energy of the nucleophile to the S_1' subsite has a larger influence on the temperature dependence of E_S than the substrate binding energy on E_H. In the temperature interval 5–45°C the maximum change in these stereoselectivities was found to be about one order of magnitude.

Ionic Strength

Equilibrium–Controlled Synthesis

Qualitatively, an increased ionic strength should stabilize isolated ions, leading to a decrease in K_{app}. Quantitatively, this can be deduced from the ionic strength dependence of the dissociation constants in the last factor of Equation 7.7. The other factors do not depend on the ionic strength. The pK values of carboxyl groups decrease and for amino groups are practically constant (or increase for amino acids with unprotected carboxyl groups) with ionic strength. This is in agreement with the qualitative prediction. Thus, for soluble condensation prod-ucts, increasing the ionic strength should reduce the yield. This implies that the use of ions in buffers, other than those involved in the condensation reaction, should be minimized.

Kinetically Controlled Synthesis

For both Case i and ii (Table 7.1) systems, the ionic strength influences the binding of the nucleophile to a charged S_1' binding site. When opposite charges are involved, as for the synthesis of β–lactam antibiotics catalysed by *E. coli* penicillin amidase, an increase in the ionic strength causes a decrease in the binding of the nucleophile and the product yield. In this case, where both the activated substrate and the nucleophile contribute to the ionic strength, the amount of ions to buffer the reaction medium should be minimized. When equal charges are involved, as in peptide synthesis with negatively charged amino acids (un-protected or as amides as nucleophiles catalysed by endoproteases with a negative charge in the S_1' or S_2' subsite, an increase in the yield with increasing ionic strength is expected. This has been verified in experimental studies.

Solvent composition

Organic solvents have been used in protease catalysed peptide synthesis to increase the solubility of hydrophobic substrates and products, to perturb K_{app} in equilibrium–controlled synthesis and to isolate the products by solvent partitioning. The use of organic solvents here should be limited to environmentally unproblematic solvents that can be easily recycled. Among these is supercritical CO_2 that has recently been extensively used for lipase catalysed reactions. Its possible use for peptide synthesis has not yet been studied in detail.

Equilibrium–Controlled Synthesis

From Equation 7.7 it follows that the addition of water–miscible organic solvents influences the activity coefficients of the uncharged solutes and H_2O. This, and their effects on the dissociation constants in the last factor, may change K_{app}. The dissociation constants for carboxyl groups have been found to decrease with the organic solvent content. This partly explains the observed increase in K_{app} with organic solvent content shown in Fig. 7.7. Only when trie water activity coefficient $\neq 1$ does it influence K_{app} in these systems. This is possible only in non–ideal mixtures of H_2O and organic solvents.

The theory of such systems is still not sufficiently developed to allow a more rational analysis of the influence of organic solvents on pK values and activity coefficients. Recently, it has been found that enzymes are more easily inactivated in water–miscible than in water–immiscible organic solvents. With the latter, two–phase systems are formed that are of practical interest, especially when the solubility of the peptides is low in an aqueous system. The reaction rates in these systems are, however, reduced by orders of magnitude compared to the latter system.

Kinetically Controlled Synthesis

From Table 7.1, it follows that the product yield can be increased when $[H_2O]$ is reduced by the addition of water miscible organic solvents that marginally influence the properties of the enzyme used. Alcohols are not suitable organic solvents here, as transesterification products may be formed and as they tend to decrease the stability of the proteases. For peptide synthesis (Case ii), increase in yields with reduced water content have been observed with DMF. The data available cannot yet be used to analyse whether these organic solvents influence the P_1'–specificity of these enzymes selectively. This however, can be snidied with water–immiscible solvents. With these, indirect solvent effects are expected because of the partitioning of the substrates and products in the two liquid phases.

Proteases, as other globular proteins, have a surface area of which about 40% consists of hydrophobic groups that may bind organic solvent molecules in the aqueous phase that is saturated with the water–immiscible solvent. (This is the basis for hydrophobic interaction chromatography used to purify proteins.) This gives rise to a direct solvent effect due to bound organic molecules that may change the specificity constants, $(k_T/k_H)_{app}$ and the stereoselectivities E_H and E_S. Large effects due to such direct effects have been observed in the P_1' specificity of chymotrypsin with bound hexane and octane. For hydrophobic amino acids (Leu and Phe) it increased, and for Arg a decrease was observed, whereas for Ser, Val and Ala the changes were within the experimental error. The same applied for the specificity constants for the hydrolysis of N–Acetyl–Try–OEt. No direct solvent effect could be observed for endoproteinase K and penicillin amidase.

Synthesis in Suspensions

Recently, protease–catalysed equilibrium and Idnetically controlled peptide syn-thesis has been carried out in suspensions with solid substrates and/or products as an alternative to the use of organic solvents. In these systems, high total substrate contents (up to *30%* w/v) can be obtained. The reaction rate is limited by the rate of mass transfer from the solid to the aqueous phase, where the protease–catalysed reaction occurs. The latter is increased when the size of the solid substrate particles is decreased. In such systems for equilibrium and Idnetically controlled peptide synthesis, greater product–formation rates (space–time yields) can be obtained than in systems with organic solvents or in aqueous systems with dissolved substrate. Its use is demonstrated in Protocol 1.

Protocol 1

Protease–catalysed synthesis of the dlpeptide N–acetyl–L–tyrosine–L–arginlne amide in aqueous two–phase system in suspension with solid substrate (N–acetyl–L–Ty–OEt, ATEE) and soluble nucleophlle L–Ar–NH_2 using free or Immobilized α–chymotrypsln (CT).

Equipment and reagents

- HPLC–apparatus, pH–meter, thermostatted reaction vessel that can be stirred with a hanging stirrer to avoid mechanical damage of the immobilized enzyme
- Stock solution of CT (10 mg/ml) prepared in 10^{-3}–10^{-2} HCl to avoid autolysis (Section 6, 2). It is stable for years when frozen. CT is immobilized in a suitable support as Eupergit C (≈ 1 mg CT/ml settled gel)

- Stock solutions of the nudeophile L–Arg–NH_2 and dissolved substrate ATEE in carbonate buffer, pH 9.0 (I = 0.2 M), the latter are used for calibration in the HPLC analysis

Method

1. Selection of optimal *protease*. Prom Fig. 7.4, it follows that CT is the only enzyme with both the desired P_1 and P_1' specificity. In cases where several enzymes can be used, the one with the highest P'_1 specificity $(k_T/k_H)_{app}$ value in Hgurc 5 should be selected. A suitable choice is to use CT. For first studies, free enzyme can be used. To avoid disturbance in the HPLC analysis, the free enzyme used must not adsorb to the capillaries or stationary reversed–phase in the columns. Such adsorption has been observed with β–lactamase but, fortunately, not for CT, trypsin, penicillin amidase, endopeptidase K and thermolysin. Commercial bovine CT samples are homogeneous and can easily be immobilized to different supports,
2. Substrate activation and protection of carboxyl group in P'_1. An ester should be used as activated substrate, as CT is a much better esterase than amidase. It follows that a P'_1 amide gives higher product yields than a P_1' ester.
3. *Buffer, pH, ionic* strength. An inert buffering system at a pH above the pK for L–Arg–NH_2 as carbonate buffer (I – 0.2 M) of pH 9.0 is a suitable choice. The ionic strength (in a homogeneous systems) should not be too high, for immobilized systems it must be selected to minimize possible pH gradients in the support.
4. HPLC analysis. Prepare solutions of activated substrate, hydrolysed substrate, nucleophile (and if available of condensation product). These solutions can be used to identify reactants and products and to calibrate the HPLC peaks for quantitative measurements. Reference samples and samples from the reaction mixture (≈ 50 µl) are analysed after dilution (≈ 1:10) with the mobile phase using an RP 18 column and as eluent, 70% (v/v) 0.067 M KH_2PO_4, pH 4.7, and 30% (v/v) MeOH). The peak areas at 280 nm (for aromatic amino acids) and 220 nm (for all amino acids) are used to quantify substrates and products.
5. Synthesis. Preparing the reaction mixture (1–5 ml): dissolve the nucleophile (400–800 mM) in carbonate buffer, pH 9, (I = 0.2 M), then add the activated substrate (400–750 mM); adjust the pH of the suspension to 9.0 with 2 M NaOH and stir the suspension vigorously to disintegrate particle aggregates of the solid substrate. Add enzyme (10–20 µg free or 100 µg immobilized enzyme per ml reaction suspension) to the thermostatted (25°C) reaction mixture to start the synthesis. The product peptide and hydrolysis product N–acetyH–Tyr are soluble in this system. Withdraw liquid samples at different times and analyse them as described in step 4, The pH in the suspension should be kept at 9.0 during the reaction with 2 M NaOH or 3 M Na_2CO_3. The latter should be used when the enzyme or products are unstable at pH > –10.
6. From the results in *Figure 7.8* it follows that this synthesis, as expected, can be described by Case ii (Table 7.1 and *figure* 7.2). Both the results in solution and in suspension can be used to determine $(k_T/k_H)_{app}$. No peaks other than these shown were observed and only one condensation product involving the α–amino group

($pK \approx 8$) is formed. Condensation products involving other amino groups (ε–amino or guanidino of Lys and Arg, respectively) have been observed in other studies (own unpublished data).

7. Yields and reaction *rates*. The $(k_T/k_H)_{app}$ values calculated from *Figure 7.8* show that the rates and the product peptide yield are reduced when the enzyme is immobilized. This is in agreement with the predictions that can be made from the reactions in Table 7.1. In the aqueous phase ($[NH]_0 >> [AB]_0$) the yield then increases with $(k_T/k_H)_{app}$. That this ratio and the rates decrease when the enzyme is immobilized is expected from quantitative analysis of such systems, and reduction of pH in the im-mobilized enzyme system cannot be avoided even at high buffer capacities. At constant $[NH]_0$ the yield can be increased. The peptide product yields (as % of the activated substrate) in Figure 7.8 are 87 % for free CT and 83% for immobilized CT. This, and the high rates and space–time yields demonstrate the potential of enzyme–catalysed reactions in suspensions.

Planning a Protease–Catalysed Synthesis of a Peptide Bond

What enzyme?

The proteases that may be used as catalysis can be selected from the data in figures 7.4 & 7.5. Where several proteases maybe used, the selection of the most suitable protease depends on whether a dipeptide or an oligopeptide should be synthesized. If an oligopeptide, the protease should have a low specificity for the hydrolysis (and synthesis) of the peptide bond(s) in the reactant peptides. This is especially important in equilibrium–controlled synthesis. For kinetically controlled synthesis with amino acid (peptide) esters as activated substrates, these undesired side–reactions are minimized by the use of proteases with large ratios of esterase to amidase activity.

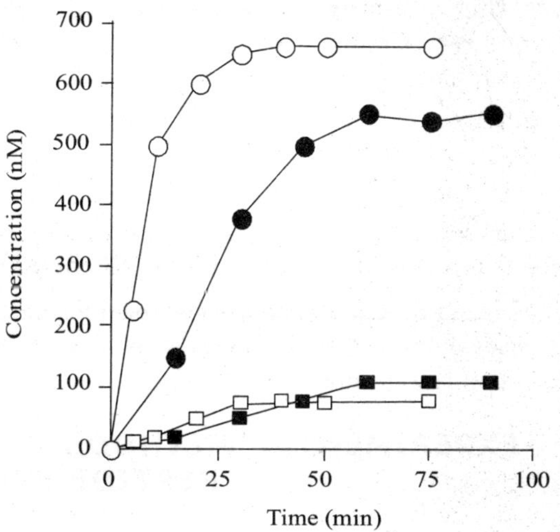

Fig. 7.8. Kinetically controlled synthesis of soluble ATM (circles) from ATEE as a suspension and soluble L-Arg-NH_2 with free (open symbols) and immobilized (filled symbols) CT according to *Protocol* 1. Conditions: 25°C, pH 9.0 (carbonate buffer, I = 0.2 M), the pH is kept constant during the reaction. Free enzyme: enzyme content 10 µg/ml; starting concentrations: 800 mm L-Arg-NH_2 and 750 mM (total content) ATEE as a suspension. Immobilized enzyme: 200 µl CT immobilized in Eupergit C (0.5 mg CT/ml settled gel) per ml reaction suspension; Starting concentrations were 700 mw L-Arg-NH_2 and 660 mM (total content) ATEE as a suspension; circles, ATM; squares, the soluble hydrolysis product N-acetyl-L-Tyr. The ATEE content in the aqueous system during the reaction with free enzyme was <1 mM, with immobilized enzyme it decreased from 25 mw at time 0. Without enzyme the dissolved ATEE concentration was 25 mM.

Equilibrium–Controlled or Kinetically Controlled Synthesis?

All proteases can catalyse the equilibrium–controlled synthesis. The kinetically controlled synthesis can only be catalysed by proteases that form covalent acyl–enzyme intermediates. Since much less enzyme is needed, and the reaction times required to reach maximal yields are shorter in the latter case, kinetically controlled synthesis should be used when possible. They should be performed under Case ii conditions (Fig. 7.2). This requires the use of esters as activated substrates. In this case, and with an excess of nucleophile, the maximal yield can be estimated using the relations in Table 7.1.

Free or Immobilized Enzyme?

For first experiments, low concentrations of free enzymes should be used. With proteases, autolysis cannot be neglected. It is then advantageous to use immobilized enzymes. This also reduces enzyme costs and the costs of separating enzyme from products. The immobilization cannot influence the yields in the equilibrium–controlled synthesis. In kinetically controlled synthesis, the maximal yields and steric purity of the products can be reduced compared with the free enzyme, at low (< 100 mM) nucleophile concentrations because of mass transfer effects. However, this difference is reduced at the higher concentrations (> 100 mM) that are of practical interest. Changes in enzyme properties due to immobilization may, however, influence the yield and steric purity (Table 7.1).

Peptide synthesis involves the formation of acids and bases. With immobilized proteases this may give rise to considerable pH–gradients (several pH units) in the supports with the enzyme. This changes the rates and may also lead to pH–induced denaturation of the protease. These negative effects can be minimized when sufficient buffer capacity (I > 0.2 M) is used and the system is well stirred to obtain maximal mass transfer rates.

Immobilized proteases can also be used for peptide synthesis when the prod-uct precipitates. In such cases a small pore support, where the product cannot precipitate in the support, must be used. Then the immobilized protease can be reused.

EXPERIMENTAL METHODS FOR PROTEASE–CATALYSED PEPTIDE SYNTHESIS

Enzyme Purity and Purification

Commercial protease preparations are generally not pure. They contain different forms of one protease (such as α– and β–trypsin in bovine trypsin) and/or other proteases as impurities. In the former case (enzyme heterogeneity) this only influences the yield in kinetically controlled synthesis. Foreign proteases as impurities may cause undesirable hydrolysis of products and reactants. To obtain reproducible results in kinetically controlled peptide synthesis, pure and homogeneous protease preparations are required. It is thus essential to check the purity of the preparations. A convenient and rapid method is IEF with substrate overlay techniques to visualize the protease. For fundamental studies, only pure homogeneous protease preparations should be used.

Enzyme Immobilization

Suitable techniques and supports for enzyme immobilization are described in ref. When dissolving proteases, buffers that minimize autolysis must be used. They should ideally be dissolved in buffers at a pH removed from the optimum for activity. During immobilization, the concentration of proteases must be kept below 1 mg/ml to avoid autolysis during the

Substrates and Buffers

In equilibrium–controlled synthesis, amino (P_1) and carboxyl (P_1') protected amino acids should be used (Figure 7.1). In kinetically controlled synthesis, the amino group in P_1 must be protected. To obtain the desirable Case ii condition (Figure 7.2) for the synthesis, esterification is better than amidation for the activation of the P_1-carboxyl group. This requires the use of proteases that are better esterases than amidases. The P_1' amino add carboxyl must be protected (Section 4.1). For the stepwise synthesis of peptides, P_1–P_1'. P_1'–esters are better than amides; then, the resulting P_1-P_1' peptide is also activated.

Buffers that have been shown to have minimal influence on activities should be used at the lowest possible ionic strength. Their pK–value should be close to the pH that is optimal for the enzyme catalyzed synthesis. Tris must not be used as a buffer component because it can deacylate acyl–enzymes.

Monitoring the Synthesis; Purification of Products

High–pressure liquid chromatography is the best analytical technique to monitor reactants and products in protease–catalysed peptide synthesis. Simultaneous detection at several wavelengths (280 nm for aromatic amino acids and about 220 nm for all amino acid) simplifies the identification of peaks. Detection at 220 nm is especially important to identify products from side–reactions such as the hydrolysis of carboxyl–protected P_1' amino acids or formation of P_1'–P_1'–X. For preparative isolation of products, preparative HPLC columns can be used under the same conditions as the analysis.

Optimizing the Yield

Once the optimal system for the synthesis—Case ii—has been selected, the yield can be further increased by changing yield–controlling factors discussed in Section 4. From the determination of $(k_T/k_H)_{app}$ and the relations in Table 7.1, the P_1' content required for quantitative transfer of P_1 to P_1' in the kinetically controlled synthesis can be calculated.

For equilibrium–controlled synthesis, the yield can be increased by the use of organic solvents. In these cases it is essential to determine the enzyme stability under the conditions used for the synthesis. Only proteases whose activity changes marginally should be selected as catalysts.

PROTEASES IN PEPTIDE SYNTHESIS: LIMITATIONS AND PERSPECTIVES

For practical purposes, more peptides are synthesized by chemical peptide synthesis than in protease–catalysed processes. For some peptides a combination of chemical and enzymatic condensation steps is used. The automation of chemical peptide synthesis using peptide synthesizers, and newer developments that reduce the amounts of problematic solvents, still favour this procedure over the enzymatic approach. However, the latter has advantages, including prevention of racemization during the synthesis, reduced use of problematic (toxic) solvents and possible re–use of the catalyst (as an immobilized protease). Every possible peptide can also be synthesized. Finally, proteases can be used for simple amidation of P_1'–carboxyl end groups in peptides (hormones) by transpeptidation with P_1'–NH_2 as a nucleophile.

Since the appearance of the first edition of this handbook some of the main developments in protease catalysed peptide synthesis are:

- the extension of the systems used to aqueous suspensions and frozen systems where higher rates and yields can be obtained than in systems with organic solvents;
- more systematic studies on the stereoselectivity of enzymes;
- new data for proteases including those from extremely thermophilic organisms;
- the introduction of enzyme catalyzed deprotection in peptide synthesis;
- the increase in the number of recombinant proteases, especially those of medical interest.

These advantages have not yet led to a more extensive use of proteases in peptide synthesis, owing to several factors. These include the lack of systematic studies on the stereospecificity of proteases. Most proteases are not strictly stereospecific (Figure 7.6). Thus D,L mixtures of P_1' cannot generally be used for the synthesis of sterically pure peptides. They can, however, be applied to synthesize L–D (D–L) peptides with L (P_1' or P_1)–specinc proteases. There is a paucity of information and sources of (cheap) proteases that can be used for the synthesis of Pi–Pi bonds where data are missing or the specificity is low (Figures 7.4–7.6). We also lack data on enzyme properties and stability in organic solvents. Finally, there have been no systematic studies on the use of protected and activated P_1 and P_1' amino acids that can be used both for chemical and enzymatic synthesis.

Even when this information is available, one should not expect that protease–catalysed peptide synthesis would eventually replace chemical peptide synthesis. The protease catalysed synthesis cannot be as readily automated as chemical synthesis, and the latter will remain the first choice for laboratory–scale peptide synthesis. For preparative synthesis of commercially important peptides, protease–catalysed steps will probably increase in importance.

The results from studies on the protease–catalysed peptide bond synthesis have been used to analyse the factors that influence product yield. However, they may also be used to analyse more basic enzymological problems. Based on the kinetic data on P_1' specificity reviewed here, systems for direct studies on the P_1'–S_1' interaction by X–ray crystallography can be selected. Such investigations can be used for comparative analysis of the sequence (P_1') specificity of proteinases using natural systems or enzymes designed by site–directed mutagenesis. The results could provide basic information on interactions between substrates and the active site on enzymes that influence their properties as catalysts.

Survey and Summary

There are three main groups of restriction endonucleases (REases) called Types I, II, and III. Since 1973, REases and DNA methyltransferases (MTases) have been named based on an original suggestion by Smith and Nathans. They proposed that the enzyme names should begin with a three–letter acronym in which the first letter was the first letter of the genus from which the enzyme was isolated and the next two letters were the first two letters of the species name. Extra letters or numbers could be added to indicate individual strains or serotypes. Thus, the enzyme Hindll was one of four enzymes isolated from *Haemophilus influenzae* serotype d. The first three letters of the name were italicized. Later, a formal proposition for naming the genes encoding REases and MTases was adopted. When there were only a handful of enzymes known, these schemes were very useful, but as more enzymes have been found, often from different genera and species with names whose three–letter acronyms would be identical, considerable laxity in naming conventions has appeared. In addition, we now know that each major type of enzyme can contain subtypes. This especially applies to the Type II enzymes, of which more than 3500 have been characterized. In this paper we revisit the naming conventions and outline an updated scheme that incorporates current knowledge about the complexities of these enzymes. We describe a set of naming conventions for REases and their associated MTases. Since the homing endonucleases have been named in an analogous fash-ion, we propose that similar guidelines be applied to that group of enzymes. Finally, it is important to realize that the aim of this document is to provide a nomenclature for these enzymes, not to provide a rigorous classification.

GENERAL RULES

First, we introduce a number of general changes, standard abbreviations and definitions that are recommended for use.

1. 'Restriction enzyme' and 'restriction endonuclease' should be regarded as synonymous and the abbreviation REase (or in some cases, R) is preferred. However, the abbreviation ENase, which has been used extensively, may also be used. Alternative names such as restrictases should be avoided. The abbreviation R–M should be used for restriction modification. Homing endonuclease should be abbreviated HEase.

2. Methyltransferase is the preferred name, since it correctly describes the activity. Methylase, while in common use, is not strictly accurate and should be avoided in print. The abbreviation MTase (or in some cases, M) should be the standard.
3. Italics will no longer be used for the first three–letter acronym of the REase or MTase name. Many journals already avoid italics and retaining the italic convention is not easily translated to computers and serves no essential purpose. The convention of naming different enzymes from the same isolate of the same organism with increasing Roman numerals will continue.
4. Restriction enzyme names should not include a space between the main acronym and the Roman numeral. This practice, which has been employed to avoid the inelegant look caused when characters in italic fonts are juxtaposed next to characters in a regular font, is incorrect. Now that italics will no longer be used in names there is no reason to continue this practice.

   The previous scheme of using a raised dot after the prefix will be abandoned and a normal dot (period) should be used. Furthermore, except for the single period or hyphen (in homing endonucleases) that is used to separate the prefix from the main part of the name, no punctuation marks, such as parentheses, periods, commas or slashes, should be used in REase or MTase names. Only alphanumeric characters should appear. Already the enzymes from *Nostoc* species C have been changed from their original Nsp(7524)I to Nspl and many others have also changed. The most recent is Bst4.4I, which has changed to Bst44I.
5. The designation of the three main types of REases as Type I, Type II, and Type III will continue, with the capital *T' preferred. However, they will be divided into subtypes as indicated below. One new type of REase will be added. This is Type IV, which will include those systems that cleave only methylated DNA as their substrate and show only weak specificity, such as the McrA, McrBC and Mrr systems of *Escherichia coli.*
6. The sequence databases contain many genes that are excellent candidates to encode DNA MTases and REases, based on sequence similarity. These will be named according to the same guidelines as are used for biochemi-cally characterized enzymes, but will carry the suffix P to indicate their putative nature. Once they have been characterized biochemically and shown to be active, the P will be dropped and their names will be changed to a regular name with the next Roman numeral that is appropriate.
7. The current convention of naming R–M enzymes with a prefix M, R, etc. will be expanded to include the protein products of related genes such as the controlling proteins (e.g., C.BamHI) and the nicking enzymes that cleave G/T mismatches (e.g., V.Hpall for the vsr–like enzyme associated with the Hpall system) and N.BstNBI for the regular nicking enzymes. In addition, up to two characters will be allowed in the prefix. This will enable enzymes, such as Eco57I, with both REase activity and MTase activity fused in a single protein to be designated RM.Eco57I. Its accompanying MTase would remain as M.Eco57I. Note that the current convention of permitting the REase to be named either with or without the 'R' prefix will be continued. Thus, R.EcoRI and EcoRI will be considered synonymous as will RM.Eco57I and Eco57I.

For certain nicking enzymes that have been obtained from the Type IIT enzymes, where one of the two heterodimeric subunits has been inactivated the resultant mutant nicking enzymes should be called Nt.BbvCI or Nb.BbvCI, where the 't' and the 'b' indicate cleavage of the top or bottom strand of the normal recognition sequence.

8. When two REase or MTase genes are present and associated with a single R–M system, they should be referred to with the second character of the prefix being an Arabic 1 or 2. Thus, the two M gene products of the HphI R–M system would be Ml.HphI and M2.HphI.
9. The standard abbreviations for methylated bases should be 5–methylcytosine (m5C),N4–methylcytosine (m4C) and N6–methyladenine (m6A). It is not necessary to use a superscript for the number.
10. Isoschizomers are REases that recognize the same sequence. The first example discovered is called a prototype and all subsequent enzymes that recognize the same sequence are isoschizomers of the prototype. Neoschizomers are that subset of isoschizomers that recognize the same, sequence, but cleave at different positions from the prototype. Thus, Aatll (recognition sequence: GACGT–↓C) and Zral (recognition sequence: GAC↓GTC) are neoschizomers of one another, while Hpall (recognition sequence: C↓CGG) and Mspl (recognition sequence: C↓CGG) are isoschizomers, but not neoschizomers. Analogous designations are not appropriate for MTases, where the differences between enzymes are not so easily defined and usually have not been well characterized.
11. The solitary MTases (i.e., not associated with an REase) such as the Dam and Dcm MTases of E.coli and the eukaryotic MTases such as Dnmtl and Dnmt3a will be named systematically in accordance with the general rules established for the prokaryotic enzymes. Thus, the systematic name for the Dam MTase of *E.coli* K12 will be M.EcoKDam and the murine maintenance MTase will be M.MmuDnmtl. However, it will be acceptable to refer to them by their more commonly used trivial names, Dam, Dcm, Dnmtl, etc., but it will simplify automated searching and cross–referencing of the literature if the systematic name, including the M prefix, also appears at least initially in a publication.

    Solitary MTases that are phage or virus borne are also named with the prefix M and the name of the phage or virus that carries them. Optionally, the host name may be included. Thus, the MTase encoded by phage SPR of Bacillus subtilis is named M.SPRI and the MTase encoded by the archaeal virus (ϕCh1 of *Natrialba magadii* is named M.NmaPhiChll.
12. The rules for naming genes of Type II R–M systems should adhere to the proposals of Szybalski et al. with the obvious extensions to accommodate C, V and N genes. Thus, the entire name should be italicized, the first letter will be lower case and the capital letter(s) used as a prefix for the protein will become the suffix for the gene. The gene for EcoRI thus becomes *ecoRIR* and that for its MTase is *ecoRIM*. In the case of genes with two prefixes in the protein name the gene name would incorporate both letters of the prefix.

Thus, the gene for RM.Eco57I would become *eco57IRM*. In the case of Type I enzymes, an acronym for the source organism should be followed by the traditional gene designations, hsdS, hsdR and hsdM. Thus, the three genes of the EcoKI restriction system would be *ecoKIhsdM, ecoKIhsdR,* and *ecoKIhsdS*. However, it will be acceptable to omit the *ecoKI* where appropriate.

13. It sometimes happens that two genes are required for a single enzyme activity, effectively encoding two subunits. In these situations the two genes and their products should carry a suffix A and B. For example, BbvCI is a heterodimeric REase. The two gene products would be called R.BbvCIA (or just BbvCIA) and R.BbvCIB (or just BbvCIB),and the active holoenzyme would be BbvCI. Note that the two separate MTases of this system would be Ml.BbvCI and M2.BbvCI. For enzymes like Eco57I, which have both endonuclease and MTase activity in the same polypeptide chain, the endonuclease would be referred to as RM.Eco57I, but the second MTase activity associated with this system would be called M.Eco57I. For MTases, an example is M.AquI, which has one gene encod-ing the N–terminal region up to the middle of the variable region of this m5C MTase and a second gene encoding the remaining C–terminal region (9). In this case, the two parts of this protein should be referred to as M.AquIA and M.AquIB and the genes as *aquIAM* and *aquIBM*.

DETAILS OF TYPES AND SUBTYPES

TYPES I, II, III, AND IV

The original subdivision of Types I, II, and III will be maintained and a new Type IV added to accommodate a class of methyl–dependent restriction enzymes. The previously proposed candidates for new types, such as Eco57I and Gsul, will be incorporated as subtypes of existing Type II enzymes.

Type I

The key characteristics of the Type I R–M systems are that these enzymes are multisubunit proteins that function as a single protein complex and usually contain two R subunits, two M subunits and one S subunit. The symbol for Type I systems is *hsd,* thus the genes are *hsdR, hsdM,* and *hsdS,* and their protein products are HsdR, HsdM, and HsdS, respectively. The protein products can be abbreviated by omitting Hsd. The S subunit is the specificity subunit that determines which DNA sequence is recognized. The R subunit is essential for cleavage (restriction) and the M subunit catalyzes the methylation reaction: in all known cases the methylated base formed is m6A. When Type I enzymes act on unmethylated substrates, they function mainly as REases (they may also methylate unmodified sites with a low probability) and have an absolute requirement for ATP during cleavage. They cleave the DNA at variable positions away from their recognition sequence.

The location of the cleavage sites is determined by either the collision and stalling of two such complexes during translocation along a DNA chain, or the stalling of a single enzyme on a single–site circular substrate following DNA translocation. The biochemical nature of the termini produced upon cleavage is unknown and the enzymes do not turn over in the cleavage reaction. In contrast, when these complexes encounter a hemimethylated substrate, in which one strand of the recognition sequence is methylated, as would occur immediately after DNA replication of a fully methylated substrate, then the complex functions as a DNA MTase, using S–adenosylmethionine (AdoMet) as the donor of the methyl group. A complex of two M subunits and one S subunit is fully functional as an MTase. Probably the best known Type I enzyme is EcoKI. The REase is referred to as R.EcoKI or EcoKI, but it is important to remember that it is also an MTase.

The MTase complex of two HsdM and one HsdS is referred to as M.EcoKI. When referring to phenotypes the preferred convention is r_K + m_{KI} + etc. Four subcategories of Type I enzymes (A, B, C, and D) are in common use. These are based on genetic complementation and their use will be continued. If experimental evidence defines new subtypes, then additional letters may be used as suffixes to describe them. A number of artificially created hybrid enzymes have been described, which often include those with new specificities. These should be named as deemed appropriate, but without a Roman numeral at the end.

Type II

The Type II REases recognize specific DNA sequences and cleave at constant positions at or close to that sequence to produce 5'–phosphates and 3'–hydroxyls. Usually they require Mg^{2+} ions as a cofactor, although some have more exotic requirements (see below). They may act as monomers, dimers or even tetramers and usually act independently of their companion MTase. The MTases usually act as monomers and transfer a methyl group from the donor S–adenosyl–L–methionine directly to double–stranded DNA and form m4C, m5C,or m6A. Because of the interest in these Type II REases for recombinant DNA technology, more than 3500 have been characterized.

Given the assay that is used to find them, which detects any activity yielding a consistent DNA fragmentation pattern, it is no surprise that they come in a large variety of 'flavors'. Early on it was recognized that while then–normal Type II enzymes recognized palindromic sequences and cleaved symmetrically within them, the Type US enzymes cut outside their normally asymmetric sequences and differed in other interesting ways. We now know of additional enzymes that cleave on both sides of their recognition sequence (e.g., Bcgl), are activated by AdoMet (e.g., Eco57I), interact with two copies of their recognition sequence (e.g.,EcoRII) or have unusual subunit structures (e.g., BbvCI). These additional kinds of enzymes will be considered subdivisions of Type II. It should be recognized that for the purposes of nomenclature some enzymes would fall into more than one subdivision.

Specifically, some of the criteria are based on the sequence cleaved and others on the structure of the enzymes themselves, so not all subdivisions are mutually exclusive, e.g., Bcgl is both Type IIB and IIH. Type US enzymes, originally designated as enzymes with cleavage sites shifted away from their recognition sequence, will be retained, but a new Type IIA will be defined that includes all Type II REases that recognize asymmetric sequences. A new Type IIP will be used to desig-nate the enzymes that recognize symmetric sequences (palindromes). The overriding criterion for inclusion as a Type II enzyme would be that it yields a defined fragmentation pattern and cleaves either within or close to its recognition sequence at a fixed site or with known and limited variability.

In general, the Type II REases and their associated MTases are separate, inde-pendent enzymes, but in several classes (e.g., IIB, IIG, and IIH) the R and M genes are fused into a single composite gene. The nomenclature for the subtypes of the Type II enzymes currently known is shown below. It should be noted that these designations are not intended to be exclusive, but rather to permit enzymes with common characteristics to be referred to as a group. Conservation of structural domains with associated enzymatic activities is observed between different classes of Type II enzymes and also between other types of R–M enzymes. The Type II subdivisions are summarized in Table 1 and described in more detail below.

Type IIP

This would be used as a generic description for all enzymes that recognize symmetric sequences, often termed palindromes, and cleave at fixed symmetrical locations either within the sequence or immediately adjacent to it. The recognition sequences and cleavage sites of these enzymes should be represented as in the following example: EcoRI: GlAATTC. In full double–stranded form this corresponds to:

5′ G↓AATTC

3′CTTAA↑G

Note that enzymes such as SinI (recognition sequence: GGWCC), Bgll (recognition sequence: GCCNNNN↓NGGC) and Hindll (recognition sequence: GTYRAC) belong to Type IIP because the recognition mechanism still involves a symmetric homodimer.

Table 8.1. Subtype of Type II REases

Subtype	*Defining Feature*	*Examples*	*Recognition Sequence*
A	Asymmetric recognition sequence	Fok I	GGATG (9/13)
		AciI	CCGC (–3/–1)
B	Cleaves both sides of target on both strands	BcgI	(10/12) CGANNNNNNGC (12/10)
C	Symmetric or asymmetric target. R and M f unctions in one polypeptide	GsuI	CTGGAG (16/9)
		HaeIV	(7/13) GAYNNNNNRTC (14/9)
		BcgI	(10/12) CGANNNNNNTGC (12/10)
E	Two targets; one cleaved, one an effector	EcoRII	↓CCWGG
		NaeI	GCC↓GGC
F	Two targets, both cleaved coordinately	SfiI	GGCCNNNN↓NGGCC
		SgrAI	CR↓CCGGYC
G	Symmetric or asymmetric target. Affected by AdoMet	BsgI	CTGCAG (16/14)
		Eco57I	CTGAAG (16/14)
H	Symmetric or asymmetric target. Similar to	BcgI	(10/12) CGANNNNNNTGC (12/10)
		AhdI	CACNNN↓NNGTC
M	Subtype IIP or IIA. Require methylated target	DpnI	Gm6 A↓TC
P	Symmetric target and cleavage sites	EcoRI	G↓AATTC
		PpuMI	RG↓GWCCY
		BslI	CCNNNNN↓NNGG
S	Asymmetric target and cleavage sites	FokI	GGATG (9/13)
		MmeI	TCCRAC (20/18)
T	Symmetric or asymmetric target. R genes are heterodimers	Bpu10I	CCTNAGC (–5/–2)
		BslI	CCNNNNN↓NNGG

[a] Note that not all subtypes are mutually exclusive. For example, BsII is of subtype P and T.

[b] The abbreviation indicates double-strand cleavage as shown below :

5′CC↓TNAGC

3′GGANT↑CG

Type II A

This would be used as a generic designation for any Type II enzymes that recognize asymmetric sequences irrespective of whether they cleave away from the sequence or within the sequence. Typically these systems have one REase gene and two MTase genes, one to modify each strand of the asymmetric recognition sequence. However, occasionally two R genes are found as with Bpu10I, or both M genes are fused as with M.FokI. When more than one R or M gene is present the genes and their protein products should be named with either an Arabic 1 or 2 in the prefix of the name. Thus, the two MTases of the Sapi system would be named Ml.SapI and M2.SapI if the proteins are being referred to, or *sapIMl* and *sapIM2* for the genes. However, the two subunits of the BpulOI REase would be designated R.BpulOIA and R.BpulOIB and their genes *bpulOIAR* and *bpulOIBR.* The recognition sequences and cleavage sites of the Type US REases should be represented as in the following example:

HphI: GGTGA(8/7) where the first numeral in the parentheses indicates the position of cleavage on the strand written and the second numeral indicates the cleavage position on the complementary strand. In full double–stranded form this corresponds to:

5′GGTGANNNNNNNN↓

3′CCACTNNNNNNNt↓

Note that when recognition sequences are assigned, the convention is to write the single–stranded sequence such that cleavage lies downstream of the sequence. If cleavage takes place within the sequence, then the single–strand designation is always written so that the sequence of the strand is first alphabetically.

Type IIB

This would be used for enzymes that cleave on both sides of the recognition sequence. At present there are many well defined members of this class (Alol, BplI, Bst44I, Bael, Bcgl, BsaXI, Bsp24I, Cjel, CjePI, HaelV, Hin4I, and Ppil). In this case the recognition sequence and cleavage sites should be represented as exemplified for Bcgl:

Bcgl– recognition sequence: (10/12)CGANNNNNNTGC(12/10)

Here, the (10/12) preceding the recognition sequence indicates that cleavage occurs 10 bases in front of the sequence on the strand written and 12 bases before the sequence on the complementary strand. The (12/10) following the recognition sequence indicates cleavage 12 bases after the recognition sequence on the strand written and 10 bases after the sequence on the complementary strand. In double–stranded form this would be written:

↓NNNNNNNNNNCGANNNNNNTGCNNNNNNNNNNNN↓

↑NNNNNNNNNNNNGCTNNNNNNACGNNNNNNNNNN↑

Type IIC

This would be used as a generic term for all enzymes that have a hybrid structure containing both cleavage and modification domains within a single polypeptide. Examples include all of the Type IIB, IIG and some Type IIH enzymes.

Type IIE

This would be used for enzymes that interact with two copies of the recognition sequence, one being the actual target of cleavage, the other being the allosteric effector. The best studied examples are EcoRII and Nael. FokI, MboII, and SauSAI were found to exhibit similar properties. Other enzymes such as Acc36I, AtuBI, Bsgl, BpmI, Cfr9I, Eco57I, Hpall, Ksp632I, Narl, SacII, and SauBMKI are likely to be members of this group because they are reported to be stimulated by oligonucleotide duplexes containing the specific recognition site.

Type IIF

This would be used for enzymes that interact with, and cleave coordinately, two copies of their recognition sequence. Examples include BspMI, CfrlOI, NgoMIV, SfiI and SgrAI.

Type IIG

This would be used for enzymes that have both R and M domains fused to form single polypeptides and that may be stimulated or inhibited by AdoMet, but otherwise resemble Type II enzymes. These include Bce83I, BseMII, BseRI, Bsgl, BspLUllIII, Eco57I, Gsul, Mmel, and Tthlllll. The recognition sequences may or may not be asymmetric. Thus, both Type IIA and Type IIP enzymes may be of Type IIG.

Type IIH

This would be used for enzymes that contain genetic features resembling Type I enzymes, but biochemically behave as Type II enzymes. At present three examples have been characterized: Ahdl and PshAI, both of which comprise a three gene system akin to that of a typical Type I enzyme (G.G.Wilson, unpublished results), and Bcgl, which is a two gene system. Several hypothetical systems have gene organizations that resemble that of Bcgl.

Type IIM

This would be used for Dpnl and similar enzymes that recognize a specific methylated sequence in DNA and cleave at a fixed site. Note that the methyldependent enzymes such as McrA, McrBC are not considered members of this subclass, because they do not have well defined recognition sequences and cleavage sites. They are included within the Type IV enzymes.

Type IIS

This would be used for Type IIA enzymes that cleave at least one strand of the DNA duplex outside of the recognition sequence (i.e. cleavage is shifted relative to the recognition sequence). Note that for some enzymes, such as BsmI (recognition sequence: GAATGC), cleavage of the strand written takes place outside of the recognition sequence, whereas cleavage of the complementary strand takes place within the recognition sequence. This is still considered a Type US enzyme. However, in most cases both strands are cleaved away from the recognition sequence, which therefore remains intact. This was the earliest sub–class of the Type II restriction enzymes to be recognized.

Type IIT

This would be used for enzymes that are composed of heterodimeric sub–units. This subtype includes enzymes like BbvCI, BpulOI and Bsll.

Nicking Enzymes

Two types of nicking enzymes are known. One type includes those that behave functionally like REases, but cleave only one strand of the DNA substrate. These enzymes should be named with the prefix N and their recognition sequences should be written such that the strand displayed is the strand nicked. Thus, N.BstSEI has the recognition sequence: GAGTCNNNN↓ which is abbreviated to GAGTC. Similarly, the mutants of Alwl and Mlyl that have interrupted the dimerization function, and which have become nicking enzymes, are named N.AlwI and N.Mlyl. For enzymes such as Bpu10I, where the wild–type REase has two subunits, each of which nicks a different strand, the mutant nicking enzymes made by inactivating one or the other subunit should be named Nt.Bpu10I for the enzyme that nicks the top strand of the normal recognition sequence and Nb.Bpu10I for the enzyme that nicks the bottom strand.

In full double–stranded format Nt.BpulOI would recognize

5′CC↓TNAGC

3′GGANTCG

while Nb.BpulOI would recognize

5′CCTNAGC

3′GGANT↑CG

Alternatively this may be written

5′GC↓TNAGG

3′CGANTCC

A single–stranded representation of their recognition sites would be NtBpu10I (recognition sequence: CC↓TNAGC) and Nb.Bpu10I (recognition sequence: GC↓TNAGG or CCTNAT↑GC). Note that the use of T always denotes cleavage of the lower strand. A second type of nicking enzyme is found exclusively in association with m5C–MTases, where it serves to nick the G/T mismatches that can result from deamination of m5C within the recognition sequence of the MTase. The best studied of these is the Vsr protein that accompanies the Dcm MTase of *E.coli* K–12, M.EcoKDcm. Vsr recognizes the specific G/T mismatch that occurs if there is deamination of the methylated cytosine residue within the context of the CCWGG recognition sequence of M.EcoKDcm. These kinds of mismatch nicking enzymes are named with the prefix V and should be given the acronym of the MTase gene with which they are associated.

Thus, Vsr, the product of the V gene that overlaps with the gene for M.EcoKDcm, is system-atically named V.EcoKDcm. However, the trivial name Vsr, which was origi-nally designated for this protein, is an acceptable synonym. For other V genes and their products the systematic names are preferred. Thus, V.Hpall is the preferred name for the mismatch nicking endonuclease that accompanies M.HpaII.

Control Proteins

Some R–M systems are found to have an additiona gene that encodes a protein involved in the control fo expression of the R gene. The best studied examples are the PvuII and BamHI systems, wherr the products of the genes, C. PvuII and C.BamHI, serve as transcriptional activators; this prevents the expression of the R genes following transfer of the systems into naive hosts, until such time as C protein has accumulated and methylation is sufficient to provide protection against what would otherwise be the deleterious action of the REase.

Type III

These systems are composed of two genes *(mod* and *res)* encoding protein subunits that function either in DNA recognition and modification (Mod) or restriction (Res). Both subunits are required for restriction, which also has an absolute requirement for ATP hydrolysis. For DNA cleavage, the enzyme must interact with two copies of a non–palindromic recognition sequence and the sites must be in an inverse orientation in the substrate DNA molecule. Cleavage is preceded by ATP–dependent DNA translocation as with the Type I REases. The enzymes cleave at a specific distance away from one of the two copies of their recognition sequence.

The Mod subunit can function independently of the Res subunit to methylate DNA: in all known cases the methylated base formed is m6A and full modification is actually hemimethylation. This is not deleterious because of the requirement for two unmodified sites in inverse repeat orientation for cleavage. DNA replication puts all of the unmodified sites in the same orientation. The best–known examples of Type III enzymes are EcoPlI and EcoPlSI. Putative Type III R–M systems are easily recognized because of their similarity at the sequence level. When naming the genes for these enzymes the mod gene of EcoPlI would be systematically named *ecoPHmod,* but the abbreviation mod is acceptable when it does not result in confusion.

Type IV

These systems are composed of one or two genes encoding proteins that cleave only modified DNA, including methylated,hydroxymethylated and glu–cosyl–hydroxymethylated bases. Their recognition sequences have usually not been well defined except for EcoKMcrBC, which recognizes two dinucleotides of the general form RmC (a purine followed by a methylated cytosine – either m4C or m5C) and which are separated by anywhere from 40 to 3000 bases. Cleavage takes place –30 bp away from one of the sites. The best studied example at both the genetic and biochemical level is EcoKMcrBC of *E.coli,* but on the basis of sequence similarity it is likely that there are many such systems in other bacteria and archaea. As with the genes of the Type I and Type III systems, the abbreviations McrBC for the enzyme and *mcrBC* for the gene are acceptable.

Hypothetical Enzymes

Hypothetical REases and DNA MTases can often be found by similarity searching in DNA sequences or their presence may be inferred when specific sequences in plasmid or bacterial DNAs are found to be methylated. It is convenient and useful to be able to refer to such hypothetical enzymes by name. The following convention for naming these enzymes is proposed. They should be named as though they were normal R–M systems, but should carry the suffix 'P' to indicate their putative nature. Once biochemical or unequivocal genetic activity, such as phage restriction, is demonstrated the suffix 'P' and any open reading frame (ORF) designations can be dropped allowing the main element of the name to be retained.

Furthermore a Roman numeral should be included to indicate whether it is the first, second, third, etc. enzyme to be found in that organism. Note that the P extension should remain with the gene until such time as a gene product has been demonstrated to be functional. This 'P' convention is illustrated with genes from *H.influenzae* serotype d. Two Type II REases, Hindll and Hindlll, and their associated MTases had been characterized biochemically. One Type I system had been demonstrated genetically and the MTase, presumably associated with this system, had been partially characterized biochemically. In the genome there are two putative Type I systems, although only one has a complete set of intact genes. The intact system therefore carries the designation Hindi. In addition to these three systems, there was also known to be a Dam–like MTase, now called M.HindDam. However, also in the genome are putative m5C–MTase and REase genes (genes HI 1040 and HI 1041) that show high similarity to the known R–M system, HgiDI.

The MTase encoded by HI1041 leads to a functional protein with specificity identical to that of M.HgiDI (R.D. Morgan, J. Patti, and R.J. Roberts, unpubl. results). It is there-fore named M.HindV. However, the adjacent gene for the putative endonuclease is inactive and so it is named HindVP. One other R–M system can also be seen in the genome, this time encoding a Type III system. Neither the R nor the M genes have yet been demonstrated to be active and so these are named HindORF1056P and M.HindORF1056P. If they are shown to be active they would be renamed HindVI and M.HindVT. The convention here is to name the system after the ORF encoding the MTase gene. This is to ensure that the two genes are given names that indicate they are part of the same R–M system.

Homing Endonucleases

Homing endonucleases have been classified into four families according to conserved sequence motifs. These are the LAGLIDADG, GIY–YIG, H–N–H and His–Cys box families. Nomenclature of the homing endonucleases is patterned after that of REases, with a three–letter genus–species designation, followed by a Roman numeral. Whereas intron endonucleases are characterized by the prefix I– (for intron), the intein endonucleases are characterized by the prefix PI(for protein insert), and where the endonuclease is not intron– or intein–encoded, the prefix is F– (for freestanding).

The systematic nomenclature does not preclude maintaining historic names. Counter to the original conventions proposed, the above nomenclature will extend to putative homing endonucleases without demonstrated catalytic activity. As with hypothetical REases, the suffix P will be used to denote the putative nature of the assignment, and the P will be dropped once nuclease activity has been confirmed. Hybrid homing endonucleases will be preceded by the prefix H–, followed by the authors' designation, e.g., an I–DmoI/I–Crel chimera could be H–Drel, or an I–TevI/I–Bmol hybrid could be H–TevBmo. Those homing endonucleases that have been characterized biochemically will continue to be listed within REBASE.

Adherence to These Conventions and Updates

The authors of this proposal have all agreed to follow these recommendations and it is hoped that other authors and journals will also adhere to these conventions. If further changes become appropriate, then REBASE (5) should be consulted for the latest modifications and practices.

Evolution of Restriction Endonucleases

The phenomenon of restriction and modification (R–M) was first discovered in the early 1950s. It was observed that certain strains of bacteria inhibited ('restricted') the growth of bacteriophages previously propagated on a different strain. In the early 1960s, it was found that the restriction is due to the enzymatic cleavage of the phage DNA by sequence–specific endonucleases (REases), which are sensitive to covalent modification of bases in the target sequence. Some of the REases produced discrete DNA fragments upon cleavage. This property proved very useful for analyzing and rearranging DNA, which soon prompted the rapid development of genetic engineering techniques as well as the search for more REases with novel recognition sequences. It was in the mid–1970s when cloning of R–M enzymes themselves began. It was found that most of restriction enzymes are genetically linked with modification enzymes of cognate specificity, forming R–M systems, but a few solitary enzymes were also characterized. Several kinds of R–M systems have been discovered.

The differences concern enzyme composition and cofactors, the symmetry or asymmetry of the target sequence, and position of the cleavage site with respect to the target sequence. The classification and nomenclature of R–M systems have been recently revised and reviewed. Briefly, restriction enzymes have been classified into four types: I (Enzyme Commission number:), modification-directed (M–D, not classified by EC). Type II has been subdivided into numerous subtypes. Only selected relevant members of subtypes IIP, HE, IIP, and US will be discussed in this article. Enzymes of Types I, II, and III are parts of restriction–modification (R–M) systems, which comprise two opposing activities with similar specificities: a REase activity, and a methyltransferase (MTase) activity. While REase targets and cleaves specific sequences in the DNA, MTase modifies the same sequences by addition of a methyl group to cytosine or adenine, and thereby renders them invincible to cleavage by REase. Modification is required to protect the 'self DNA against the cleavage.

However, M–D restriction enzymes do not require the cognate MTase activity-conversely, they recognize and cleave sequences with modified bases, for instance those methylated by MTases from other R–M systems. The subunit composition and sequence-structure-function relationships in different types of R–M systems have been recently reviewed (Bujnicki 2001b) and will not be discussed here in too much detail. Instead, this review will be focused on phylogenetics of restriction enzymes, their origin and evolutionary basis of the diversity and variability of their catalytic domains. For specialized reviews of classification of restriction enzymes (and R–M systems in general), their biochemistry and structural biology, see other chapters of this volume and other relatively recent reviews.

GENOMIC CONTEXT OF R–M SYSTEMS

R–M systems occur ubiquitously among Eubacteria and Archaea. Closely related systems are found quite commonly in phylogenetically distant organisms, the codon usage of their genes is often different from that of the host genome, and they are known to appear in distinct genomic context in different strains of the same organism. All these features suggest that R–M systems have been frequently transmitted by horizontal gene transfer. It is commonly believed that the principal biological function of R–M systems is to protect the genome of the host organism against invasion of foreign genetic elements, such as phages or conjugative plasmids.

The DNA of the host is usually modified by the MTase, and hence resistant to cleavage, while the DNA of the invader is unmodified or exhibits 'foreign' patterns of methylation and can be shred into pieces upon its penetration into the recipient bacteria. Many phages have found it advantageous to circumvent restriction by various means, initiating an 'arms race' with the bacteria. Some phages encode their own MTases with the same specificity as the host R–M systems or other DNA modification enzymes, such as hydroxymethyltransferases or glucosylases, which make them resistant to restriction by Type I, II, and III enzymes. It has been hypothesized that development of the ability to enzymatically introduce hydroxymethylcytosine (hmC) as well as 'typical' methylated bases (such as m^6A) into the DNA by T–even phages was an evolutionary response to the presence of the 'classical' R–M systems in their hosts (Bickle and Kruger 1993). This in turn led to the development of M–D enzymes, which specifically targeted phage–like patterns of modification.

As a response, the T–even phages developed the ability to glycosylate the hmC in their DNA, rendering them resistant to all known restriction enzymes of *E. coli.* Incidentally, the first observation of the restriction phenomenon was due to M–D enzymes (later described as McrA and McrBC), which were active against T–even phages con-taining nonglycosylated hmC in their DNA, but inactive against glycosylated hmC. The most recent stage in this intriguing evolutionary scenario is probably the development of a PvuRtsII nuclease from *Proteus vulgaris,* which seems to restrict glycosylated T4 DNA more effectively than the same DNA containing nonglycosylated hmC. In addition to their function of a simple "immune system" in prokaryotes, there is evidence of functional and structural similarity of restriction enzymes to recombinases and transposases.

Recently, it has been proposed that the major function of R–M systems is to promote the frequency of genetic rearrangements, and thereby to ensure a rich diversity of genomes. Thereby, REases would act for the benefit of the biological evolution of the population rather than the benefit of individual organisms or the R–M systems themselves (Arber 2000). However, all these functions can be interpreted also in the light of the 'selfish gene' hypothesis, which requires less assumptions to be made. It states that restriction enzymes simply assure their own maintenance against all other genetic elements by cleaving 'non-self DNA and that all other aspects of their function are of "second–order". The REase activity not only includes destruction of incoming foreign DNA, but also contributes to a high frequency of horizontal transmission of R–M systems that allow them to invade new genomes (Jeltsch and Pingoud 1996; Kobayashi 2001).

When an R–M system enters a new host on a DNA fragment and the REase is expressed, its cleavage activity creates recombinogenic free DNA ends within the host genome. This creates the opportunity of the entire R–M system to integrate into the cut site at increased frequency, for instance by illegitimate recombination. In this regard, R–M systems behave in a very similar way to the homing endonucleases (Gimble 2000). R–M systems can be regarded as intracellular parasites at the verge of mutualism in the same sense as lysogenic bacteriophages, which can confer advantage in certain contexts, such as bacterial resistance to antibiotics. For a specialized review of "selfishness" of R–M systems and the evolutionary context of their interactions with the host genome, see Kobayashi (2001, and an article in this volume).

EVOLUTIONAR HISTORY

Early sequence comparisons of REases revealed a striking lack of common features between enzymes that recognize different sequences. Neither over-all sequence similarity nor common motifs could be identified. However, amino acid sequences of some isoschizomers that cleave the same sequence at the same position were found to be evidently similar; *e.g.*, EcoRI and RsrI (G/AATTC, where "/" denotes the site of cleavage and underline denotes the 'sticky end' produced after the cleavage) (Stephenson *et al.* 1989), MthTI and NgoPII (GG/CC)(Nolling and de Vos 1992), and Xmal and Cfr9I (C/CCGGGX Withers et al. 1992). Typically, if statistically significant similarity between isoschizomers was observed, it was very high (often >90 % identity), suggesting that they diverged very recently. Nonetheless, some isoschizomers were found to be dissimilar; for instance, Haelll shares only 15% sequence identity with MthTI or NgoPII, despite these enzymes are functionally equivalent with respect to the recognition sequence and the cleavage site.

Enzymes that cleave the same sequence at different positions (neoschizomers), are generally as dissimilar as typical pairs of non-izoschizomers, *e.g.*, Smal (CCC/GGG) and Xmal (C/CCGGG) share only 13 % identity. The lack of any detectable common motifs or statistically significant sequence similarity between non-isoschizomeric REases suggested that they arose independently by convergence, and not by divergence from a common ancestor (Wilson 1991). Later, overall sequence similarity was detected in a few cases between enzymes that recognize sequences that differ by one or two base pairs, *i.e.*, Muni (C/AATTG) and EcoRI (G/AATTC) (Siksnys *et al.* 1994) or SsoII (/CCNGG, where N=any base) and PspGI (/CCWGG, where W=A or T)(Morgan et al. 1998). Nonetheless, until the mid-1990s, the prevailing view was that REases evolved by convergence. The divergent evolution was recognized only within some small families that group enzymes with the same or nearly the same specificity and these fam-ilies were believed to be unrelated to each other. This belief was soon to be shattered by the results of crystallographic analyses.

CRYSTALLOGRAPHY OF TYPE II REases

The first crystallographic analysis of a restriction enzyme (EcoRI complexed with its target DNA at 3-A resolution) was published in 1986. However, the original model was found to be partly incorrect and a substantially revised structure has been released in 1990. The second published REase structure was that of EcoRV, in 1993. In the years 1990-2002, structures of 14 type-II REases have been determined. Comparisons of these structures revealed that all of them share a common structural core (four-stranded mixed p-sheet flanked by cc-helices), which serves as a scaffold for the active site, typically comprising two or three acidic residues (Asp or Glu) and one Lys residue. In a few enzymes, Lys was found to be replaced by Glu or Gin. These residues are the only shared sequence feature, forming a very weakly conserved

Table 9.1. Structurally characterized REases and related proteins with the PD-(D/E)xK fold

Name	*PDB*	*Reference*	*Specificity*
BamHI	2 bam	Viadiu and Aggarwal (1998)	G/GATCC
BglI	1 dmu	Newman *et al.* (1998)	GCCNNNN/NGGC
BglII	1 d2i	Lukacs *et al.* (2000)	A/GATCT
Bse634I*	1 knv	Grazulis *et al.* (2002)	R/CCGGY
BsoBI	1 dcl	van der Woerd *et al.* (2001)	C/YCGRG
Cfr1OI*	1 cfr	Bozic *et al.* (1996)	R/CCGGY
EcoRI	1 eri	Kim *et al.* (1990)	G/AATTC
EcoRV	1 rvb	Kostrewa and Winkler (1995)	GAT/ATC
FokI	1 fok	Wah *et al.* (1997)	GGATGN$_9$/NNNN[a]
Hindi	1 kc6	Horton *et al.*(2002)	GTY/RAC
Muni	1 d02	Deibert *et al.* (1999)	C/AATTG
Nael	1 iaw	Huai *et al.* (2001)	GCC/GGC
NgoMIV	1 fiu	Deibert *et al.* (2000)	G/CCGGC
PvuII	1 pvi	Cheng *et al.* (1994)	CAG/CTG
λ exo*	1 avq	Kovall and Matthews (1998)	5′-exonuclease
MutH*	1 az0	Ban and Yang (1998)	/GATC[b]
Vsr	1 cw0	Tsutakawa *et al.* (1999a)	C/(T:G)WGG[b]
TnsA*	1 flz	Hickman *et al.* (2000)	Tn7 transposon end
T7endol*	1 m0i	Madden *et al.* (2001)	Holliday junction
*Ss*Hjc*	1 hh1	Bond *et al.* (2001)	Holliday junction
*Pf*Hjc*	1 gef	Nishino *et al.* (2001)	Holliday junction
*Sc*RPBS*	1 dzf	Todone *et al.* (2000)	not a nuclease
*Mj*EndA*	1 2a79	Li *et al.* (1998)	tRNA introns
*Af*EndA*	NA	Li and Abelson (2000)	tRNA introns

The most representative entry from the Protein Data Bank (PDB) (http://www.rcsb.org) has been chosen for each enzyme, with the preference for protein-DNA complexes and structures solved at possibly highest resolution. Indicates the enzymes for which protein-DNA cocrystal structures are not available in PDB. *Ss, Sulfolobus solfataricus. Pf, Pyrococcus furiosus, Sc, Saccharomyces cerevisiae, Mj, Metanococcus jannaschii, Af, Archaeoglobus fulgidus, NA,* coordinates have not been deposited in PDB.

[a] The isolated catalytic domain is nospecific

[b] Nicks single strand

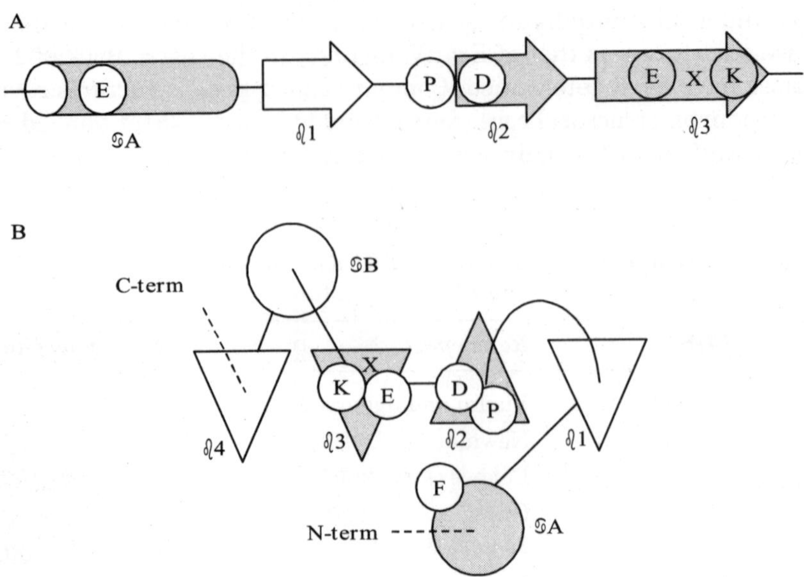

Fig. 9.1. A, B. The conserved core and the most typical active site of enzymes from the PD-(D/E)xK superfamily. Structural elements that serve as a scaffold for the conserved residues are shown in *gray*. Elements that display little or no sequence conservation but are present in nearly all PD-(D/E)xK nucleases are shown in *white*. A Primary structure of the otppp structural motif and the associated e-X_{10-30}-p-D-$x_{10\ 30}$-(D/E)-x-K sequence motif. The most typical residues are shown in *circles* (for simplicity, the 'D/E' carboxylate is represented by '£'). Common secondary structural elements (α-helices and β-strands) are depicted as *cylinders* and *arrows*. B Topology diagram depicting the spatial orientation of the conserved elements. *Circles, triangles,* and *connecting lines* represent α-helices, β-strands, and loops, respectively. Two alternative positions of the triangles correspond to the parallel/antiparallel orientation of the p-strands

motif associated with a characteristic ot(3 | 3p pattern of secondary structure elements. The common sequence pattern is e-X_{10-30}-p-D-x_{10-30}-(D/E)-x-K (where lowercase indicates partial conservation, uppercase indicates nearly universal conservation and X is any amino acid) and hereafter will be referred to as "PD-(D/E)xK" for simplicity and in accord with the tradition.

The structure-based predictions were supported by site-directed mutagenesis studies of several REases, which demonstrated that the conserved residues are indis-pensable for the DNA cleavage activity. Nonetheless, the common secondary structures harboring the active site of different REases were found to be embedded in very different peripheral elements, which sometimes constituted the majority of the protein. Moreover, the dimerization modes of different enzymes were often dissimilar. Initially, it was not clear if the few common features represent sufficient evidence for the common ancestry of all REases or only of those most similar (e.g., EcoRI and BamHI vs. EcoRV and PvuII). Traditionally, homology of genes/proteins was inferred from the similarity of DNA/amino acid sequences.

As more and more protein structures were determined, it was found that conservation of structural and functional similarities despite extreme erosion of sequence similarity is common in the protein world. Large-scale comparisons of protein sequences and structures revealed that in homologous superfamilies, protein sequences diverge most quickly, then functions, then structures. In other words, homologous proteins usually retain the common three-dimensional fold regardless of the divergence at the sequence level. Hence, structural conservation should be regarded as the most robust evidence of protein homology.

Nonetheless, evolutionary analyses of the proteins in the "midnight zone of homology" (where sequences of homologous proteins are saturated by amino acid substitutions to the level of less than 15-10% identity and any genuine similarities disappear in the random noise) are complicated by the fact that protein structures also change in the course of evolution. These changes may be as drastic as insertions, deletions, swapping or substitutions of secondary structure elements or even entire domains, invasion or withdrawal of strands in the common (3-sheet, or circular permutations of polypeptide segments. At the level of the quaternary structure, typical evolutionary changes involve the development of new oligomerization modes and swapping of structural segments between the domains. Because similar structures can arise independently in evolution (at least in theory) and truly homologous structures may become partly dissimilar due to divergence, structural similarity alone does not provide sufficient evidence of common ancestry.

In the absence of obvious sequence similarity between the proteins of interest, their evolutionary relatedness is typically established based on the presence of a common structural scaffold (even if the peripheral structures are dissimilar) and in addition: the similarity of generic biological function (e.g., DNA binding) as well as molecular function (common residues in the binding site, common mechanism of reaction), retention of unusual structural features (unlikely to be reinvented independently), a common domain organization (conserved fusions with unquestionably homologous domains), a common genetic context (homologous neighbors on the chromosome) or, preferably, a combination of the above. As mentioned above, all REase structures determined to date (March 2003) share a geometrically and topologically equivalent core, which acts as a scaffold for a similar metal-binding/catalytic site (the "PD-(D/E)xK" motif) located at the heart of the DNA-binding site.

Recent analysis of crystal structures of REases has suggested that the active site and the recognition sites are the major centers of structural stabilization preserved in the evolution. Independently of the analysis of the crystal structures, Jeltsch et al. (1995) used a statistical Monte-Carlo procedure to find a highly significant correlation between the genotype (amino acid sequence) and the phenotype (target DMA sequence) of Type II REases and came up with a conclusion that they did not arise independently in evolution, but rather evolved from one or a few primordial nucleases. From the current perspective of protein evolution, the above-mentioned results are robust evidence for the common origin of those REases, whose structures are known, which rule out the lingering hypothesis of convergent evolution beyond any reasonable doubt. Since all structurally characterized REases turned out to be related despite they share little if any sequence similarity, it was hypothesized that all restriction enzymes (including those, whose structures remained unknown) may be related, regardless if their sequences exhibit significant similarity or not.

HOMOLOGY

Crystallographic studies revealed that the PD-(D/E)xK superfamily comprises not only restriction enzymes, but also other nucleases, implicated in DNA recombination and repair, including: phage λ exonuclease, two bacterial enzymes exerting ssDNA nicking in the context of methyl-directed and very-short-patch DNA repair: MutH and Vsr, Tn7 transposase TnsA, two closely related Holliday junction resolvases (Hjc) from different species of Archaea, and a Holliday junction resolvase (endonuclease I) from phage T7. As with most of restriction enzymes, these nucleases did not exhibit easily identifiable sequence similarity to their remote homologues. That T7 endonuclease I and Hjc enzymes belong to the PD-(D/E)xK super-family was predicted earlier based on combination of bioinformatics and mutagenesis, but the relationship of other proteins to restriction enzymes came as a surprise. Remarkably, Crystallographic analyses revealed the PD-(D/E)xK fold in two proteins that do not function as deoxyribonucleases at all.

The N-terminal domain (NTD) of the RPB5 subunit of RNA polymerase from yeast *S, cerevisiae* exhibits perfect conservation of the restriction enzyme-like structure, but lacks any catalytic residues it is plausible that it functions as a nucleic acid binding domain devoid of any catalytic activity (Todone *et al.* 2000). The C-terminal catalytic domain (CTD) of tRNA splicing endonuclease EndA also bears striking resemblance to the minimal core of the PD-(D/E)xK fold. EndA-CTD also lacks the catalytic residues common to restriction enzymes and other related deoxyribonucleases. However, in the different' part of the common fold, it developed an RNase active site, whose geometric configuration is very similar to that of a His-Ty r-Lys triad in structurally unrelated RNase A. The structural similarities between these proteins and restriction enzymes have not been anticipated from sequence comparisons and were inferred only after the crystal structures were solved. ' Since EndA and RPB5-NTD lack the PD-(D/E)xK active site, their evolutionary relationship to restriction enzymes is less certain compared to Holliday junction resolvases or other *bona fide* deoxyribonucleases.

However, detailed analyses of sequence alignments generated by superposition of atomic coordinates, revealed common residues important for the formation , of the hydrophobic core. The conservation of aromatic and hydrophobic residues in the critical positions of the PD-(D/E)xK fold is particularly striking between EndA-CTD, archaeal Hjc enzymes and X exonuclease and ' between RPB5-NTD and the catalytic domain of FokI (data not shown). Moreover, PD-(D/E)xK nucleases are the only structures, to which EndA-CTD , and RPB5-NTD exhibit any significant similarity. If EndA-CTD and RPB5-NTD acquired the PD-(D/E)xK fold'by convergence', one could expect to find some lingering similarity between their structures and a non-PD-(D/E)xK 'ancestral fold.

The present version of the protein structure database contains no such fold unrelated to the genuine PD-(D/E)xK nucleases, which could be credibly identified as a potential ancestor of EndA-CTD and/or RPB5-NTD. Hence, the most parsimonious hypothesis is that EndA-CTD and RPB5-NTD are true members of the PD-(D/E)xK superfamily. However the ultimate conformation of this hypothesis (and other hypotheses based on structural similarities in the absence of sequence similarities) must await the identification of "evolutionary intermediates", whose sequences would provide links of statistically significant sequence similarity between now "isolated" members of the family. Alternatively, intermediates would be found that link EndA-CTD and RPB5-NTD with some other proteins.

NON-HOMOLOGOUS ACTIVE SITES

Solution of the crystal structure of Type II REase Cfr10I revealed the conservation of the classical PD-(D/E)xK fold, with unusual modification of the active site: the conserved "(D/E)" carboxylate has been replaced by Ser. Strikingly, another carboxylate side chain has been identified in the active site, provided by a Glu residue located in a non-homologous structural element, namely in an α-helix that immediately follows the oc(3f3(3 core. That the nonclassical (version II) E204 residue is essential for catalysis, while the Ser residue is not, has been confirmed by mutagenesis. Remarkably, the Ser-Glu swapped CfrlOI variant, in which the "classical" (version I) architecture of the active site has been restored, retained approximately 10% activity of the wild-type enzyme, suggesting that the spatial conservation of the functional groups rather than sequence conservation plays a major role in the organization of the active sites of PD-(D/E)xK nucleases.

Crystallographic analyses of enzymes NgoMIV and Bse634I, revealed that they too exhibit version II of the active site. It is noteworthy that spatial migration of catalytic residues within the structural framework of the active site has been well-documented in many different enzymes, such as the thioredoxin superfamily and the enolase superfamily). Bioinformatics analyses suggested further tantalizing examples of potentially novel versions of the PD-(D/E)xK active site in restriction enzymes and related nuclease domains. For instance, it has been found that SsoII and related nucleases possess an "alternative" carboxylate in the same position as Cfr10I and NgoMIV (version II), and an additional carboxylate, 4 residues towards the N-terminus (*i.e.*, on the same face of the α-helix). Using site-directed mutagenesis, it has been demonstrated that either of the two Glu residues in SsoII or PspGI is required to support catalysis, although the highest activity is displayed by the wild-type enzyme that retains both carboxylates.

It remains to be determined experimentally, if putative restriction enzymes such as MjaVIP or Hpy99VIIIP, which possess only version III Glu and no version I, or II acidic residues, represent yet another configuration of the PD-(D/E)xK active site.

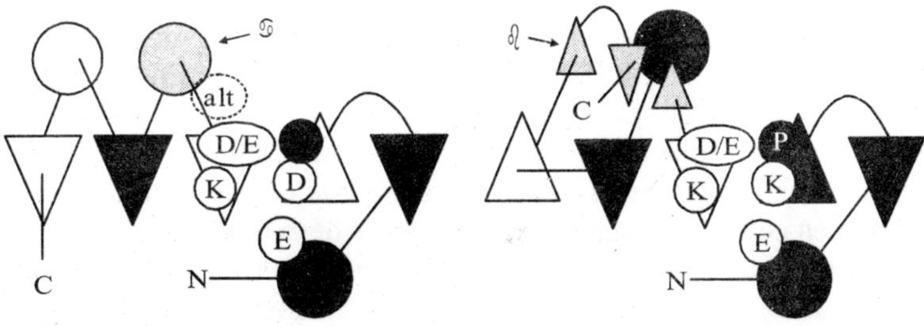

Fig. 9.2. Topology diagrams depicting major structural differences between the a (EcoRI-like) and (3 (EcoRV-like) subclasses of PD-(D/E)xK enzymes. Common secondary structures are shown in *black*. Key elements involved in DNA recognition are shown in *gray* (in α class it is a universally conserved α-helix B, in β class it is an additional small (β-sheet). Other elements specific for a and P subclasses (including the topologically variable fifth β-strand) are shown in *white*. The alternative site in α-helix B, to which the D/E carboxylate migrated in some of the enzymes from the a subclass, is indicated as '*alt*' This figure represents a slightly modified version of a diagram in Bujnicki (2003).

A fourth alternative location of the essential carboxylate has been proposed for a DNA repair enzyme Vsr and an M-D restriction enzyme Mrr. In Vsr, divergence of the active site involved replacement of the "(D/E)xK" element by "FxH", and development of an additional, unique catalytic His residues in another part of the common fold. Yet another version of the same active site was found in a PD-(D/E)xK domain found by sequence searches in R2 and other site-specific, non-long terminal repeat retrotrans-posable elements. Unlike for Vsr, whose crystal structures with and without the DNA are available, predictions of the unusual active site in Mrr and R2 have been supported only by limited mutagenesis studies, and the definite confirmation of these hypotheses must await determination of the crystal structures and detailed biochemical analyses. Careful sequence database searches carried out after the Vsr structure was solved, provided a convincing evolutionary link between the Vsr family and other nucleases with "normal" (version I) active site, comprising typical carboxylates and Lys side chains.

These analyses proved that PD-(D/E)xK enzymes can lose the "PD-(D/E)xK" signature and develop novel active sites to catalyze the phosphodiester cleavage with different mechanisms. From the mechanistic point of view it is not clear "what was the purpose" of these conversions of the active site. From the evolutionary point of view, however, an enzyme with any version of the active site can be selected (and subsequently "fine-tuned") to catalyze a given reaction, as long as it is able to "do the job" and appears "at the right time and the right place" to provide a selective advantage for the host cell.

CLADISTIC ANALYSIS

Structural and functional differences prompted classification of Type II REases into two groups, represented by the archetypal enzymes EcoRI and EcoRV. On the one hand, enzymes that cleave to produce 5′-staggered ("sticky") ends usually recognize specific bases in the DNA mainly via residues from an a-helix and have been termed EcoRI-like or the β-class. Oil the other hand, REases that cleave to produce "blunt" ends usually recognize specific bases in the DNA mainly via residues from an additional β-sheet-they have been termed EcoRV-like> or the [3-class. Fig. 3 shows the topology diagrams of the common core of oc and β-classes. It is evident that both α- and (β-class REases retain the elements that serve as a scaffold for the active site, while the peripheral elements are quite distinct.

It has been pointed out that even though all restriction enzymes comprise a 5-stranded β-sheet, the C-terminal (fifth) strand is topologically non-equivalent in the two classes-it is parallel or antiparallel with respect to the central (β-sheet in α- and β-class REases), respectively. The observation of major structural differences between α- and β-class enzymes and relatively smaller variations within each of these two classes, suggested that the measure of structural differences could be used as an indicator of evolutionary divergence to infer the tree of REases. There is a number of computational methods which allow phylogenetic reconstructions to be made based on protein structure rather than amino acid sequence. Some of these methods rely on inference of evolutionary distances using traditional superposition and calculation of the RMS distances for pairs of Ca atoms that are geometrically (and/or topologically) superimposable.

Others compare intrinsic features of individual protein structures, such as distribution of Cot-Coc distances within (not between) the structures of interest (Carugo and Pongor 2002). As with methods for sequence-based phylogenetic inference, all methods for structure-based treeing have shortcomings. Moreover, it is not always clear if the measures of structural divergence implemented in these methods have been at all tuned up to correlate with the accepted measures of sequence divergence. Hence, the utility of these methods as tools for calculation of the, absolute timing of evolutionary events is very limited. Nonetheless, the structure-based evolutionary inference can provide working models of phylogenies for such protein families, whose divergence definitely precludes'classical' sequence-based treeing. The utility of such phylogenies can be assessed by their predictive power and the ability to explain the available experimental data. Using the method of Johnson and co-workers, I have inferred the molecular phylogeny of the PD-(D/E)xK superfamily using the nine crystal structures available in the Spring of 1999. The results of that study were in excellent agreement with the function-based classification of REases into EcoRI-like β-class) "5′-end cutters" and EcoRV-like (β-class) "blunt-end cutters": these two groups formed two monophyletic branches on the structure-based superfamily tree.

One notable exception was the "3′-end cutter" Bgll found to be related to the EcoRV-like enzymes. However, the functional difference between Bgll and EcoRV could be easily rationalized by relatively minor modifications of the protein surface that resulted in formation of different dimers and hence-different distances between the two active sites in these two enzymes. The nicking enzyme MutH was found to be related to "blunt end cutters", in agreement with its overall similarity to the "blunt end cutter" PvuII reported earlier (Ban and Yang 1998), while the phage A exonuclease was found to be relatively close to the root of the tree (*i.e.*, close to the putative ancestor of all PD-(D/E)xK proteins). Fig. 9.4 shows the revised and updated phylogenetic tree of the PD-(D/E)xK superfamily, inferred for all structures available to date.

It illustrates a hypothetical evolutionary scenario, in which contemporary members of the PD-(D/E)xK superfamily evolved by the gradual accumulation of various "decorations" of the common ancestral fold. The tree has been inferred based on the cladistic analysis, with different "structural peculiarities" used as infor-mative characters (mainly the presence or geometry of certain non-universal 0-strands and a-helices). The topology of this tree is in good agreement with the consensus of several different structure-based treeing methods (calculated using the STRUCLA server). Analysis of the superimposed structures or, for simplicity, of the topological diagrams, reveals the structural elements at the periphery of the common PD-(D/E)xK fold that occupy equivalent position with respect to the core, yet originate from non-homologous parts of the polypeptide.

One example is the additional, sixth β-strand (present only in some nucleases): in Bglll and BamHI it is encoded by the N-terminus and assumes antiparallel orienta-tion with respect to the universally conserved third and fourth strands; in NgoMIV, FokI, PvuII, Nael, and Bgll it is parallel and in the C-terminus, while in Vsr it is antiparallel and in the C-terminus. An α-helix, which packs against strands β4, β5, (and β6, if it is present), also exhibits topological variability: in EcoRV, Hindi, Nael, Bgll, MutH, Bglll, and BamHI it is encoded N-terminally to the common core and assumes antiparallel orientation with respect to the universally conserved helix αA, while in the Hjc enzymes (β-lineage) and all members of the α-class except Bglll and BamHI, it is encoded C-terminally toi the common core and assumes parallel orientation with respect to αA.

In Vsr, this helix is formed by a polypeptide segment between strands β4 and β5, resulting in a left-handed β-α-p crossover, which is a very rare feature of protein structures in general. Remarkably, in FokI the geometrically equivalent element is formed by a connection between strands β5 and β6; it is actually broken into two helices connected by a kink, resulting in a geometrically similar, yet non-homologous left-handed β-α(kink)α-β crossover.

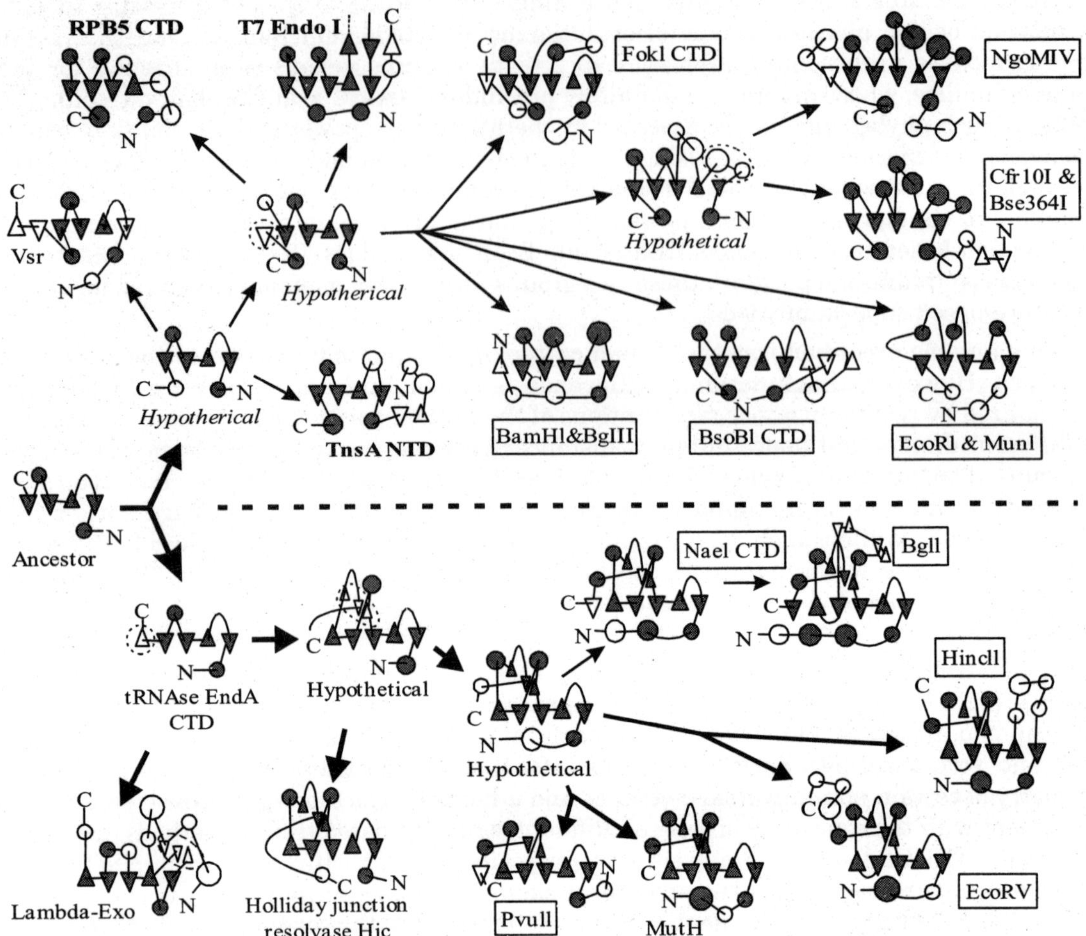

Fig. 9.3. Cladistic tree of known PD-(D/E)xK structures. Characters (features) that are either unique to one protein structure or universally conserved were ignored, because they do not provide any information to resolve inter-superfamily relationships. Evolu-tionary steps (acquisition and loss of structural elements) are indicated by *arrows*. Whenever a contemporary structure shares all major elements with the hypothetical ancestor, the contemporary structure is shown at the node instead of the ancestor, to illustrate its 'ancient' character. Elements that are conserved in a given lineage (apomorphies) are shown in *white* in this structure, in which they appeared for the first time. Elements inherited from ancestors older than the last common ancestor of the given lineage (symplesiomorphies) are shown in *gray*. The major features that allow distinction between different evolutionary lineages are depicted by *dotted circles:* (1) the directionality of the fifth β-strand (parallel in the a lineage and antiparallel in the β-lineage), (2) the additional small β-sheet in the β -lineage (note that the additional β-sheet of λ-exo is topologically different and has been acquired independently), (3) a unique tetrameriza-tion interface in the common ancestor of Cfr10I/ Bse364I and NgoMIV (formed partly by a helical insertion in one of the loops). Other notable peculiarities are the unusual left-handed β-α- β elements in the C-terminal part of Vsr and FokI, as opposed to the typical right-handed structures in other proteins, and that T7 endonuclease I forms a swapped dimer and the core of one catalytic domain is made of fragments of two different monomers (one indicated by *dotted lines*)

According to the most parsimonious evolutionary scenario, these recurrent structural variants evolved by independent extension of the common fold in different evolutionary lineages (*i.e.,* "by convergence"). Alternatively, some REase structures could be products of circular permutation or recombination between parts of catalytic domains derived from, REases with different overall structures. Nonetheless, the latter scenario seems to me unlikely. Firstly, the peripheral elements of different enzymes share no detectable sequence similarity, (hence no convincing evidence for circular permutation or recombination). Secondly, to my knowledge, active PD-(D/E)xK REases with "chimeric" catalytic domains have been obtained only by recombination of segments derived from very closely related enzymes and there have been no reports of active restriction enzymes produced by the spontaneous recombination of remote homologues in the nature. Thirdly, spatially equivalent "optional" structures in α-class and β-class REases often exhibit distinct pattern of contacts with the rest of the protein, while they tend to be more similar (and better superimposable) within each class (data not shown). In general, the natural development of functional hybrid domains is extremely rare compared to insertion or deletion of individual secondary structure elements at the periphery of the common fold.

Hence, it is reasonable to believe that different REases might have independently found ways to decorate the common core with a few similar secondary structures. Nonetheless, the structural plasticity of PD-(D/E)xK domains makes them attractive candidates for engineering protein variants with modified connectivity between the individual secondary structural elements. Studies on the activity and stability of such variants could serve as a test for the proposed evolutionary scheme of development of contemporary PD-(D/E)xK nucleases. It is noteworthy that the structures of sequence-specific restriction enzymes are more elaborate and according to the presented scenario, occupy the terminal branches of the tree, while other nucleases, usually involved in generic metabolism of nucleic acids (DNA repair, recombination etc.), are typically located close to the root of the tree.

Nucleases associated with the cellular DNA repair or recombination machinery evolve under the constant selective pressure stemming from their functionally critical interactions with conserved partners (Aravind et al. 2000). This pressure has apparently slowed down the divergence of generic nucleases, involved in recombination, and stabilized the ancestral state by limiting the development of major structural innovations. Restriction enzymes are neither essential to the cell nor involved in conserved interactions with common partners and hence are free to experiment with different extensions of the common fold. One notable exception from this rule is the MutH mismatch repair nuclease, whose ancestor could have been a type II restriction enzyme. The phylogenetic distribution of MutH is restricted to γ-Proteobacteria (Bujnicki 200la), hence it seems that the MutH proteins have become adapted to function in cooperation with the MutL/MutS system relatively recently in the macroevolutionary time scale. The extreme divergence between most of families of Type II enzymes (both at the sequence and structural level) paralleled by very close similarity between known isoschizomers can be explained as a consequence of fluctuations in the selection pressure.

On the one hand, R-M systems have been identified in nearly all prokaryotic genomes, with the probable exception of a few intracellular pathogens (Roberts et al. 2003b). On the other hand, functional analyses have demonstrated that many of the REase genes are inactivated by mutations-some bacterial genomes are practically "graveyards" full of dead REase. It seems that periods of the strong selective pressure on both high activity and strict sequence specificity of Type II REases (such as during the attack of bacteriophages on the host) are separated by long "idle" interludes. Hence, restriction enzymes may enjoy long periods of relaxation, when they proliferate by horizontal gene transfer and at the same time accumulate random mutations that render most of the copies inactive, but also (with low probability) enable generation of variants with new interesting features. Attacks of phages on a bacterial population may kill those hosts, whose restriction enzymes relaxed too much. However, if any nuclease is able to "stand up and fight" with the invading DNA and allows it host to survive, its evolution will be directed towards optimization of both the activity and the specificity.

An additional condition for survival is the presence of the functional MTase, which would protect the host against the active REase. Of course, under strong selective pressure, the specificity of the MTase may be concurrently tuned up to the specificity of the REase. Thereby, phage infections may serve as evolutionary bottlenecks for restriction enzymes, allowing selection of quite peculiar structural variants of the PD-(D/E)xK fold.

IDENTIFICATION OF PD-(D/E)XK DOMAINS

Using various bioinformatics methods, PD-(D/E)xK domains have been detected in nucleases, for which no crystal structure has been solved yet (including genuine REases, proteins implicated in DNA repair and/or recombination as well as many putative open reading frames). Most of these analyses failed to confidently identify the structures outside the common αβββ core, however in a few cases more extensive similarities to one REase structure or another have been found, suggesting a specific location where a sequence-based branch should be grown on the structure-based tree. Here, I will mention only these predictions that can be reliably integrated with the phylogeny of the PD-(D/E)xK superfamily, with the preference for those that have been validated by site directed mutagenesis.The readers are referred to previously published broad surveys for the extensive list of predicted Rease domains, most of which still await experimental confirmation.

For a recent review of bioinformatics methodology see (Bujnicki 2003). Figure 5 shows the phylogenetic tree of the PD-(D/E)xK superfamily, which combines the structural and bioinformatics data. Two notable cases of "secondary convergence", or independent invention of the same func-tion in different evolutionary lineages are illustrated. Multiple sequence alignment analysis, followed by structure prediction and identification of the characteristic otppp pattern of secondary structures and conserved residues, were used to identify a PD-(D/E)xK domain fused to a tandem of P-loop ATPase domain in of the R subunits of type-I and type-III enzymes as well as in the RecB/RecE helicase family. In agreement with these predictions, alanine mutagenesis of the suspected catalytic Asp, Glu and Lys residues abolished the nuclease activity of these multifunctional enzymes, thereby uncoupling the cleavage and ATP-dependent DNA translocase activities. Interestingly, in Types I and III REases the PD-(D/E)xK domain was found either in the N-or C-terminus of the ATPase module.

It is not clear if these fusions were generated independently or one was derived by circular permutation from the other. Independent fusions of the PD-(D/E)xK domain with different helicase domains have been identified in many enzymes implicated in DNA repair and recombination (Aravind et al. 2000). According to secondary structure predictions, the nonspecific nuclease domain in most of these proteins represents a relatively compact version of the PD-(D/E)xK fold, which appeared quite early in the evolution of the superfamily. A relatively compact PD-(D/E)xK domain was also identified in a functionally uncharacterized protein PrfA conserved in Gram-positive bacteria.

A statistically significant match between the PrfA sequence and the structure of REase Pvull was obtained using a fold-recognition method. Although the cellular role of PrfA remains to be determined, its predicted nuclease activity was confirmed by showing that it nicks supercoiled DNA. The structural model of PrfA was further supported by mutage-nesis of one of the carboxylate residues.

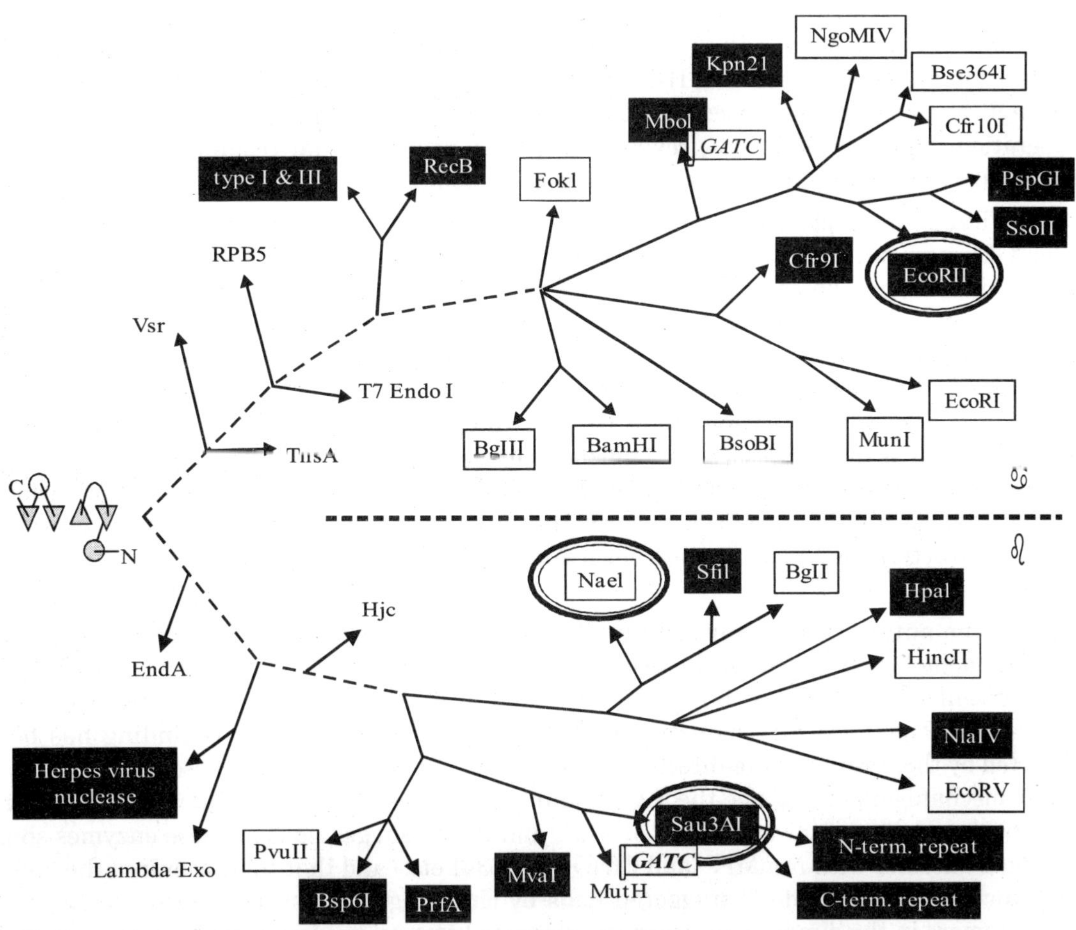

Fig. 9.4. Phylogenetic tree of the PD-(D/E)xK superfamily, based on the 'structural' tree (Fig.4) and expanded to include additional members, identified by sequence analyses and protein-fold recognition. Restriction enzymes are shown in *black frames.* PD-(D/E)xK domains identified by bioin-formatics and not by crystallography are shown in *white on gray background.* Sub Type HE enzymes from three different lineages are indicated by *circles.* Isoschizomers Mbol and Sau3AI that originated from two different lineages are indicated by a label with their recognition sequence GATC. Parts of the tree, which could not be confidently resolved based on either sequence or structural analyses, are shown in *broken lines*

Iterative sequence database searches and secondary structure prediction were used to identify the relationship between two groups of PD-(D/E)xK REases: structurally characterized Type IIP enzymes that recognize different sequences with a common core X/CCGGX[2] and cleave to produce tetranucleotide staggered ends (representative members: CfrlOI and NgoMIV), and Type IIP and HE enzymes that recognize sequences with a common core /CCXGG and cleave to produce pentanucleotide staggered ends (representative members: SsoII, PspGI, and EcoRII). This group of enzymes can be regarded as a model system for studying the evolution of three features: (1) development of novel versions of the PD-(D/E)xK active site (discussed earlier in this article); (2) development of additional protein-DNA contacts out-side the common core that change the specificity; and (3) development of quaternary interactions characteristic for different functional subtypes (IIP, HE, or IIF). Comparative analysis of the crystal structures of CfrlOI (recognition sequence: R/CCGGY (where R=G or A and Y=C or T), and NgoMIV (recognition sequence: G/CCGGQ revealed that they utilize similar residues and mechanism to recognize the central CCGG bases (Deibert et al. 2000).

The "RxDR" motif present in both Cfr10I and NgoMIV was implicated in direct recognition of the "GG" segment. However, NgoMIV apparently developed a different N-terminal extension of the common core, which provided some of the key residues involved in recognition of the outer GC pair, including D34. Analysis of superimposed structures of Cfr10I and NgoMIV reveals that other residues that participate in direct recognition of the outer bases in NgoMIV are substituted in Cfr10I by different side chains (including R227 vs D240 and Q63 vs. K64). The availability of several isoschizomers of Cfr 101 (including the crystallographic structure of Bse634I) as well as their T/CCGGA-recognizing homologues such as Kpn2I, suggests that careful examination of conservation patterns in this family of REases could help identify variable residues that control the specificity towards the outer base pair. Such "evolutionary hot spots" could be attractive targets for mutagenesis aiming at engineering of REases with non-native specificities.

However, author is not aware of any attempts to engineer new specificities in Cfr10I or any of its homologues based on the evolutionary information. The equivalent of the "RxDR" motif present in NgoMIV has been also identified in SsoII (recognition sequence /CCNGG), EcoRII, and PspGI (both recognizing /CCWGG, where W=A or T). This finding has been supported by the, results of site-directed mutagenesis, suggesting that all these enzymes utilize similar mechanism to recognize the 'GG' module (Pingoud et al. 2002). Sequence analysis and theoretical modeling of the structures of SsoII and PspGI suggested that these enzymes share a common ancestor with NgoMIV (and CfrlOI, Bse634I etc.) and that their acquired the ability to produce pentanucleotide 5′-staggered ends by changing the respective orientation of the two monomers in the dimer, to increase the distance between the two catalytic sites compared to NgoMIV.

A similar reorientation of monomers with respect to each other and to the DNA has been demonstrated by crystallographic analysis and used to explain the evolution of the Bgll enzyme, which produces trinucleotide 5′-staggered ends, from EcoRV-like enzymes, which produce blunt ends (Newman et al. 1998). Thus, it appears that reorientation of monomers in a dimer is a common evolutionary mechanism for development of REases with new specificities. Analyses of structural elements appended to the common PD-(D/E)xK fold as extensions or insertions in regions outside the common fold of SsoII, PspGI, NgoMIV, and EcoRII, revealed that they are responsible for development of features characteristic for different functional subtypes (IIP, IIP, or HE). First, the catalytic domain of PspGI and SsoII retained the "orthodox" features of sub-Type IIP. Second, the ancestor of the NgoMIV/Cfr10I lineage developed longer insertions in the loops between strands β1/ β2 and β2/ β3, which formed a novel tetramerization interface required to produce a sub-Type IIP enzyme. Third, the catalytic domain of EcoRII was N-terminally fused to a novel DNA-binding domain, which allowed the enzyme to simultaneously bind two identical sites, a feature typical for sub-Type HE enzymes (Reuter et al. 1999).

Remarkably, it was shown that deletion of the N-terminal domain of EcoRII restores the "orthodox" characteristics of subType IIA. It is noteworthy, that the archetypal Type HE enzyme, Nael (Huai et al. 2000), belongs to the β-class of REases and shows no particular similarity to Type HE enzyme EcoRII, the a class member. The subType HE features of Nael were also found to result from fusion of the PD-(D/E)xK domain with a DNA-binding domain, but in this case the additional domain was found C-terminally to the catalytic domain. Moreover, the auxiliary DNA-binding domains of EcoRII and Nael show no relationship to each other (data not shown). Another Type HE enzyme, Sau3AI was also shown to be a member of the PD-(D/E)xK superfamily.

It was found that Sau3AI comprises a tandem duplication of two PD-(D/E)xK domains, of which both exhibit statistically significant sequence similarity to the DNA repair enzyme MutH,but only the N-terminal copy retains the hallmark catalytic residues. The C-terminal copy of the PD-(D/E)xK domain in Sau3AI has apparently lost the active site, but retained the ability to bind the same sequence (GATC) as the N-terminal domain. Hence, both domains of SauSAI and the sole catalytic domain of MutH share similar sequence specificity. It is striking that the catalytic domains of three subType HE enzymes (Nael, EcoRII, and Sau3AI) show different relationships with other PD-(D/E)xK nucleases and their auxiliary DNA-binding domains are unrelated. The only reasonable explanation is that different sub Type HE enzymes have independently (*i.e.,* "convergently") found a similar solution to the same problem of binding two identical DNA sites.

Another evidence of convergent evolution of a similar function is presented by enzymes Mbol and Sau3AI, which both recognize /GATC, yet apparently originated from two different lineages, *i.e.,* α and β, respectively. It will be of interest to find, if these isoschizomers use similar or different mechanisms for recognition of the same substrate. Since they are homologous, it could be expected that they interact with the DNA in a similar manner. Nonetheless, as mentioned earlier, α-and β-class REases utilize different structural elements as the scaffold for recognition of the target sequence, hence major differences should not be surprising. Based on the results of protein fold-recognition analysis carried out via our "meta-server" I have provisionally classified several Type II REases as members of the β-class, based on their extensive similarities to known structures: NlalV to EcoRV, Hpal to EcoRV and Hindi, Sfil to Bgll, Bsp6I to PvuII, and Mval to MutH.

The 'new' members of the β-class recognize and cleave different sequences than their structural templates, yet share with them all key structural and functional elements. Theoretical models of NlalV and Bsp6I were validated by site-directed mutagenesis of residues involved in dimerization, DNA-binding and catalysis (M.Radlinska and J.M. Bujnicki, manuscripts in preparation), other predictions remain to be confirmed experimentally. A large-scale comparative analysis of REases has been launched in my laboratory, aiming at detection of as many evolutionary links between different REases as possible, to use this information as a guide for experimental analyses.

Sequence Analyses

Despite the notion of 'secondary' convergence in the terminal branches of the PD-(D/E)xK superfamily, all crystallographic studies and sequence analyses of many nuclease families substantiated the hypothesis of primary common origin of all REase domains in restriction enzymes and many DNA repair and recombination enzymes. Ironically, at the turn of the millennium, soon after the compelling hypothesis of homology between all Type II enzymes started to be generally accepted as a general working model, new studies have! demonstrated convincingly that some *bona fide* Type II REases are in fact evolutionary unrelated and structurally dissimilar to the PD-(D/E)xK super-family. First, sequence analysis of a novel Type US restriction enzyme Bfil revealed that its N-terminal part exhibits low, statistically insignificant similarity to an EDTA-resistant nuclease (Nuc), a member of the phospholipase D superfamily (PLD).

(Sapranauskas et al. 2000). The intriguing identity of several key residues between Bfil and Nuc prompted the Siksnys group to carry out further experiments. In accordance with the hypothesis of remote homology of the two enzymes, Bfil turned out to share the peculiar features of Nuc, namely to be active in the absence of Mg^{2+} as well as to hydrolyze an artificial substrate bis-(β-nitrophenyl) phosphate. Recently, site-directed mutagenesis and biochemical analyses demonstrated that Bfil, exactly like Nuc, forms a dimer with a single active site composed of residues from both subunits. Second, bioinformatics analyses of the HNH endonuclease superfamily, revealed their relationship to many genuine REases, including M-D enzyme McrA, Type IIA restriction enzymes Hpyl, Nlalll, SphI, SapI, NspHI, Nspl and Kpnl, as well as Type US enzyme MboII and its homologues from *Helicobacter pylori* (Bujnicki et al. 2000; 200la; Aravind et al. 2000). To my knowledge, no crystallographic or mutagenesis studies have been published to support this prediction, but a few are currently in progress. Third, closely related Type II enzymes Eco29kl, NgoMIII, and Mral were found to be remotely related to the GIY-YIG superfamily of nucleases.

At the time of that analysis, no crystal structure of a GIY-YIG superfamily member was available. In order to provide a structural platform for experimental validation of our prediction, we attempted to predict the fold of an archetypal GIY-YIG nuclease, the N-terminal (NTD) catalytic domain of I-TevI, using ab initio simulations based on sparse NMR restraints published by the Belfort group. Recently, a crystal structure of the I-TevI-NTD has been determined, confirming the three-dimensional fold predicted by our preliminary model. The refined model of Eco29 kl was corroborated by the results of site-directed mutagenesis of the predicted catalytic residues (our unpubl. data).

Relationships to Other Polyphyletic Groups of Nuclease

Recent analyses reinforce the original suggestion that restriction enzymes as a whole are polyphyletic and that they have been re-invented multiple times. There are many protein folds that served as a scaffold for the development of various nuclease active sites. Most of them are α/β proteins, with the catalytic residues located at the C-termini of β-strands, but nucleases with all-α and all-β folds are also known. It is remarkable that the majority of known

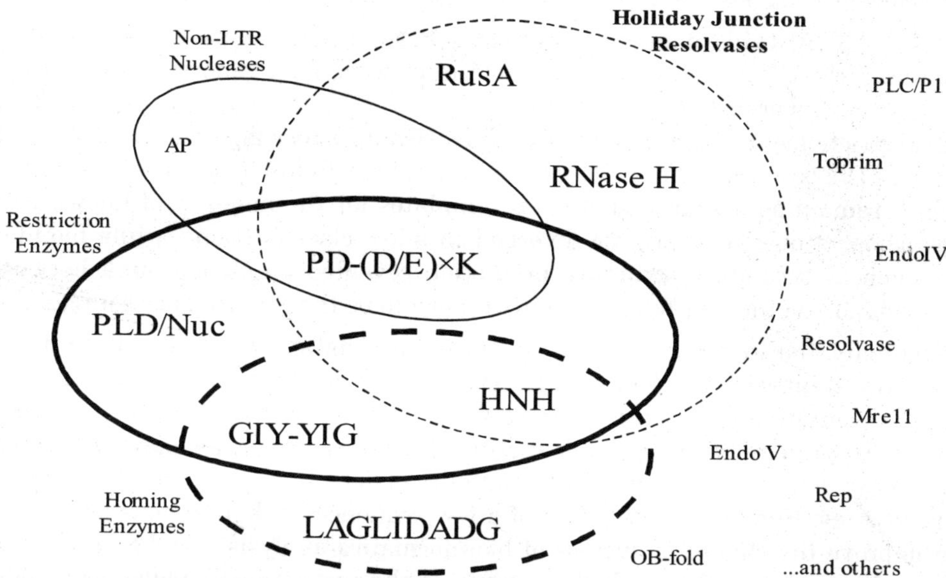

Fig. 9.5. Polyphyletic groups of nucleases. Restriction enzymes evolved from at least four different folds. Homing enzymes evolved from at least three different folds. Holliday-junction resolvases originated from at least four different folds. The PD-(D/E)xK domain found in the non-LTR transposable is very often replaced by a nuclease domain related to a superfamily grouping DNase I and apurinic/apirymidinic (AP) DNA repair endonucleases. There are other protein folds, which comprise nuclease members. It remains to be seen, if any of them gave rise to restriction enzymes.

REases originated from the PD-(D/E)xK superfamily and relatively few have been found to belong to other nuclease families. In this context it is noteworthy to mention the key difference between two large families of 'selfish' nucleases: restriction enzymes and homing endonucleases. Whereas the formerl typically recognize short, well-defined sequences and cleave them with remarkable specificity, the latter prefer very long and degenerate targets. Both homing enzymes and restriction enzymes evolved from several different folds: either types of nucleases were derived from the GIY-YIG and HNH superfamilies, while the PD-(D/E)xK and PLDl superfamilies are so far unique to restriction enzymes, and the LAGLIDADG, fold is unique to homing enzymes.

On the one hand, the LAGDLI-DADG fold is a saddle-shaped structure compatible with the major groove of DNA, which provides several β-turn structures that can be extended to form a 'very long binding site, as observed in known structures of homing enzymes). On the other hand, adaptation of the PD-(D/E)xK fold to cover larger fragments of the DNA requires adding new strands to the central beta sheet or fusions with additional domains. Nonetheless, the PD-(D/E)xK fold seems to be better suited for grasping the target DNA both from the major and the minor groove site,] thereby increasing the number of possible protein-DNA contacts. Incidentally, the EndA nuclease, which has been mentioned earlier in this article, is a fusion of PD-(D/E)xK-like and LAGLIDADG-like domains, of which neither exhibits the functional deoxyribonuclease active site (Bujnicki and Rychlewski 200Id).

Other LAGLIDADG-like domains devoid of the REase active site have been found in various transcription factors, suggesting that this fold may serve as a generic nucleic acid-binding module. The same is probably true for the PD-(D/E)xK fold. It remains to be seen if there are indeed no restriction enzymes related to the LAGLIDADG super-family or no homing enzymes from the PD-(D/E)xK superfamily. There are many superfumilies of nucleases with distinct folds, which remain as potential structural templates for the majority of known restriction enzymes, whose sequences were too diverged to allow classifications, while bioinformatics analyses such as secondary structure prediction revealed no obvious link between their sequences and any of the four folds currently known to be the evolutionary sources of REases.

Undoubtedly, the increasing size of the structural database will facilitate detection of remote relationships with the fold-recognition methods. However, some restriction enzymes may be representatives of extremely rare folds or entirely new folds and therefore not amenable to the protein-fold recognition analysis. In both cases, the availability of homologous sequences may considerable improve the reliability of detection or dismissal of similarities to known structures. Hence, identifica-tion of a few sequences of homologous isoschizomers in order to improve the chance of success of bioinformatics analysis may be an alternative to costly crystallographic analyses. To date, many of the restriction enzymes were thought to represent distinct folds, until the structures were actually solved and "more of the same old PD-(D/E)xK" was identified.

From the analyses of PD-(D/E)xK nucleases and many other proteins it is now clear that the criterion of sequence dissimilarity is worthless as an indicator of a lack of evolutionary and structural relationships. Guided by the working model of the REase evolution, bioinformatics analyses may help to focus the search of new, unusual restriction enzyme structures as well as the candidates that either seem homologous to other nuclease superfamilies or do not exhibit significant structural similarity to any protein in the database. As mentioned above, fold-recognition and homology modeling methods have been very useful in identification of potential active sites of REases. In the age of structural genomics more and more structures, including those of REases and related proteins, are expected to be solved by X-ray crystallography and NMR.

It is anticipated that bioinformation methods will serve as tools to fill in the gaps on the landscape of sequence-structure-function relationships, providing a useful platform for experimental studies on protein function in the structural and evolutionary context.
Acknowledgements. I am grateful to Aravind, Alfred Pingoud, Rich Roberts, and Virgis Siksnys for numerous stimulating discussions on various aspects of evolutionary and structural biology of restriction enzymes. My research on R-M enzymes is supported by EMBO, the Howard Hughes Medical Institute (Young Investigator Programme award), and by KBN. The Fellowship for Young Scientists from the Foundation of Polish Science is also gratefully acknowledged.

Structural Variation of Restriction Endonucleases

The Type I and. III restriction endonucleases are large, multimeric protein complexes with four enzyme activities; DNA methyltransferase, DNA endonuclease, ATPase and DNA translocase. It has been demonstrated that ATP-dependent protein motion along DNA is necessary for endonuclease activity. Studies have shown that Type I enzymes remain bound to their I recognition sites whilst simultaneously translocating adjacent non-specific] dsDNA past a stationary complex. This occurs bi-directionally so that two DNA loops are extruded. An equivalent unidirectional mechanism has been suggested for the Type III enzymes. DNA cleavage generally results when the enzymes stall against another restriction enzyme complex. Both the HsdR subunits of the Type I enzymes and the Res subunits of the Type III enzymes) carry amino acid motifs characteristic of superfamily 2 helicases. In this review, the structural and mechanistic implications of this relationship are discussed and models suggested for how the ATP-dependent restriction enzymes might couple chemical energy to mechanical motion on DNA.

Energy-Dependent DNA Processing

The cellular control of complex genomes requires a fair degree of molecular I gymnastics from large consortiums of proteins. In this regard, enzymatic dsDNA cleavage may seem a rather straight forward affair, involving the, recognition of a specific sequence and the alignment of a pair of active sites proximal to a pair of scissile phosphodiester bonds. Glancing through commercial catalogues at lists of restriction enzymes may certainly give thisi impression. However, it is becoming increasingly clear that even here what, may seem like simple DNA recognition and cleavage may in fact involve more complex molecular communication events such as DNA looping (*e.g.,* see A.J./Welsh *et al.,* this Vol.).

The NTP-dependent restriction endonucleases, which comprise the modification-dependent restriction (MMR), Type I and Type III systems, are less well known than the Type II enzymes principally because they have not been exploited commercially. However, these enzymes are widely distributed in nature with almost half of the current sequenced microorganism genomes containing one or more putative NTP-dependent syste. Their dependence on an external energy source is key to their mechanism-communication between sequence-specific DNA recognition and non-specific DNA cleavage requires one-dimensional DNA translocation, Nearly every genetic processes including replication, recombination, transcription and repair relies on enzymes which consume NTPs. In doing so, these proteins convert chemical energy into mechanical energy and hence they have come to be considered as “molecular motor proteins”.

The work done could be the unwinding of a DNA duplex, backtracking of a disrupted replication fork (Mahdi et al. 2003), remodelling of chromatin, or ousting of a stalled RNA polymerase. In a great many cases these proteins share common amino acid sequences and folds, such as helicase or AAA+ motifs, that are also found in the NTP-dependent restriction enzymes. Hence, a relatively straightforward process-DNA cleavage-is, in many cases, being undertaken by complex molecular machines. Much of our understanding of restriction-modification enzymes arose from pioneering genetic work carried out on NTP-dependent systems. This culminated in the purification of the first restriction endonuclease, the Type I enzyme.

With the subsequent purification of Type III enzymes (Haberman 1974) there has followed over 35 years of elegant work on these systems. A number of recent reviews have covered the history, biology and mechanism of these enzymes in detail. Nonetheless, a great many questions remain unanswered, principally in the details of how the chemical energy from ATP is converted into mechanical work during DNA processing. Given the depth of coverage in the earlier reviews, the aim of this article is to focus on the structural features of the “motor” domains from the Type I and III restriction endonucleases to help elucidate mechanistic features of these enzymes. We will also discuss how this information can be informed by, and inform, our understanding of helicases, AAA+ ATPases and related motor enzymes.

Motor Enzyme Architecture of the ATP-Dependent Restriction Endonucleases

For all NTP-dependent restriction endonucleases, DNA cleavage is carried out by a large macromolecular complex built from multiple copies of specific sub-units encoded by genetically-linked genes. Nonetheless, once assembled these complexes can undertake their reactions without the necessity for other protein factors (it should be noted, however, that other proteins may play a role in vivo - for instance, topoisomerases may be important in relieving topological strain induced by DNA translocation). The ATP binding and hydrolysis func-tions are carried out in each case by one particular subunit: the HsdR subunit in Type I systems and the Res subunit in Type III systems.

The catalytic centre that utilises ATP is constructed from a series of discontinuous amino acid motifs that are common to DNA helicases and to related AAA+ ATPases (Gor-balenya and Koonin 1991; Murray et al. 1993; Titheradge et al. 1996; Caruthers and McKay 2002). Because of their self-contained activity and relative simplicity, the Type I and III restriction endonucleases are attractive test systems with which to explore helicase mechanism.

Motor Enzyme Motifs in the Type I and III Restriction Endonucleases

A wide variety of proteins from bacteria to man can be classified on the basis of characteristic amino acid motifs into one of four helicase superfamilies-SF1, SF2, SF3 or SF4

(Hall and Matson 1999). Whilst some members are undoubtedly *bonafide* helicases (*i.e.* many SF1 enzymes unwind dsDNA into the corresponding ssDNA), there is mounting evidence that other members, particularly those in SF2, do not *directly* unwind DNA (Singleton and Wigley 2002). Instead, these enzymes couple chemical energy from NTP hydrolysis to other mechanical events such as processive dsDNA translocation, remodelling of protein-DNA complexes such as chromatin or stalled polymerases, etc. The HsdR and Res subunits fall into this SF2 class. With ongoing genome sequencing programmes, putative Open Reading Frames (ORFs) have been identified in a broad range of microogranisms which may encode HsdR or Res subunits as part of restriction endonuclease operons. In combination with characterised ORFs, we undertook an alignment of 143 HsdR sequences and of 39 Res sequences. Using ClustalW, we re-checked the consensus sequences for the previously identified helicase motifs and identified a number of new motifs which may be involved in ATP binding and hydrolysis and/or DNA translocation (Table 10).

Gross *Organisation of the Type I HsdR Subunits*

The HsdR subunits are typically 900-1050 aa in length and have theoretical molecular weights of 100-120 kDa. These values compare well with the characterised values for enzymes such as EcoKI, EcoR124I, EcoAI, etc. The subunit domain organisation was similar throughout. There are two regions of significant sequence homology; the N-terminal Region X characteristic of endonuclease activity and the centrally located collection of helicase motifs. Some variation in distance between helicase motifs la and III and between IV(Y) and V was observed (Fig. 1A), but this can be easily accommodated in the helicase fold. A new, putative helicase motif, the Q-tip helix, was also identified N-terminal to Motif I. There was no significant sequence homology elsewhere.

Moreover, the biggest variations in protein length were seen outside of the conserved regions. Proteolytic digestion of the HsdR subunit of EcoKI identified that interactions with the HsdS subunit occur within the C-terminal domain, but no sequence homology was observed in this region. However, secondary structure conservation was observed. Alignments of subsets of HsdR sequences from phylogentically related organisms did not reveal any additional motifs. Some ORFs lack the N- or C-terminal regions which suggest that they would not be able to cleave DNA or assemble into active complexes. However, the majority of the sequences appears intact and could potentially participate in active Type I systems.

Gross Organisation of the Type III Res Subunits

The Res subunits are typically 950-1000 aa in length and have theoretical molecular weights of -110 kDa. These values compare well with the characterised values for enzymes such as EcoPIl and EcoPlSI. As with the Type I subunits, there were two regions of significant sequence homology. In this case, however, the relative order was switched; the helicase motifs were located in the N-terminal domain whilst Region X was located near the C-terminus. Initial attempts to align all 39 sequences were unsuccessful but convincing alignments were obtained by splitting the sequences into two subsets which we have termed IIIA and I1IB. The IIIA sequences were very closely related; in addition to the seven classical helicase motifs, four other regions of homology were identified which may play roles in the motor mechanism. In

Table 10.1. Helicase motifs in HsdR and Res proteins. Amino acid sequence alignments were carried out using ClustalW (Higgins et al. 1994) at the European Bioinformatics Institute (http://www.ebi.ac.uk/clustalw/index.html). HsdR and Res Protein sequences were obtained from REBASE (Roberts et al. 2003). 143 Type I HsdR sequences were aligned, of which 121 were translations of uncharacterised ORFs (this includes some dupli-cation between strains). For the Type IIIA enzymes (see below), 21 Res Sequences were aligned, of which 16 were translations of uncharacterised ORFs. For the Type IIIB enzymes (see below), 18 Res sequences were aligned, of which 15 were translations of uncharacterised ORFs. In all cases, sequences were discarded which had <20% identity, or which were truncated, across the helicase region. The name and consensus sequences of the SF2 helicase motifs (Hall and Matson 1999) are shown in the left-hand column. Regions which are likely to be characteristic of the restriction endonuclease family alone are indicated in italics. The order of the motifs are as shown; note, however, that the relative spacing will vary. For the consensus sequences, black shaded abbreviations indicates >80% identity, uppercase abbreviations indicate >60% identity, a "+" indicates a hydrophobic residue with >80% identity, an "o" indicates a hydrophilic residue with >80% identity, and an "x" represents a variable residue. Underlined amino acids are those residues mutated and shown to affect ATPase, endonuclease and translocation activities (see main text). [Type IIIA Res: EcoPII (Humbelin et al. 1988), PhaBI (Ryan and Lo 1999), StyLTI (Dartois et al. 1993), BceSI (Hegna et al. 2001), HpyAXI (Lin et al. 2001), *Bacteroides fragilis* ORFC143 and ORFC196, *Burkholderia pseudomallei* K ORFC742, *Chlorobium tepidum* T ORF908, *Clostridium thermocellum* ORFS4, *E. coli* CFT073 ORF5372, *Fusobacterium nucleatum Knorr ORF416, Helicobacter pylori* 26695 ORF1522 and J99 *ORF*1411, *Magnetococcus* species MC-1 ORFC186, *Neisseria gonorrhoeae* FA1090 AX, *Neisseria meningitidis* Z2491 ORF1467 and B strain MC58 *ORF*1261, *Pasteurella multocida* Pm70 ORF698, *Salmonella typhimurium* CT18 ORF388 and LT2 ORF357. Type IIIB Res: LlaFI {Su et al. 1999), Pstll (G.G.Wilson and M.D.S, unpublished observations), Hpy1061 (De Vries et al. 2002), *Chlorobium tepidum* TLS ORF1729, *Ferroplasma acidarmanus* ORFC158, *Geobacter metallireducens* ORFC20, *Helicobacter pylori* 26695 ORF1370 and J99 ORF1296, *Mycoplasma pulmonis* UAB CTIP ORF3960, *Nitrosomonas euro-peaea* ORFC219, *Neisseria gonorrhoeae* FA1090 ORFC707, *Neisseria meningitidis* Z2491 ORF1590, *Neisseria meningitidis* B strain MC58 ORF1375, *Rhodospirillum rubrum* MORFS3P, *Thermosynechococcus elongatus* BP-1 ORF1481, *Thermoplasma volcanium* ORF1464 and ORF1476, *Xan-thomonas campestric* C ORF1068]

Q-Tip, α_Q Helix	Type IIIA	LXOQXXA + OA + XX + FOG	Recognition and stabilisation of the base in ATP (*e.g.*, Theis et al., 1999)
	Type IIIB	XQOOA + OO +	
	Type I	RXXQXXA	

Motif	Type	Sequence	Comments
Motif I (Walker A) GxGK(T/S)	Type IIIA	+D+oI-IETGTCHTYpoYoxoTxF H	Highly conserved sequence throughout helicases and related NTPases. Interaction with phosphates of MgATP/MgADP and coordination of the Mg^{2+} ion.
	Type IIIB	+XFOMATGXGKTX+MAX+I	
	Type I	+++XHOOGSGKSTXT	
Motif a +++xPoo	Type IIIA	GXXKI++VVPSXAIKEG$(X)_{9-:1}$ EHF	Possibly involved in contacting DNA? For the type IIIA group, Motifs I and Ia are contiguous. This sequence is weakly conserved within the IIIB group
	Type IIIB	OF+++OOX	
	Type I	++++XDRXOLOOQ	
Motif II (Walker B) +++DexH	Type IIIA	(C) P+OX+XXXRP++I+DEPHO	Highly conserved sequence throughout helicases. Coordination of Mg^{2+} ion and ATP catalysis. Interactions with Motif III?
	Type IIIB	O+V++XDEAHH	
	Type I	(C) XX++++DEAHR	
Motif III +x+(T/S)(A/G)(T/G)	Type IIIA	(S H) PXXXLRYGATF	Role of Motif III in SF2 enzymes yet to be resolved fully.
	Type IIIB	(T) LOFSATX	
	Type I	+GFTGTP	
Region III_a	Type IIIA	OLOA+DAXNOXLVKX+X+	Region IIIa is immediately adjacent to Motif III. Possibly involved in contacting DNA?
	Type IIIB	LXO+XOXGXVK	

Region V	Type IIIB	FXXXXXXYXXXXXXXD (E) FGOX+OXYO+OOA+OD (81/121)	Alternative to Motif IV in type 1 enzymes (Titheradge et al, 1996) and some of the type IIIB enzymes. Possibly involved in contacting DNA?
	Type I	F(X)$_{7-25}$ YOXOOAIOD (E) (40/121)	
Motifa IV +++(F/Y) xxoxo	Type IIIA	IKXLSLFFIDO + XXYR (T)	Possibly involved in contacting DNA? No significant alignment was observed in the type I or type IIIB enzymes (see Region Y).
Region IV$_a$	Type IIIA	VHXGYFSODOO	Between Motifs IV and V but not observed in the B group; may represent a non-essential sequence?
Region V$_a$	Type IIIA	ILKOKEKLLS +	Immediately adjacent to motif V. Possibly involved in contacting DNA?
	Type IIIB	XXXOOOOLLOX	
Motif V +x(T/S)xxxxxG+o+xo+	Type IIIA	PXRFIFSOOALREGWDNPNVFXICOL	Possibly involved in contacting DNA and ATP? These sequences are strongly conserved amongst the type I and II enzymes but do not clearly fit into a pattern with other SF2 enzymes.
	Type IIIB	P + R + I + OVOLOEGWDVXN + XOIVXL	
	Type I	XXX + _ IVXOX + LTGFD + PXXOXXX + DK	
Motif VI **(arginine finger)** QxxGRxxR	Type IIIA	KXQEVGRGLRIXVNOCGOR	Highly conserved sequence thoughout helicases and related NTPases; implicated in stabilisation of the transition state during ATP hydrolysis
	Type IIIB	XOQ + + GRG + R	
	Type I	+XQX + XRXNR	
Region VI$_a$	Type IIIA	F + + OOLQOEI	An extended region of homology beyond Motif VI which may represent an additional DNA contact loop.

comparison, the I1IB sequences were less well conserved and with greater variability in the spacing of the motifs.

Intriguingly, the II1B sequences have a greater sequence identity across the helicase region with the HsdR sequences than with the IIIA Res sequences-in particular, Region Y appears conserved in some Type II IB enzymes. A structural characterisation also suggests that Type I and IIIB may be related. However, representatives of both A and B groups have DNA cleavage activity typical of Type III enzymes rather than Type I enzymes (L.J. Peakman, G.G.Wilson and M.D.S. The Q-tip helix motif was also identified in both groups, but was located further from Motif I than in the Type I enzymes.

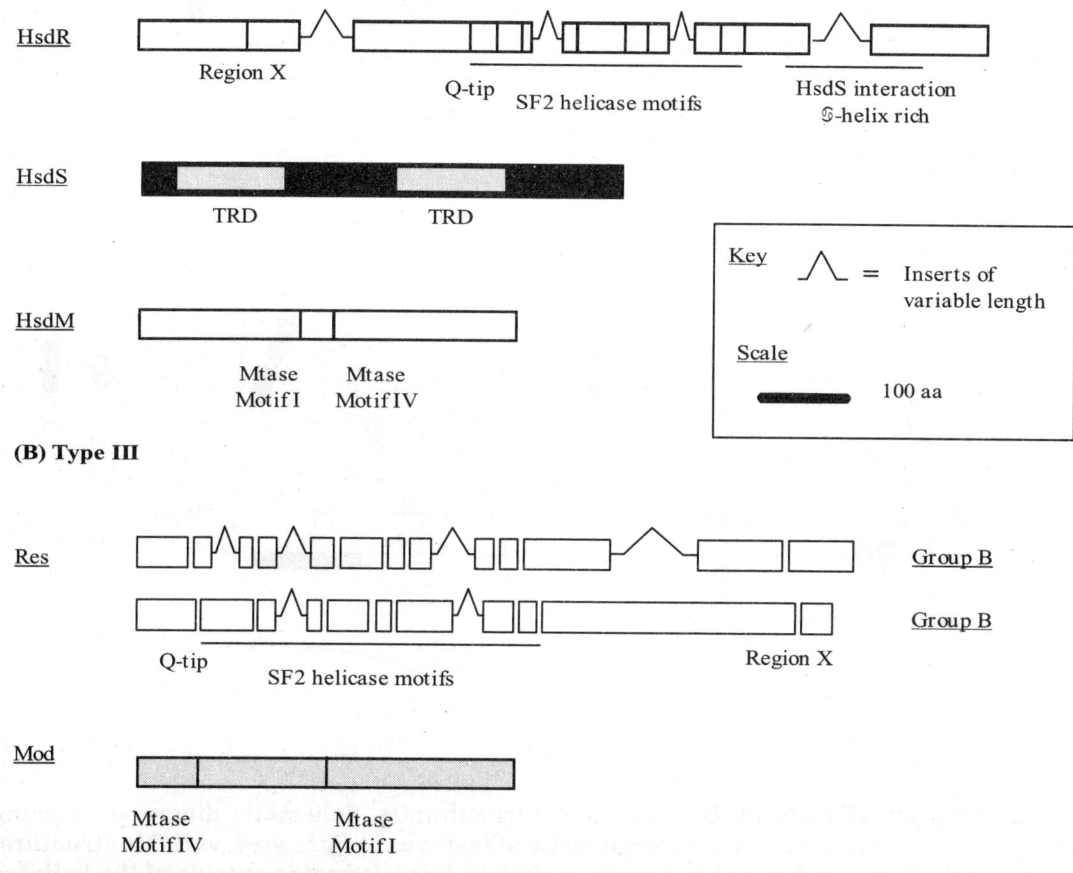

Fig. 10.1. Subunit domain architecture of the ATP-dependent restriction endonucleases. Proteins are represented as rectangles approximately to scale with inserts of variable length indicated (see key, *inset*). *Shaded rectangles* or *vertical lines* indicate regions with a defined mechanistic function (see main text) which may or may not show sequence conservation. A *horizontal line* indicates a collection of motifs which form a specific catalytic site or fold. A The HsdR, HsdS and HsdM proteins of the Type I enzymes. B The Res and Mod proteins of the Type IIIA and IIIB enzymes. TRD Target recognition domain.

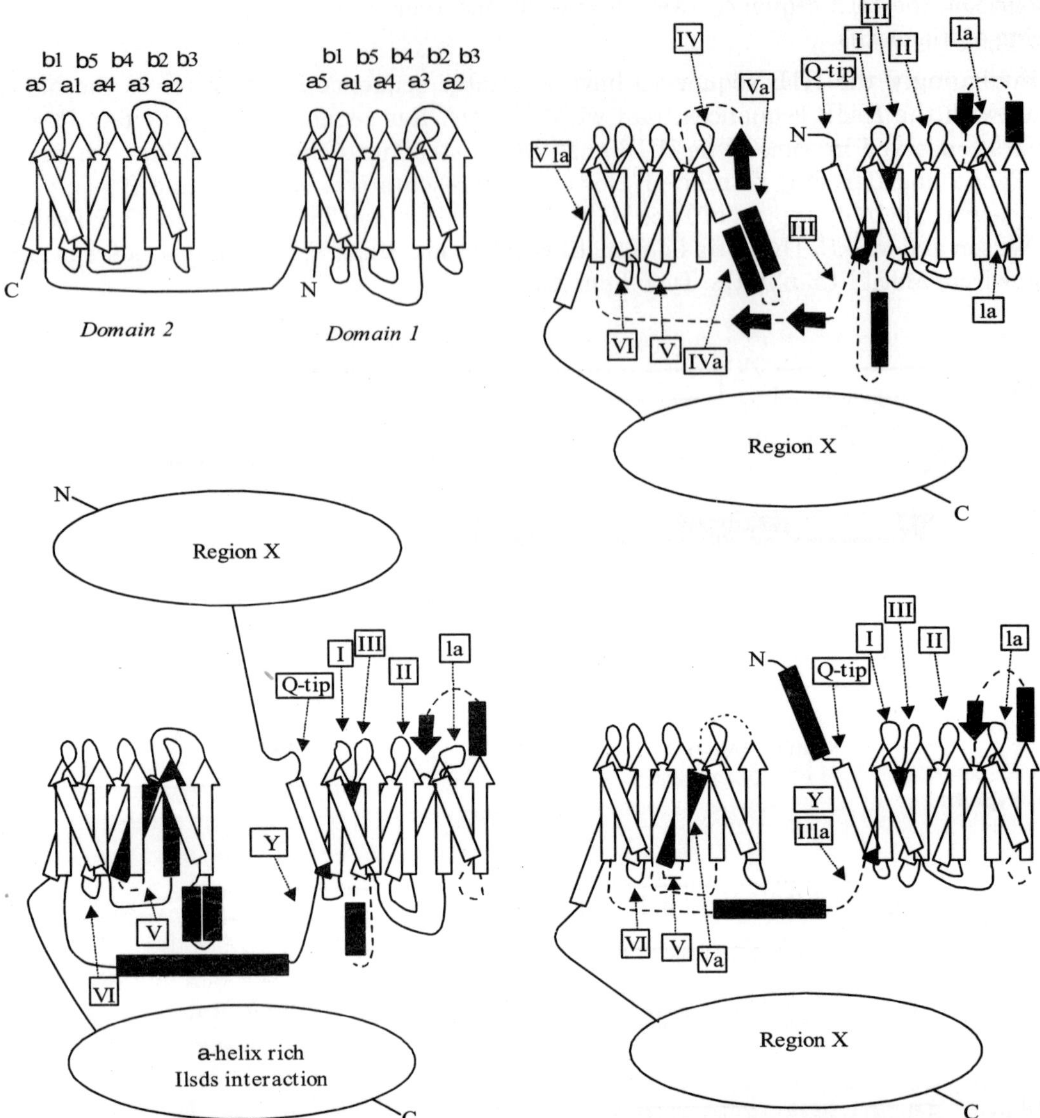

Fig. 10.2. Topological folds in the HsdR and Res subunits. Schematic diagrams showing α-helices and β-strands, with conserved structural features in *light grey*, variable structural features in *dark grey*, and variable inserts as *dashed lines*. Domains outside of the helicase motor are shown as ellipses. The locations of conserved motifs are indicated. Topologies of the HsdR and Res subunits were determined with PSIPRED using at least four representative sequences in each case. A Representative arrangement of the fold and structural domains in a helicase. The structural elements are named above and below the diagram. B Representative arrangement of the fold and structural domains in HsdR. C Representative arrangement of the fold and structural domains in Type IIIA Res. D Representative arrangement of the fold and structural domains in Type TUB Res.

Core Helicase Motifs in ATP Binding and Catalysis

The crystal structures of a number of SF1 and SF2 DNA and RNA helicases have revealed that the superfamily motifs cluster into an NTP binding and catalysis pocket between two structurally-related domains. A similar structural fold is also seen in other superfamilies and in related AAA+ ATPases such as RuvB. The central core of the NTP pocket comprises Motifs I and II [the so-called Walker A and B boxes first identified in ATP synthase (Walker et al. 1982)], and Motif VI. These motifs coordinate the magnesium ion and phosphate groups of MgATP/MgADP. Accordingly, these motifs are the most strongly conserved in the HsdR and Res subunits (Table 10.1) and it is highly likely that these core regions function in a very similar manner to many other ATPases.

In all helicase structures to-date (Singleton and Wigley 2002; Caruthers and McKay 2002): The amino group of the conserved lysine of Motif I [GxGK(T/S) for SF2] interacts with the mononu-cleotide phosphates whilst the hydroxyl group of the conserved serine/threonine [GxGK(T/S) for SF2] coordinates the Mg^{2+} ion; the conserved aspartate of Motif II [DExH for SF2] also coordinates the Mg^{2+} ion whilst the conserved glutamate [DExH] is a catalytic residue; and, the first conserved arginine in Motif VI [QxxGRxxR for SF2] is implicated in binding they phosphate group (note however that in some helicase structures the side chain is not ideally placed to bind the phosphate; Theis et al. 1999; Caruthers et. Mutations at some of these key residues affect the coupling of ATP hydrolysis to DNA translocase and cleavage by the Type I enzymes and the ATPase and nuclease activities of the Type III enzymes. A number of other contacts beyond Motifs II and VI may play additional structural roles. Residues immediately C-terminal to the DExx sequence of Motif II may be involved in forming salt bridges with other structural domains, *e.g.*, with motif V and/or with Motif III.

The "Q-Tip Helix" - A New Helicase Motif?

In aligning the HsdR subunits around Motif I, it was noted that a highly-conserved glutamine and small hydrophobic residue were present in 99% of the sequences, located ~30 aa N-terminal to the first conserved G residue of Motif I. A similar motif was also found in the Type IIIA and IIIB enzymes, albeit with more variable spacing to Motif I. In structures of the PcrA, rep, UvrB, and RecG helicases and the AAA+ helicase Ruv, a Qxx + motif is also present at the N-terminal end of an α-helix which packs against the recA fold in domain 1. A similarly-located conserved Q-residue has been implicated in ATPase activity of DEAD-box RNA helicases. In the UvrB and RecG structures the glutamine side chain points into the ATP binding pocket, packs against the adenine and forms hydrogen bond with the cyclic N7 and exocyclic N6 positions of the base.

In other ATPase structures the residue is located close to the ATP binding pocket but is not ideally oriented to contact the base. However, mutants at the glutamine residue in RuvB and the BLM helicase have been identified as deleterious to function (Iwasaki et al. 2000; Bahr et al. 1998). Applying secondary structure. prediction algorithms to the Type I and III sequences in this region indicate that the conserved glutamine also resides at the end of a helix. We have noted that this residue and helix are common to a broad range of SF1 and SF2 helicases and some AAA+ ATPases (Szczelkun, manuscript in preparation) and thus *represents a new* helicase motif we suggest that the motif is named the 'Q-tip helix' as it shows both sequence and structural conservation. Further testing will be required to probe its importance to the Type I and III restriction endonucleases.

The DNA Binding Motifs - Family Specific Deviations

Many of the remaining helicase motifs are implicated in DNA/RNA interactions and in domain-domain interactions that both stabilise the motor structure and communicate conformational changes during ATP binding/hydrolysis. Analysis of the Res and HsdR subunits reveals that whilst some of these motifs are conserved, others are either weakly conserved or absent altogether. In some cases, alternative conserved sequences can be identified that may undertake similar roles to the 'missing' motif. *Motif la:* In most helicases, the amino acid backbone of motif la primarily makes contact to the sugar-phosphate backbone of the polynucleotide sub-strate.

The close packing of motifs la, II and III suggest that changes in the ATP/ADP binding can be communicated to the DNA via la and III. In UvrB, a salt bridge is also made to motif V. In the Type I and III enzymes, this region is poorly conserved. Since the DNA contacts may be made via amino acid backbone contacts, it follows that Motif la could accommodate variations in amino acid sequence. Strong sequence conservation is observed in the Type IIIA enzymes, but as Motifs I and la are contiguous this may simply reflect closely related sequences rather than a conserved motif *per se*. Conversely, this region is not well conserved in the Type IIIB group. Because Motif la is a less well-defined consensus sequence overall (+++xPoo, Hall and Matson 1999), it has not been explored as fully in structure-functions studies as some other motifs.

However, mutations in this motif in the Type I enzyme EcoKI revealed its importance in coupling ATPase activity to DNA cleavage and translocation. *Motifs HI and* III_a In all SF2 helicase structures to-date, the serine/threonine residues in either position of the conserved (T/S)(A/G)(T/G) sequence of Motif III make hydrogen bond contacts to motif II. In the SF1 enzymes, the unrelated Motif III sequence also makes contacts to the phosphates of ATP/ ADP and to the polynucleotide substrate - the same contacts have yet to be implicated for the SF2 enzymes. Unlike the contacts made by motif la, Motif III of the SF1 enzymes makes hydrogen bond and stacking interactions with the DNA bases. Although the Type I and III enzymes carry helicase motifs, DNA unwinding activity has never been observed (Szczelkun 2000) and a growing number of SF2 enzymes have been designated dsDNA translocases as opposed to unwindases.

Consequently, the DNA contacts made by HsdR and Res may differ from those seen in SF1 helicases. Nonetheless, the importance of Motif III has been proven in the Type I enzymes. Motif Ilia was identified in the Type IIIA and B groups, immediately adjacent to Motif III (except in *Salmonella typhimurium* sequences - not shown). Given the possible role of Motif III, this conserved region could also be involved in contacting the DNA. *Motif IV,* IV_a *and Region Y.* In SF1 enzymes, Motif IV is implicated in mononucleotide binding, whereas in SF2 enzymes a role in polynucleotide binding is suggested. However, this motif is not always identified unambiguously in SF2 enzymes.

Moreover, the consensus is sufficiently accommodating that many alignments have inappropriately assigned functions to random sections of sequence. In fact, Motif IV from the SF2 helicase NS3 was shown to have a high sequence similarity to a region of the SF1 helicase Rep revealed to be in contact with. Therefore, there may be considerable family-specific variation in how the polynucleotide substrate is contacted. No significant similarity to Motif IV was observed in alignments of the Type I or Type IIIB enzymes. For the Type I enzymes, an alternative region of similarity, Region Y, is the more likely candidate. A subset of the Type I enzymes have a highly variable non-specific insert in this region - intriguingly, 12 of this subset are archaeal sequences. We also found evidence that at least some of the Type IIIB enzymes contain a Region Y-like sequence. This putative motif is in a similar position to region III_a so may not be relevant.

However, Type IIIB sequences showed stronger sequence and structural conservation across the helicase domain with Type I sequences than with IIIA sequences. An additional region of homology in the IIIA sequences was also identified. This sequence was located between motifs IV and V in a region of variable structure.

***Motif V and Region V_a* :** The sequence similarity in Motif V was strong across the Type I and III sequences. Generally, this region is more divergent in sequence and length amongst the helicase families. Motif V has been implicated in forming a network of interactions which ligate MgATP/MgADP and which stabilise domain contacts (*e.g.,* contacting Motif II) and in making contacts to the polynucleotide. Mutations in this motif in EcoKI affect ATPase, nuclease and translocase activities. An additional region of strong homology N-terminal to Motif V called Region VJ was observed in the Type IIIA group. A weakly-conserved lysine-rich region was also observed in some of the IIIB group. These sequences may simply be extensions of motif V in the Type III enzymes.

***Family-Specific DNA Binding Loops* :** Outside of the helicase motifs identified by amino acid sequence homology are a number of other structural motifs that are involved in DNA binding: *e.g.,* the QxxR motif of DEA(D/H) box RNA helicases, the TxGx ssDNA binding motif, the TRG motif of RecG (Mahdi et al. 2003) and Mfd (N. Savery, pers. comm.). These regions do not show the same conservation of sequence between families as seen in the classical motifs but instead relate to structural elements found in specific locations within the conserved topological fold. It has been suggested that the dsDNA translocation mechanism of RecG and Mfd may be a prototype for other SF2 enzymes. However, the TRG motif which is found C-terminal to Motif VI following a helix-turn-helix structure is not found in either the HsdR or Res subunits (R. Lloyd, personal communication).

An additional region of homology, Motif Via, was also found in the Type IIIA enzymes. This sequence is located in ot_5 of domain 2, but does not have the same relative location as the TRG in RecG. Aligning the sequence motifs onto a structural fold gives some indications of alternative regions that may be important, but it remains to be seen if these structurally-conserved features can be confirmed as critical parts of the restriction enzyme motors.

A Motor Enzyme Fold in the Type I and III Restriction Endonucleases

The growing number of motor enzyme structures available in the PDB have revealed that for helicases and many related ATPases, the core sequence motifs occupy common positions within a shared protein fold. In every case, an α-β domain is found that is similar in topology and structure to the RecA recombination protein. In many SF1 and SF2 enzymes, a second domain is also present which is essentially a repeat of the first, but which lacks the classical

ATP binding motifs. It is the combination of the two domains which arrays the helicase motifs into an ATP/ADP binding pocket. The arrangement of each domain is such that the amino acid sequence can make excursions into separate topological folds or domains without disrupting the RecA fold.

The result is a core motor enzyme architecture onto which any other functionality can be 'bolted' - in the case of the Type I and III enzymes, domains involved in DNA cleavage and protein-protein contacts. This modular design makes the helicases highly adaptable. Secondary structure predictions using PSIPRED were undertaken on a selection of HsdR and Res sequences. In combination with previous analyses, it is clear that the core β-sheet components of the motor fold are largely present in the HsdR and Res subunits and that they are arranged into a two-domain structure. Based on the topology diagram for a representative helicase proposed by Caruthers and McKay (2002), putative model architectures were constructed for the HsdR and Res subunits.

The positions of the motor motifs and α-helices were matched according to the positions of the predicted β-sheets and by reference to the structure of the SF2 helicase RecG. For the Type I enzymes, the N-terminal third of the protein contains region X but no conservation of structure was observed in this domain. The C-terminal third is proposed to interact with the HsdS subunit and in this case an α-helix rich structure was conserved which may be important in the protein-protein contacts. Within the central domains 1 and 2 of the representative helicase fold, a number of variable regions exist, some of which have distinct structural elements. α_Q was positioned according to the packing of similar structures in RecG, and UvrB.

The region between α2 and α3 is particularly changeable; the alignment as shown puts the variable loop between β3 and α3, but it is equally likely that this could be rearranged with the variable loop positioned between α2 and β3. For some enzymes, such as EcoAI, this region is truncated which may downplay its importance (EcoAI is smaller overall and may be a more compact structure). Nonetheless, it is worth noting that the region between β3 and β3 coincides with the TxGx motif implicated in RNA interactions by eIF4A. An additional helical domain may be present between α4 and β5. Region Y is located after Motif III but appears to have mixed structure from the predictions. However, following this motif is a conserved, α-helix rich region which we suggest forms the link between domains 1 and 2. The positions of the conserved motifs in domain 1 relative to the α-β fold are exactly as predicted for a SF2 helicase. Whilst the repeated α-β structure of HsdR domain 1 appears remarkably well conserved, there is more variability in domain 2.

Core structural components such as β2 and α3 are absent in some cases and additional structures, mainly helical in nature, may exist between α2 and β3. It should be noted that domain 2 is less well conserved in other helicases; for example, α3 is absent in RecG such that β3 and β4 are linked by an extended loop carrying Motif V. Nonetheless, Motifs V and VI are positioned in HsdR as expected. In many heli-cases, the β2 and α3 structural elements position Motifs IV and/or QxxR proximal to the DNA. The variability seen here suggests that alternative parts of the structure may be used instead, possibly region Y located closer to domain 1. For the Type III enzymes, the N-terminal halves of the Res sub-units comprise domains 1 and 2 of the helicase fold whilst the C-terminal halves are less well conserved α-β structures containing region X.

The arrangement of the Q-tip motif in the III enzymes is somewhat different to the Type I - a variable loop structure with conserved α-helical elements is inserted between α_Q and β1, such that the Q-tip might be part of a helix-turn-helix. We assumed that α_Q still packs against the RecA fold as observed in other helicase structures. Beyond that difference, both Type III structures follow a very similar topology in domain 1 to that observed for the Type I enzymes, with a similar extended variable region present between β3 and α3. Thus, this region may play a conserved role in the restriction endonucleases. In contrast, the topology of domain 2 was distinct in each case. For the Type IIIA enzymes, an extended loop with some β-sheet structure connects domains 1 and 2. Following Motif IV, a variable α-helical region was observed which, in some cases, lacked β3. The additional motifs IVa and Va could be found in this region.

Motif Via was located in α_5 of domain 2. For the Type IIIB enzymes, the overall topology of domain 2 appears more similar to that of the Type I enzymes with the absence of α3 most notable. Domains 1 and 2 are linked by an α-helix rich stretch containing the putative, Region Y/Motif IIIa in a similar manner to HsdR. As the region encompassing Motif IV is less well defined, it may be that region Y plays a similar role to the Type I enzymes. Given the results of this exercise and results from similar studies on EcoKI, it is highly likely that the Type I and III motor subunits share the same topological fold and arrangement of amino acid motifs as seen in many other helicases and related motor enzymes. We await a crystal structure of either HsdR or Res to confirm these predictions.

Macromolecular Assembly of the Type I and III Restriction Endonucleases

Neither the HsdR nor the Res subunits have appreciable DNA-binding, ATPase or motor activity in isolation. Instead, the subunits must assemble as part of a larger macromolecular complex. Similarly, many helicase show relatively poor enzyme activity when separated from other protein cofactors. The Type I and III endonucleases must remain bound, or at least close to, their recognition sequences whilst also translocating expanding loops of non-specific DNA. For neither system has the architecture of the complex that undertakes this con-voluted DNA manoeuvring been resolved. For the Type I enzymes, a single HsdS subunit is required for DNA recognition. This subunit contains two variable Target Recognition Domains (TRDs) within a conserved protein structure with two-fold rotational symmetry.

Each TRD has a motif comprising two loops and a α-stran which is likely to recognise, via the major groove, one half of the bipartite Type I recognition sequence. This subunit forms a scaffold for the addition of two HsdM subunits (Dryden et al. 1997) which contain motifs characteristic of adenine methyltransferases. There is some evidence that this interaction occurs through the C-terminal region of HsdM. Two HsdR subunits are then required for DNA cleavage and translocation activities although complexes with a single HsdR can translocate DNA in a non-processive manner. There has been a great deal of debate in the helicase field as to whether SF1 and SF2 enzymes operate as monomers.

One way to assemble a Type I endonuclease is with each HsdR placed at opposite ends of the complex, attached to non-specific DNA. This allows for the bidirectional DNA translocation from both sides of the site. There is at least some evidence for this arrangement from DNA footprinting studies. In this case the monomers are likely to act independently of each other. However, translocation studies using complexes with a single HsdR suggest that when a subunit releases DNA during motion, it can bind and translocate the DNA on the *opposite* side of the complex. Therefore, an alternative proposition is that the HsdR sub-units are cheek-by-jowl.

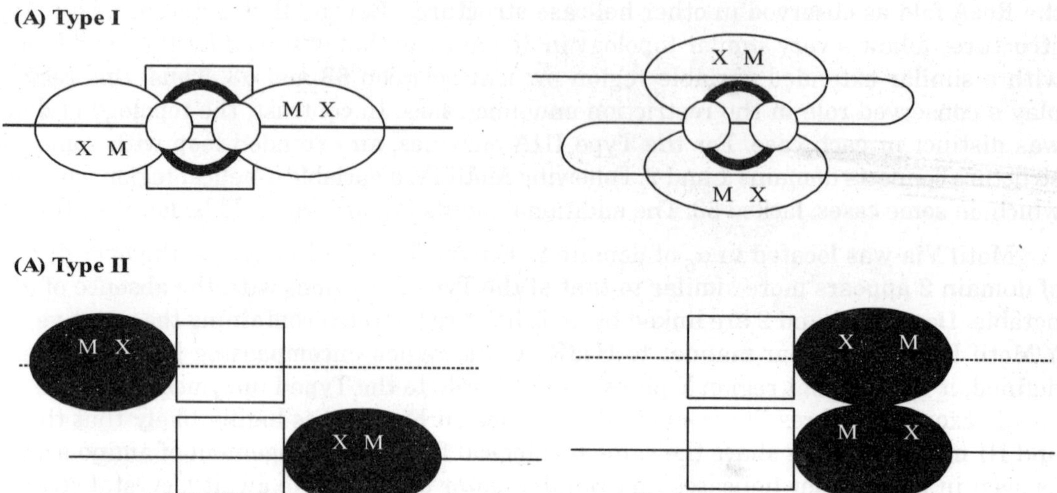

Fig. 10.3. Subunit assembly of the Type I and III restriction endonucleases. A Assembly of the Type I enzymes. The HsdS subunit is illustrated as *two grey circles* (TRDs) connected by a *black circular line.* The HsdM subunits are illustrated as *rectangles* filled with dots. The HsdR subunits are illustrated as *ellipses* with the motor *(M)* and region X *(X)* domains indicated. DNA is indicated by a *black line.* Two arrangements are shown: *left* the HsdR subunits are independent and bound at opposite ends of the complex; *right* although still bound at opposite ends of the DNA, the HsdR subunits are in contact with each other and can cooperate. This requires a wrapping of the DNA around the complex. B Assembly of the Type III enzymes. The Mod subunits are illustrated as *two grey rectan-gles.* The Res subunits are illustrated as *black ellipses* with the motor (M) and region X *(X)* domains indicated. DNA is indicated by a *black line.* Putative DNA contacts are indicated bya *dotted line.* Two arrangements are shown: *left* the Res subunits are independent and i bound at opposite ends of the complex. This might allow bi-directional translocation although this is less likely with Type III enzymes; *right* the HsdR subunits are cooperative. Either both Res subunits translocate one DNA, or the second Mod and Res may be involved in non-specific contacts to stabilise the complex

In this case the non-specific DNA may need to wrap around the complex to contact the motors. DNA distortion during binding may assist in this process. This alternative arrangement could give several advantages: Firstly, the expanding loop would be preformed. This may assist in the initiation of motion when under-twisting needs to be accommodated in the loop; Secondly, the HsdR subunits could directly communicate, which may explain why two subunits produce processive motion but one alone produces non-processive motion; and thirdly, cleavage of circular DNA with a single recognition site would be facilitated as the HsdR subunits could assist in producing a double strand break. In conflict with this model is recent evidence from translocation studies on circular DNA (S.E.McM. and M.D.S., unpubl data) which suggests that DNA twisting accumulates at much less than one turn per 10 bp, so loop topology may not be as limiting as previously thought. Furthermore, there is some evidence that a single HsdR can cut both strands so that cooperation is not required. To-date, there is no unambiguous experimental data that maps where the HsdR subunits start translocating DNA, although we do know that it cannot occur over a long distance via DNA looping. Nor is there any clear evidence as to whether processive translocation occurs as part of an HsdR dimer or is undertaken by a single HsdR. What is clear is that processive DNA translocation by this class of helicase motor occurs as part of a larger complex and therefore the role of the complex *as a whole* must not be ignored.

DNA sequence recognition by the Type III enzymes is undertaken by a single subunit, Mod, which contains motifs characteristic of adenine methyltransferases. Surprisingly, this occurs as a Mod_2 complex. Unlike the Type I complexes where two target adenines are methylated, only one base is targeted for methylation in Type III sites. Thus this arrangement appears to result in a spare Mod sub-unit. Similarly, when Res binds to form the endonuclease, two subunits attach. Two alternative arrangements are suggested. Given the stoichiometry of the complex, each Mod-Res unit could interact with a separate cognate DNA site. However, on a two-site substrate two Res_2Mod_2 complexes are required for maximum DNA cleavage (L.J. Peakman, unpubl. data). It is possible that non-specific DNA binding by the auxiliary Mod stabilises the complex.

A similar recognition subunit redundancy has been observed with the Type US enzyme BspMI, which is a tetramer that binds two asymmetric sites.The translocation model for Type III enzymes is based largely on that proposed for the Type I enzymes and it has yet to be proven unambiguously whether translocation is uni- or bi-directional, or as efficient as Type I motion. It is possible that each Res subunit in a Res_2Mod_2 complex could translocate DNA. Alternatively, the Type III enzymes could be examples of dimeric motors with the two Res subunits cooperating to translocate DNA unidirectionally. It has been long appreciated that the relative arrangement of the DNA motor and DNA cleavage domains in HsdR and Res determines whether the enzymes cut non-specific DNA at distant or proximal sites.

If both motors share the motor polarity common to SF2 enzymes and move along dsDNA using one strand in a 3′-5′ direction (*e.g.*, Whitehouse et al. 2003), then for the Type I enzymes Region X would be ahead of the motor, near to distant DNA, whilst for the Type III enzymes Region X would be behind the motor, near to the proximal DNA.

FUTURE DIRECTIONS

Coupling Chemical Energy to Mechanical Motion

The Type I and III restriction endonucleases contain motifs and structural elements typical of SF2 DNA helicases and, like many of the family (Delagoutte and von Hippel 2003), they are active as part of a larger macromolecular machine. Therefore, how does DNA translocation occur? What has become increasingly clear from helicase studies is that conserved motifs and structure do not define helicase activity per. Instead they define a mechanocoupling device that, upon binding and hydrolysing NTPs, can undergo domain motions and rearrangements which can then drive specific processes. For the helicases, the minimal motor fold produces translocation of a polynucleotide: In some cases, extra domains are involved in unwinding dsDNA such that the polynucleotide transported is ssDNA. In other cases it is more likely that the polynucleotide transported is intact dsDNA and that the extra domains play alternative roles. DNA translocation has been proven unambiguously for the Type I enzymes and there is convincing evidence for a similar overall mechanism for the Type III enzymes. However, in neither case has any evidence for strand separation been obtained. Therefore, it is more likely that, in common with many other SF2 enzymes, the Type I and III endonucleases are dsDNA translocase in which the motor domains are coupled to an endonuclease/protein-protein interaction domain(s). A model to demonstrate how DNA motion may occur is illustrated. This scheme is based around recent evidence from other SF2 helicases, in particular the RecG helicase.

Despite the mechanistic complexities of the Type I and III enzymes, it is likely that this simple model will be the underlying basis for DNA translocation in both systems. The model shows an HsdR subunit attached via HsdS to a Type I complex (a similar mechanism is likely to occur for a Res subunit in a Type III complex.) The HsdR subunit contacts non-specific DNA adjacent to the recognition site. Upon binding of ATP and Mg^{2+} ions into the cleft between

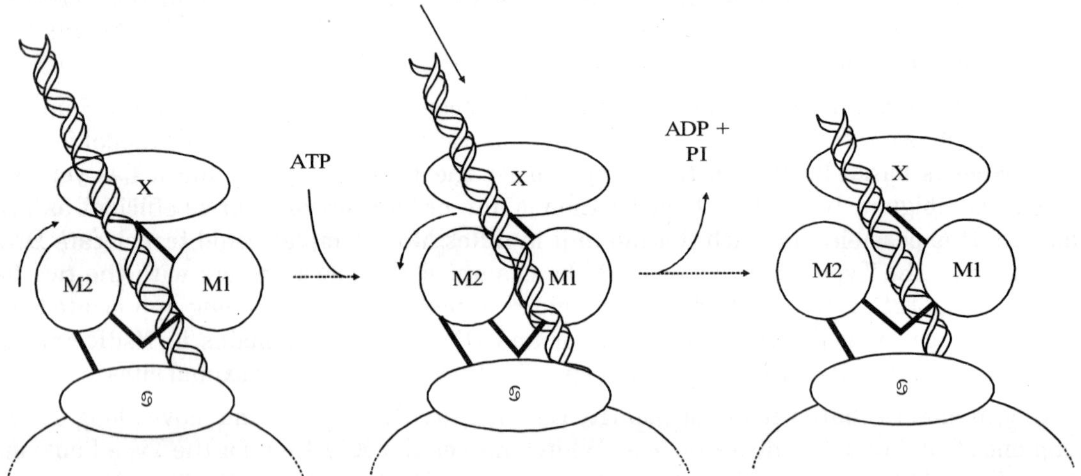

Fig. 10.4. Model for DNA translocation by a restriction endonuclease. A model is illustrated for Type I enzymes, but a similar model can be envisaged for Type III enzymes. Only one side of the bi-directional motor is shown for clarity. The HsdR subunit is illustrated as three domains with the connectivity. N-terminal region X *(X)*, a helicase domain *(M1* and *M2)*, and a C-terminal protein-protein domain (a). HsdS is shown in *grey* (the rest of the complex is omitted). Upon binding ATP, the motor domains 1 and 2, come together. This causes conformation rearrangements which transports the DNA past HsdR into the expanding loop. The hydrolysis of ATP and release of products causes the motor domains to relax and reset to a new position further along the DNA. An alternative view might be that the same domain motions occur but that the DNA is passed back-and-forth between two closely arranged HsdR subunits. The direction of the DNA has not been confirmed for SF2 enzymes, and may be more horizontally-placed across M1 and M2.

motor domains 1 and 2, the cleft closes and a conformational change occurs which transports the DNA from ahead of the complex into the expanding loop. Binding of S-adenosyl methionine, most likely by the methyltransferase domains, also produces conformational changes which stabilise the translocation state. To allow the system to repeat the mechanical transport of DNA, the motor domain must reset. This is achieved by hydrolysing the ATP, releasing ADP and inorganic phosphate and re-opening the cleft.

To prevent the DNA simply sliding back, it must be held in contact at all times. There are a number of putative loops and motifs in the HsdR and Res structures which could undertake the role of a 'molecular ratchet'; for instance Motif Y in the Type I and IIIB enzymes. However we should not discount the possibility that the HsdS, HsdM, or Mod subunits may also be involved in these processes. The model assumes a monomeric motor, but an alternative possibility is that the second HsdR or Res in the complex is also involved. In this case the DNA could be passed back-and-forth between the two motor subunits (the rolling model of helicase motion proposed by Wong andLohman 1992). This is a less likely scenario given the current data on the Type I enzymes (Bourniquel and Bickle 2002), but nonetheless it needs to be considered.

The DNA intact duplex, but contacts need not be made to both strands. Indeed, other SF2 helicases which translocate intact DNA do so using one strand with a defined polarity (*e.g.*, Whitehouse et al. 2003). Furthermore, there is growing evidence that the SF2 enzymes utilise sugar-phosphate backbone contacts rather than the bases and there is evidence from DNA crosslinking studies using psoralen that this is likely to be the case for the Type I enzymes. The step size, or amount of DNA transported per physical step, is not defined by our model. From studies of SFl helicases, this could be anything from 1-2 bp (Dillingham et al. 2000) to 3-5 bp (Ali and Lohman 1997) or even 23 bp (Bianco and Kowalczykowski 2000). Consequently, the coupling efficiency-the number of nucleotides translocated per NTP consumed-is also not defined. This has been estimated for SFl helicases as between 3-0.3 nucleotides/NTP. One complication of ascertaining the latter has been that the Type I enzymes have at least two ATP hydrolysis states; one during translocation and another that occurs after collision and DNA cleavage.

The size of the step and the coupling efficiency will, of course, have important implications for the mechanism in terms of how a domain attaches to *a forward* section of DNA and draws it past the enzyme. Nevertheless, the general features of the model likely to remain. In our Type I model, the domain carrying Region X is sited in front of the motor such that, following multiple steps, DNA cleavage will occur at a distant site. DNA cleavage during translocation may be prevented by: (1) Region X being unable to catalyse a phosphoryl transfer reaction on a 'moving' substrate (the DNA translocation rate being several orders of magnitude faster than the cleavage rate); (2) Region X being inactive until a second HsdR from a separate complex conies into contact after collision between two translocating enzymes; or, (3) Region X being disconnected from the DNA until collision between two translocating enzymes brings it back into contact.

In contrast, the Res subunits lack an N-terminal endonuclease domain. Instead Region X is located in the C-terminal domain which also makes protein-protein contacts with the rest of the complex and is thus behind the motor. In this situation, DNA cleavage would occur close to the recognition site. Similar mechanisms may apply to prevent promiscuous DNA cleavage by the Res subunits during translocation.

Tools for Nanotechnology Rather than Biotechnology?

As illuminated elsewhere in this volume, the molecular biology revolution was built on the precision and reliability of the Type II restriction enzymes. In turn their usefulness as laboratory tools has driven the search for new activities. At first sight the NTP-dependent enzymes may seem like poor relations—their complex activities and relatively imprecise DNA cleavage loci have found only limited use in the laboratory. On the other hand, the complexity of these reactions may find a wider role in the burgeoning field of nanotechnology. The goal of nanotechnology is to design microscopic apparatus that can carry out specific tasks. A framework for such technology is provided by biological materials such as proteins and DNA and theATP-dependent restriction enzymes maybe ideal candidates for exploitation. Given the modular nature of the motor domains illustrated can readily see how the motors could be adapted; for instance, by replacing Region X, HsdR subunits could carry an alternative enzyme activity or an attached cargo.

The Type I enzymes can translocate tens of thousands of bp without dissociation at speeds in excess of 450 bp/s. This corresponds to a motor which can travel >3.4 μm - a vast distance in nanomachine terms-in just over 20 s. They can also translocate DNA of any topology without loss of speed or processivity (unpubl. data) and can displace molecular roadblocks such as DNA-binding proteins. As we learn more about the biology of the ATP-dependent restriction endonucleases, in particular their motor mechanisms, we will also be well-placed to harness their potential as miniaturised machines.

EcoRV Endonuclease

The homodimeric, orthodox Type II restriction endonuclease EcoRV cleaves double-stranded DNA with high specificity at hexameric GATiATC sites in a blunt-ended fashion. Like most of the restriction enzymes used as tools in recombinant DNA technology it can be considered as a simple hydrolytic enzyme of relatively small size with precisely defined substrate specificity and the rathe r common requirement for Mg^{2+} ions. Yet, these apparently simple enzymes become fascinating objects for structural and functional studies, and their molecular simplicity becomes an advantage once we start asking detailed mechanistic questions. For example, it is quite obvious that these enzymes have been engineered by evolution to deal with the problem of efficiently locating a specific site on DNA, which is immersed in a huge molar excess of other, partly very similar sites. Initially, the exceptionally high and diverse sequence specificity of these enzymes was the major driving force for their study, and they were considered as model systems for gaining general insights into DNA recognition by proteins.

The large number of protein-DNA complexes analyzed to date has shown that there is no simple code to classify or predict such interactions. Nevertheless, structural studies with a number of restriction endonucleases, in particular with EcoRI and EcoRV, have become a rich source for analyzing the diversity and detailed stereochemistry of such interactions. To date, the crystal structures of 14 Type II restriction endonucleases have been determined in one or more states, unliganded or bound to double-stranded DNA fragments. Their characteristic functional and structural features have been discussed in several reviews. In the course of these studies, many new and more precisely defined questions have become apparent. They relate to the full functional cycle of these enzymes which involves binding to nonspecific DNA, searching for target sites, formation of the productive complex with Mg^{2+} ions at the target site, double-strand cleavage and product dissociation.

The chemical rate of phosphodiester hydrolysis has turned out to be very sensitive to DNA substrate-enzyme interactions remote from the active site. The understanding of the structural basis of this remarkable coupling between recognition and catalysis and of the details of the divalent ion catalyzed hydrolytic mechanism have become of particular interest. Considerable adaptability in tertiary and quaternary structure is a hallmark of EcoRV endonuclease and appears to be needed to proceed through the stages of this cycle. In the first part of this chapter, we will give an overview of the crystallo-graphic work on EcoRV endonuclease, and we will describe the salient structural features that have emerged. These rather comprehensive and detailed studies have been aided greatly by the readiness of this enzyme to crystallize in several different crystal forms. This is perhaps best illustrated by the fact that 18 structures from 7 different crystal forms have been analyzed and deposited. EcoRV is a telling example on how structural and func-tional studies, carried out largely in parallel, can inspire each other.

In the second part, the emphasis will be on the biochemical characteristics of this enzyme and on the mechanistic insights that have emerged from combining the structural and biochemical data. It illustrates the pitfalls of our usually simple mechanistic interpretations that tend to assign functional properties to single functional groups or amino acid residues and neglect the complexity of networks of interactions. While not all secrets have been revealed and some issues remain controversial, the mechanistic insights that have been gained certainly bear relevance beyond the specific features of EcoRV endonuclease.

Characteristics of EcoRV Endonuclease

The first three crystal structures of the EcoRV endonuclease, jointly published in 1993 (Winkler et al. 1993), were that of the free enzyme containing 244 residues per subunit, that of a complex with a cognate decamer DNA fragment and that with a nonspecific DNA fragment. The three structures represent significantly different structural states, quite obvious at the quaternary structure level. Many additional crystallographic studies have since be reported and are summarized in Table 11.1. They have addressed a variety of specific structural questions, such as the interactions with different DNA substrates, the structural changes in mutant enzymes and the binding sites of different divalent ions. Moreover, they have revealed further structural stat whose characteristics have been systematically analyzed and compared. The asymmetrically shaped EcoRV monomer has a mixed structure and shows no subdivision into distinct globular folding domains.

However, comparison of the monomer structure observed in the differed structural states suggests that if is useful to discriminate two domains which we will term dimerizatioiil domain (DIM-domain) and DNA binding domain (DB-domain), respective. These two subdomains are linked by four chain segments (L1-L4) that show a variable degree of disorder. In view of the multitude of observed relative orientations between the DIM and DB domain of one monomer thi variable order in these linking segments is not surprising. As the DNA lies in a deep cleft between the two subunits that form a U-shapedf homodimer. Two separate parts of the protein-protein dimer interface can be discriminated.

The first one, formed between the two very small DIM-domains at thet bottom of the U-shaped dimer, consists of a short and rigid four-stranded antiparallel β-sheet interaction and is supplemented by a number of hydrophobic side chain interactions. This dimeric substructure serves as a useful reference for superposing and analyzing the different quaternary structure states. The second component of the dimer interface is variable in its detailed interactions. It involves hydrophobic side chains from the N-terminal half of the ocB helix (Nt-aB) of each subunit and some residues of the DIM domain of each subunit. Analysis of aB in the different structures shows that Nt-aB bends away from the C-terminal portion to different degrees.

Table 11.1. Crystal Structure sof EcoRV Endonuclease

Structure-type	*Oligonucletodie*	*Bound ions*	*Mutation*	*Space group*	*Resolution (Berman et al. 2000)*	*PDB code*
—	—	—	—	—	—	—
Unliganded	—	—	wt	$P2_12_12_1$ (form A)	2.50	1RVE
Unliganded	—	—	wt	($P2_12_12_1$ (form B)	2.40	1AZ3
Unliganded —	—	—	T93A	$P2_12_12_1$ (form B)	2.40	1AZ4
—	—	—	—	—	—	—
Nonspecific	CGAGCTCG	—	wt	$P2_1$	3.0	2RVE
—	—	—	—	—	—	—
Specific (I)[a]	GGGATATCCC	—	wt	$C222_1$	3.0	4RVE
Specific (III)	AAAGATATCTT	—	wt	P1	2.0	1RVA
Specific (III)	AAAGATATCTT	Mg^{2+}	wt	P1	2.1	1RVB
Specific (III)	AAAGAT/Patctt	Mg^{2+}	wt	P1	2.1	1RVC
Specific (III)	AAAGATATCTT	Ca^{2+}	wt	P1	2.0	1AZ0
Specific (III) (1998a)	CGGGATATCCC	—	wt	P1	2.0	1BGB
Specific (III)	AAAGATATCTT	Ca^{2+}	T93A	P1	2.0	1BSS
Specific (III)	AAGA (5mC[b]) (I[c]) TCTT TTCT (I)(5mC)AGAAA	Ca^{2+}	wt	P1	2.0	1BSU
Specific (III)	AAAGAC(I)TCTT	Ca^{2+}	wt	P1	2.15	1BUA
Specific (III) (1998b)	AAAGAT/ATCTT	—	wt	P1	2.10	1RV5
Specific (III)	AAAGATATCTT	Ca^{2+}	wt	P1	1.90	1B94
Specific (III)	AAAGATATCTT	_	WT	P1	2.05	1B95
Specific (III)	AAAGATATCTT	—	Q69E	P1	2.30	1B96
Specific (III)	AAAGATATCTT	—	Q69L	P1	1.90	1B97
Specific (III)	CAAGA(Tsp[d])ATCTT	Mg^{2+}	wt	P1	2.00	1EO3
Specific (III)	CAAGA(Tsp) ATCTT	Mn^{2+}	wt	P1	1.90	1EO4
Specific (III)	CAAGA (Tsp) A TCTT TTCT A (Tsp) AGAAA	Ca^{2+}	wt	P1	1.60	1EON
Specific (II)	GAAGATATCTTC	—	wt	$C222_1$	2.16	1EOO
Specific (IV)	AAGATATCTTA	—	wt	$P4_12_12$	2.60	1EOP

a Numbers in brackets indicate the crystal forms of specific complexes according to (Horton and Peerona 2000)
b 5mC: 5-methyl-cytosine
c I: Inosine
d Tsp: 3'-thio-thvmidine

To first approximation, the EcoRV dimer can be described as consisting of three rigid parts, the dimeric substructure at the bottom of the U-shaped ilimer and the two DB-domains. The relative orientation and position of the latter two with respect to the former correlates with the functional state (unliganded, bound to specific or nonspecific DNA). The deviations from twofold symmetry are small for all complexes with specific DNA fragments but are substantial in the nonspecific complex and in the iftiiganded state. The DB domain contains all the catalytic residues and all structural elements interacting with DNA. It has three segments that are partly or completely disordered at least in some of the crystal structures.

Two of these segments, the short Q and R surface loops, are disordered in the unliganded enzyme but form well-defined interactions with cognate DNA in the minor and major groove, respectively. The third segment comprises the 25 C-terminal residues 221-245. In solution, this segment is possibly completely disordered in the unliganded state and becomes ordered only upon binding to DNA. Its partial ordering seen in the crystals of the unliganded enzyme may thus be induced by crystal contacts. Furthermore, the degree of order observed for this segment in the complexes with DNA depends on the length of the bound DNA fragment. The combination of flexible and rigid structural elements in EcoRV appears well suited to cope with the functional requirement to effectively screen nonspecific DNA for the target DNA sequence and cleave it with high specificity.

DNA-Protein Interactions in Specific and Nonspecific Complexes

The steps involved in DNA recognition and cleavage by restriction endonucleases require that these enzymes interact in various ways with DNA. At least two generic binding modes, specific and nonspecific, are usually discriminated. The latter is thought to be functionally important for initial binding to large DNA substrates and for efficient location of the target sequence through facilitated diffusion on DNA. Knowledge of at least these two binding modes is needed to understand the structural basis of the high sequence specificity of these enzymes. With sub-strate DNA, complexes are only stable in the absence of the essential cofactor Mg^{2+}. Such complexes with short cognate DNA fragments have been crystallized for 12 different restriction endonucleases.

The observed interactions with the diverse target sequences are extensive and tight in all these cases indicating that the absence of the metal cofactor does not interfere with the formation of the specific complex. It has been much more difficult to obtain crystals of nonspecific complexes, probably because many different nonspecific sites can be occupied at roughly equal stoichiometries even in short duplexes. Thus far, the only two examples are EcoRV (Winkler et al. 1993) and *BamHl*. Both complexes serve as good models for an enzyme conformation competent for facilitated diffusion but incompetent for cleavage.

The Specific DNA Binding Mode

The specific interaction of a double-stranded DNA sequence with a protein is usually divided into direct readout interactions with the bases in the major and minor groove and indirect readout interactions with the DNA backbone. In most cases, the direct amino acid-base interactions in the major groove are of primary importance for discrimination. The general aspects of such interactions have been analyzed using more than 100 protein-DNA complex structures. The direct protein-DNA hydrogen bonding interactions and selected, base-specific hydrophobic contacts that are observed in the specific complexes of EcoRV are summarized. One striking feature of the specific binding mode, first revealed in the complex with the decamer oligonudeotide GGGATATCCC (4RVE), is that the protein dimer tightly embraces the recognition part of the DNA fragment that is bound in a strongly kinked conformation.

The highly localized central bend of about 50° is accomplished by a positive roll into the major groove. A compa-rable bend angle of about 50° has also been derived from solution studies with longer DNA fragments. The bound DNA further deviates from canonical B-DNA in being unwound by about 45° between the central four base pairs and in having two sugar rings 5′ to the cleavage site in C3′-endo-like rather than C2′-endo-like conformation. The distorted DNA conformation makes the major groove narrower and deeper while the minor groove becomes shallower and wider. EcoRV binds the DNA from the minor groove side and contacts the bases in the major groove through the R-loops that reach into the groove from both sides.

The relative orientation of DNA and protein along the twofold symmetry axis appears to be one distinctive feature between the two main branches proposed on the basis of structural comparisons for the restriction endonuclease superfamily. An EcoRI-like and an EcoRV-like family had been proposed before, partly based on the functional distinction of producing 5′-overhangs and blunt ends respectively upon cleavage. The two R-loops, comprising residues 182-187 of each subunit, are responsible for all base-specific interactions and are only ordered in the specific complexes. Each R-loop forms six specific hydrogen bonds to the outer two base pairs of one half-site (GAT) of the hexameric recognition sequence. Three hydrogen bonds involve main-chain donor and acceptor groups, the other three small polar side chains (N185 and T186).

The formation of a highly cooperative hydrogen bond network is a characteristic feature of protein-DNA interactions in restriction endonucleases. However, their detailed nature and architecture is found to be highly variable. The central two base pairs are not read out by hydrogen bonds to the protein, yet they are also shielded from solvent access by the tips of the two recognition loops. The domed arrangement of these two base pairs, resulting from the large positive roll at this step, is such that the N6 nitrogen of each adenine can form two hydrogen bonds, one to its Watson-Crick base-paired thymine, the other to the O4 oxygen of the intrastrand thymine on its 5′ side. Although this hydrogen-bonding arrangement does not appear optimal, it certainly helps to reduce the free energy cost of desolvation.

Deformations of bound DNA from the canonical B-form are observed in many of the specific complexes of restriction endonucleases, yet EcoRV represents the most extreme case known. Likewise, the absence of direct major groove readout for two of the six recognition base pairs is unique. The importance of individual major groove interactions has been probed extensively using both mutant enzymes (Thielking et al. 1991; Vermote et al. 1992) and oligonucleotide substrates containing star sites (differing in only one base pair from the cognate sequence) (Alves et al. 1995) or modified bases.

As will be discussed later, no or strongly reduced catalytic activity is observed for nearly all modifications that directly perturb the recognition interface. Like the R-loop, the Q-loop (residues 68-71) is disordered in the unliganded enzyme and becomes better ordered upon DNA binding. It inserts into the minor groove at about the second base pair of one recognition half-site and makes van der Waals contacts with the backbone of both DNA strands. Some conformational flexibility of this loop is observed even in the specific complexes. In particular, the conformation of the turn appears to easily switch between Type I and Type II in response to small perturbations. The side chain of Q69 extends across the minor groove and interacts with T37 of the opposing subunit in different ways.

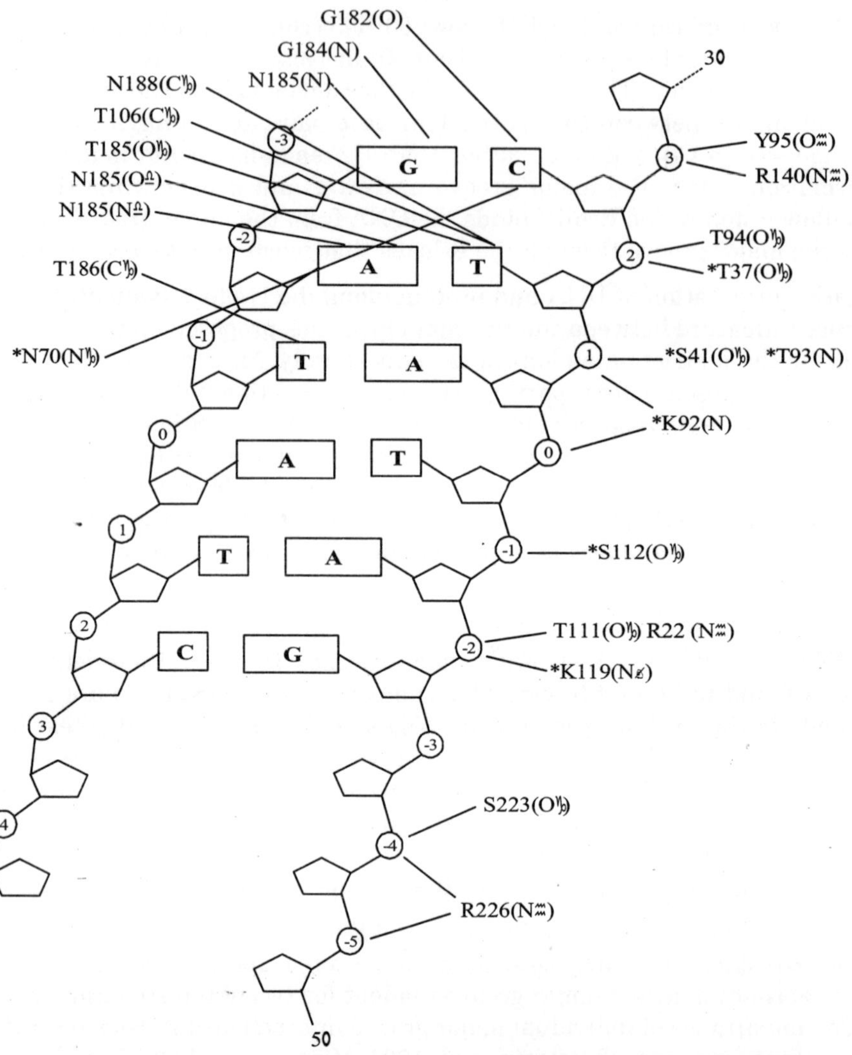

Fig. 11.1. Schematic representation of the direct protein-DNA interactions observed in the specific complexes (hydrogen bonds, salt bridges, selected hydrophobic contacts). The interactions with the DNA bases are shown in *bold* for one recognition half-site (GAT), and those with the DNA backbone in *italics* for one of the two strands. Residues of the two subunits A and B are in blue and red, respectively. Hydrophobic interactions are only shown for the two thymine 5-methyl groups. One R-loop reads the outer two base pairs of one half-site and contacts the 5-methyl group of the inner thymine of the other half-site. N70 forms the only direct hydrogen bond to a base in the minor groove. The interactions of residues marked with an asterisk (*) are not conserved in all structures. The scissile phosphodiester group at the TA step is marked as and *numbers* are increasing in the 5' to 3' direction

The side chain of N70 is observed in two conformations that correlate with the Q-loop turn conformation. In the Type I turn conformation, N70 forms a hydrogen bond to the recognition site cytosine in the minor groove. In the central part of the widened minor groove there are no direct hydrogen bonds between the bases and the protein. A number of water molecules and the side chains of the two symmetry related K38 residues occupy this loosely packed part of the protein-DNA interface.

The side chain of T37 forms a hydrogen bond to P_{+2} from the minor groove side in all high resolution structures. As discussed later, this hydrogen bond is not observed in the decamer complex where the methyl group of T37 packs tightly against the deoxyribose moiety on the 5′ side of P_{+2}.

A third chain segment, which becomes more ordered in the specific complexes with undecamer and dodecamer DNA fragments, comprises residues 221-229. Interactions with the DNA backbone are made by residues R221, S223, and R226. R226 forms salt bridges with the P_{-4} and P_{-5} phosphodiester groups. The absence of P_{-5} in the shorter decamer duplex explains why such an ordering is not observed in this complex. Further important contacts to the sugar phosphate backbone are made by a number of residues belonging to the rigid part of the DB-domain.

Most importantly, the chain segment 90-95 of β-strand pE runs almost parallel to, the DNA backbone and contacts four phosphodiester groups (P_0 to P_{+3},). On the other side of the scissile phosphate, Till and SI 12 make direct contacts with P_{-2} and P_{-1} respectively. The hydrogen bonding and salt bridge contacts are only one aspect of the DNA-backbone-protein interface. On top of the direct interactions there is a network of hydrogen bonds mediated by highly ordered water molecules. In the case of $P_{-2}/P_{-1}/P_0$ this network interconnects the three phosphates and it joins them to residues of the recognition loop, to catalytic residues, and bound metal ions. The protein-DNA-backbone contacts have been probed by site-directed mutagenesis (Wenz et al. 1996), by using dodecamer oligonucleotides substituted by either an R_p- or S_p-phosphorothioate group from P_{-4} to P_{+4} (Thorogood et al. 1996a, b) and by ethylation interference footprinting (Engler et al. 1997).

Overall, the results confirm that important interactions occur with the six phosphodiester groups from P_{+2} P_{+3}. Of the mutant enzymes probing these contacts (Wenz et al. 1996), the T37A mutation shows the most dramatic effect with a rate reduction of nearly 10^3. Kl 19, R140, and R226, which all make salt bridges to DNA phosphates, reduce binding but do not affect the steady-state turnover rate (k_{cat}). In summary, the majority of direct protein-DNA interactions in the specific complexes are provided by four chain segments of the protein, three of which undergo order-disorder transitions upon binding. The detailed analysis of this interface with a contact area of 2200-2400 A^2 (Winkler et al. 1993; Pingoud and Jeltsch 1997) reveals that about 55% are polar contacts and that each protein monomer interacts more extensively with the strand it cleaves (70%).

The Nonspecific DNA Binding Mode

The structure of EcoRV complexed with two duplexes of the self-complementary octamer CGAGCTCG, stacked end-to-end at the twofold axis, can be considered to first approximation as a complex with a 16 base pair duplex with missing central phosphodiester linkages and has been assumed to be repre-sentative for nonspecific binding. The pseudohexadecamer has no base pair in common with the recognition sequence (5′-TCG/CGA-3′ as compared to GATATC). The helical axes of the two octamer duplexes are dislocated at the center by about 4 A and there are significant deviations in this complex from twofold symmetry. This and the noncontinuous nature of the phosphodiester backbone between abutting octamers has raised some doubts on the validity of the above assumption.

However, several other observations support it: (1) the undistorted B-like DNA structure makes sense, (2) many of the protein-sugar phosphate back-bone contacts are similar to those in the specific complex and would permit a smooth transition to the specific binding mode (3) a model with continuous B-DNA bound in a symmetric fashion can be generated without stereochemical problems (Winkler et al. 1993) and (4) missing scissile phosphates do not disrupt the structure of the specific complex either. Perhaps most importantly, the structure yields a very plausible explanation why no cleavage can occur in this binding mode. The well-defined contacts between the protein and the sugar-phosphate backbone on the 5′ side of the missing scissile phosphate suggest that these contacts would be similarly formed with a continuous strand. Modeling the 3′ extension of a strand fixed in this way shows that the scissile phosphodiester group is located at least 3 A away from that observed in the cognate complex.

The nonspecific complex reported for BamHI the closest phosphate more than 6 A displaced from the scissile position in the specific complex. Here, the structural change needed to go from nonspecific to specific binding appears more dramatic than for EcoRV and involves primarily an adaptation of the protein conformation. The protein-DNA interface in the nonspecific complex is about 800 A^2 smaller than in the specific complex and the interface is much less tightly packed as indicated by a high gap volume index. One might therefore expect nonspecific binding to be considerably weaker than specific binding. Despite a number of experimental studies on this question this has remained a controversial issue as will be discussed later.

THE STRUCTURE OF THE ACTIVE SITE

The Structure in the Absence of Divalent Cations

Sequence comparison and preliminary structural results had indicated that EcoRV shares an active site motif with EcoRI, which is characterized by PD...(D/E)XK. The two acidic side chains of D74 and D90, belonging to this motif, are close to the scissile phosphate, while that of another acidic residue E45, not part of this conserved motif, is somewhat more remote. The clustering of the conserved acidic residues near the scissile phosphate suggested them to be involved in binding the essential Mg^{2+} ion. The amino function of the conserved basic residue of this motif (K92) is positioned between P_0 and P_{+1} and is assumed to be protonated. The conformation of the two acidic side chains is very similar in all structures in the absence of divalent cations because each forms a conserved hydrogen bond to a main chain N-H. In contrast, the side chains of K92 and E45 show conformational flexibility.

In the different specific complexes, the position of the DNA backbone relative to that of the rigid part of the protein active site varies up to 1 A. This explains why the hydrogen bonding network that involves the scissile phosphate, the nearby acidic and basic groups and bridging water molecules also shows variation. For example, K92 is found within hydrogen bonding distance of either P_0 or P_{+1}, or in between in some cases. The adaptability of this network suggests that rearrangements upon metal binding or during catalysis can proceed rather smoothly. It appears possible that D90 is protonated in these structures as all were crystallized at slightly acidic pH (6.5-7.0). In most structures, three water molecules are observed within hydrogen bond distance of the nonbridging oxygens of thescissile phosphate. Two of them connect to the side chain of N188 at the C-terminal end of the recognition loop which represents the most direct hydro-gen-bond-mediated contact between the R-loop and the scissile phosphate.

The Location of Bound Divalent Metal Ions

Three distinct divalent metal ion sites involving the active site residues have been observed in the crystallographic studies. Soaking of the triclinic crystals with the catalytically active cations Mg^{2+}, Mn^{2+} or Co^{2+} does lead to metal binding in the active site without cleavage. Two binding sites were identified and, on the basis of the ligating acidic residues, they were characterized as the 90/74 and the 74/45 site. Mg^{2+} and Mn^{24} ions, at 30 mM concentrations, only bind in one of the two active sites of the noncrystallographic dimer and the DNA backbone at this site shifts by about 1 A in the 5′ to 3′ direction. With Ca^{2+}, both active sites bind one metal at the 90/74 site and no DNA shift is observed. However, in complexes of the mutant enzyme T93A with the same DNA fragment, two different Ca^{2+} binding sites have been observed.

One is very similar to the 74/45 site and a new one, labeled site I, is ligated by D74, by the main-chain carbonyl oxygen of 191 and by three water molecules. It is located more towards the 3′-side of the scissile group and has been implicated in catalysis. Two similar sites are occupied in complexes with the refractory phosphorothiolate 5′-AAAGATsATCTT. Soaking studies with crystals of the complex with AAAGAUATCC, which has deoxyuridine in place of the first thymine, showed predominantly site I occupied in one subunit but sites II (the 74/45 site) and III (the 90/74 site) in the other (Winkler and Kostrewa, unpubl.). All these observations indicate that there are at least three distinct divalent cation binding sites in the active site region, and that small perturbations introduced by a mutation, by a modified substrate or by crystal packing constraints can significantly shift their occupancies. In no case, however, more than two sites have been seen occupied at the same time.

The Structure of the Enzyme Product Complex

With large DNA substrate molecules, product dissociation from EcoRV endonuclease has been found to be some 200-fold slower than cleavage, while comparable rates for the two steps are observed with small oligonucleotide substrates. It was not a priori obvious whether the DNA binding mode in a product complex would be specific-like *or* nonspecific-like. From a concentrated reaction mixture that contained enzyme, undecamer fragment 5'-AAAGATATCTT-3' and $MgCl_2$, crystals isomorphous to the substrate complex (1RVA) were obtained. The product DNA is bound in a very similar fashion as the substrate, and the structural adaptations are confined to the vicinity of the cleavage site although the 5′-phosphate is nearly 5 A away from its previous position.

Two metal ions are bound and directly coordinated by oxygens of the doubly charged 5′-phosphate group. The site involving the main-chain carbonyl oxygen of Q69 requires a Type I turn of the Q-loop. The location of both of these metal ions is different from that observed with uncleaved DNA. The other interesting structural change is a reorientation of the two central adenine bases provoked by the dislocation of their 5′ sugar-phosphate moieties upon cleavage. The two bases become involved in a cross-strand stacking at the twofold symmetry axis and the corresponding two A-T base pairs become strongly buckled. A very similar, product-like complex that lacks the 5′-phosphate of the cleaved pentamer fragment, has been crystallized in the same crystal form.

The DNA structure at the active site is intermediate between substrate and product complex but is much closer to that with substrate. The fact that the same specific binding mode is observed in this noncontinuous, 5′-phosphate and metal-free complex suggests that metal binding cannot be critical for stabilizing the product complex in this binding mode.

Biochemical Characteristics of *EcoRV* Endonuclease and Structure-Function Relationships

EcoRV endonuclease is probably the most thoroughly studied of the Type II restriction endonucleases. Many of the thermodynamic and kinetic studies employing modified substrates and mutant enzymes were carried out after the first crystal structures had become available, and there has been a very fruitful and inspiring interaction between the structural and functional investigations. In the presence of Mg^{2+}, Type II restriction endonucleases cleave DNA at their respective recognition sequence with extremely high specificities. DNA sequences differing from the recognition site by just one base pair are typically cleaved over 1 million times more slowly. This is much more than one would expect from the loss of two to three hydrogen bonding interactions with a single base pair in a substrate enzyme complex.

Restriction endonucleases exercise sub-strate specificity both in the DNA binding step (K_D) and in the subsequent chemical rate of hydrolysis (k_c). This has first been systematically investigated for EcoRI endonuclease with cognate and noncognate substrates, and both steps were found to contribute similarly to total discrimination quantified as the k_c/K_D value relative to that for a cognate substrate. Studies with a number of other restriction endonucleases have shown that total discrimination is similarly large, but the relative contribution of binding and catalysis can vary substantially. To understand the structural and energetic basis of this high discrimination one would ideally want to compare the structures of the corresponding enzyme-substrate-cofactor complexes at the transition state.

As this is experimentally impossible, the question is to what extent the observed ground state complexes can serve as substitutes or as starting points for extrapolation. At least for good substrates, it appears reasonable to assume that most of the interactions observed in the large interfaces of the ground state complexes are maintained at the transition state. The observed and thermodynamically characterized specific complexes are thus likely to be on the path to the transition state. However, we must keep in mind that the ground state structures can never reveal all the interactions of the transition state and that small changes in the details of critical attractive or repulsive interactions may have profound effects on the energetics.

Thermodynamics and Energetics of DNA Binding

In general, we must be happy if we can account in a qualitative or semi-quantitative way for larger differences in the binding constants of different sub-strates, except perhaps in cases where the changes in interactions are highly localized. Apart from the intrinsic difficulties in estimating free energy differences between different structural states, the observed structures (mostly from crystal structure analysis) are often not strictly comparable to the systems or conditions for which the thermodynamic data have been measured.

Specific and Nonspecific Binding to DNA

The determination of the specific and nonspecific binding constants for EcoRV has led to conflicting results in the absence of Mg^{2+}. Mainly three experimental techniques have been used for determining thermodynamic binding constants of restriction endonuclease-DNA complexes: (1) gel mobility shift, (2) filter binding, and (3) steady-state fluorescence anisotropy. The first systematic studies using the gel shift method led to the conclusion that this enzyme binds to all sequences with essentially equal affinity of about 10^6 M^{-1}. This surprising finding suggested that the energy difference between the two structurally characterized, distinct binding modes is very small. It was also in strong contrast to the situation with EcoRI endonuclease, which had been found to discriminate 10^3 to 10^4-fold in binding against sequences with one incorrect base pair. EcoRV was suggested to exercise sequence specificity solely in the catalytic step in the presence of Mg^{2+} and thus in a fundamentally different way than EcoRI. This conclusion was later challenged based on filter binding and gel retardation experiments carried out with shorter DNA fragments (22 base pairs) under a range of pH and salt conditions.

Specific binding was found to be strongly pH-dependent but even at pH 7.5 the selectivity value (ratio of specific to nonspecific binding constant) was postulated to be as high as 120.As neither gel retardation nor filter binding techniques are done under true equilibrium conditions the issue was reexamined by binding studies at pH 7.5 in solution using fluorescently labeled oligonucleotides for FRET (fluorescence resonant energy transfer) and FD (fluorescence depolarization) measure-1 ments. The selectivity value found in these experiments was below 10 with little variation between different oligonucleotides, apparently confirming the previous reports on no or only weak discrimination of EcoRV in DNA binding in the absence of cations. Similarly low specificity values were also observed in a later study using the gel retardation method modified in a way to minimize protein-DNA dissociation during the experiment.

The majority of studies and particularly those less prone to experimental artifacts show only weak selectivity at pH 7.5. Its increase at lower pH, which is not controversial, appears structurally plausible in view of the energetically unfavorable clustering of negative charges at the active site that would be reduced by protonation of one of the acidic residues. EcoRV and some other restriction endonucleases which show a similar behavior may thus be considered as examples with a larger unfavorable accumulation of negative charge in the specific binding mode rather than representing a fundamentally distinct mechanistic class. In accordance with such a view, EcoRI mutants like K130A, K130Eor R131E make the specific binding of this enzyme Ca^{2+} dependent. From a structural point of view, it is against our intuition that the specific and nonspecific binding mode of EcoRV should hardly differ in binding energy (in the absence of divalent cations).

A large interface with many tight interactions showing almost ideal stereochemistry stands against a much smaller and in detail much looser interface. A number of factors may help to explain this apparent discrepancy although their absolute or relative contribution is difficult to estimate. One important such factor is entropic in nature and is usually termed configurational entropy. In our case, it concerns the ordering of several structural elements associated with the specific binding mode, in particular the R-loop, residues from the C-terminus and to a lesser extent the Q-loop. In addition, the looser interface of the nonspecific complex is much less restrictive for the local and global mobility of both DNA and protein, which further increases the relative configurational entropy cost of the specific complex. In contrast, the favorable entropic contribution to binding through the release of water molecules from nonpolar surfaces should be considerably larger for the specific complex.

There are two enthalpic contributions that may destabilize the specific complex and can thus be considered as strain. One is the abundance of negatively charged groups near the scissile phosphate that, as discussed before, makes specific binding strongly pH and metal cation dependent. The other is the deformation of the DNA with an almost complete unstacking of the central two base pairs which may cost 5 kcal/mol or more. A detailed analysis of enthalpic and entropic factors contributing to site-specific DNA binding based on data for 10 different proteins, but not including EcoRV, has been presented by Jen-Jacobson et al. It shows that isothermal enthalpy-entropy compensation is important in these systems.

A direct assessment of the effect of Mg^{2+} on binding is not possible with normal DNA substrates because of rapid cleavage. A number of studies have addressed this question indirectly through investigating catalytically inactive mutants, using Ca^{2+}, which does not support catalysis, or using poor or noncleavable substrate analogues. At pH 7.5 large increases in binding affinity up to 10^4-fold are observed with Ca^{2+} for DNA fragments containing a cognate site.

Smaller increases are seen when the cognate site is substituted by cleavable sites containing modified bases (for example GAUATC) or when it contains base analogues conferring resistance to hydrolysis. Only marginal increases are observed with nonspecific DNA fragments. These results are consistent with the view that optimal binding of Ca^{2+} takes place at cognate sites and that this binding is perturbed directly (impaired metal coordination) or indirectly (impaired protein-DNA interface) with modified sub-strates. At noncognate sites, differing in just one base pair from the cognate site, the specific binding mode is already so strongly disfavored that Ca^{2+} binding cannot overcome this energy penalty. For some analogues that contain very slowly cleaved sites and for hydrolysis resistant analogues gel shift binding assays in the presence of Mg^{2+} were possible. For the former, the binding constants measured with Mg^2+ were nearly as strong as with Ca^{2+} suggesting that Ca^{2+} can be a useful substitute.

However, Mg^{2+} hardly improved the binding affinity for the noncleavable, sulfur-modified substrates. There are clearly metal specific effects whose magnitude depends on the particular nature of a perturbation and how it affects the formation of the metal binding sites. The structural interpretation of the binding experiments carried out with Mg^{2+} and Ca^{2+} ions is complicated. Both the occupancy and the detailed coordination of the binding sites may differ depending on the metal ion used and on whether a mutant enzyme or modified substrate is used. In wild-type enzyme-cognate DNA complexes, site III has been seen occupied with each of the metals Mg^{2+}, Mn^{2+}, Co^{2+}, and Ca^{2+}, and crystallization in the presence of Ca^{2+} has yielded the same binding site as soak-ing experiments.

Both site III and site 1 require the kinked DNA conformation for their formation and do not exist in the non-specific binding mode. Most likely, site III is primarily responsible for the increase in specific binding seen in the presence of divalent cations. With the soft metal ions Mn^{2+} and Co^{2+} the situation maybe even more complex due to the presence of additional binding sites distal from the active site.

The Energetics of Direct and Indirect Readout Interactions

The detailed analysis of the energetic basis of specificity in restriction endonucleases has been pioneered for EcoRl endonuclease by the Jen-Jacobson group. EcoRl shows strong specific binding in the absence of divalent metal ions. This has permitted to accurately assess the energy penalty of perturbing selected interactions on binding and catalysis by using suitable base analogs. Most single or double hydrogen bond deletions did hardly reduce the cleavage rate and thus have about equal effects on the ground state and transition state complex. In contrast, large reductions in both binding and cleavage rate occurred with noncognate DNA that differed in one base pair.

An altered conformation of the ground state complex was indicated from phosphate ethylation interference experiments in these cases.

For EcoRV, early studies of the steady-state kinetics using self-complementary dodecamer oligonucleotides that contain modified bases within the recognition sequence revealed large reductions in cleavage rates but not much change in apparent affinity (K_m). All nine major groove hydrogen bonding functions were probed with suitable base analogs. Rate reductions up to one hundred fold were also observed for some analogs carrying modifications in the minor groove. Possible explanations include the perturbation of water mediated minor groove contacts but these are all rather speculative. More recently, the major groove contacts of the outer two base pairs have been reinvestigated in a much more thorough way. Dissociation constants (K_D) in the presence of Ca^{2+} and single turnover rate constants (k_c) using Mg^{2+} or Mn^{2+} were determined for heteroduplexes and homoduplexes formed from oligonucleotides of length 19 and 22 carrying a base modifica-tions in one or both DNA strands.

The results are completely consistent with the previous findings but reveal new and more quantitative aspects. With the A:T base pair, binding of DNA with a single modification was reduced by 0.9-1.7 kcal/mol as would be expected for the simple loss of a single contact About twice the effect is measured with DNA modified in both strands and a similar behavior is observed for the reductions in the catalytic rates. With the G:C base pair, the binding penalty is about 2 kcal/mol for the single site $d^{4H}C$ modification while it amounts to 3.5-4 kcal/mol for the single dG modifications. In contrast to the A:T base pair, binding is similarly reduced with DNA carrying a G:C base pair modification in both strands. The $d^{4H}C$ modification is unique in having no effect on the catalytic rate. Thus, a single perturbation in one of the two G:C base pairs appears to cause symmetric structural adaptations, but only for the dG modifications this structural change propagates to the active site. A unique aspect of specific recognition by EcoRV is the lack of direct hydrogen bond-mediated readout of the two central base pairs.

Discrimination at these base pairs is as high as for the outer base pairs, as confirmed with an oligonucleotide substrate containing all possible single base pair substitutions (Alves et al. 1995). Cleavage rates were also determined for all possible mismatch substrates. Very heterogeneous double strand cleavage rate reductions ranging from 40 to several million-fold were observed. The rates for first and second strand cleavage of a given such substrate varied up to 1000-fold. This demonstrates that the relaxation of strain is dependent on the detailed structural context and may introduce significant asymmetry. The structural and energetic origin of the readout at the central TA step has been investigated in great detail via functional analysis of a series of symmetric base analogs.

Four modified versions of the unde-camer AAAGATATCTT with the central TA step replaced by UA, MI, CI and CG were investigated. The TA-UA and MI-CI pairs probe the loss of the only direct readout interaction, the hydrophobic contact between the methyl groups of the thy mine base and the T186 side chain, in two different contexts. The relative changes in k_c/k_d transform into very similar AAG values of 4.7 and 4.9 kcal/mol but they are quite differently partitioned into K_d and k_c. The crystal structure of the MI complex is essentially identical to that of the cognate complex. A number of distinct structural differences are, however, observed for the CI complex (1BUA) although its binding is only marginally weaker. The small but numerous structural adaptations of the CI analog suggest that considerable enthalpy entropy compensation may be involved. However, the structural relaxation in the ground state is not on the path to the transition state and the corresponding gain in binding energy must be paid back during catalysis.

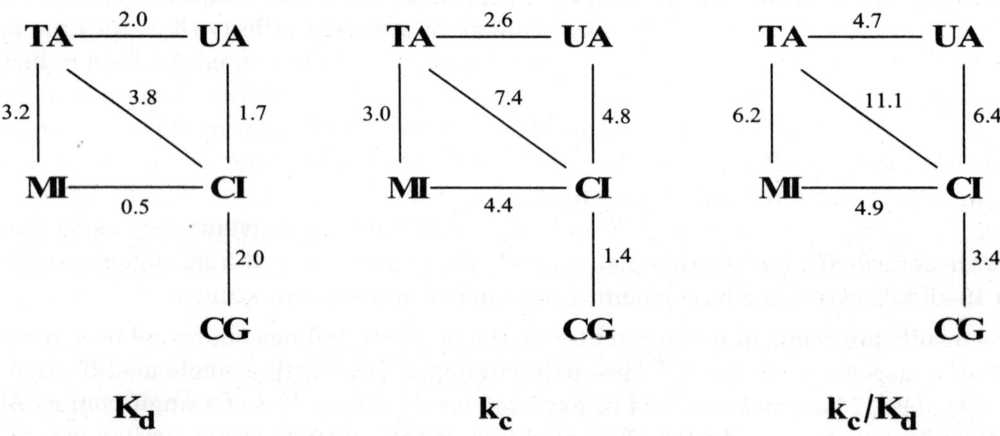

Fig. 11.2. Free energy component analysis for DNA substrates with a central TA, UA, MI, CI and CG step. (U=deoxyuracil, M=5-methylcytosine, I=inosine a gua-nine-like base lacking the minor groove 2-amino group). Thermodynamic dissociation constants (K_d) were determined in the presence of Ca^{2+} by gel shift analysis, and single-turnover reactions were used to determine the chemical rate constants (k_c) for phospho-diester hydrolysis. The ratios of these values have been converted to free energy differences AAG given in kcal/mol.

Most of the total discrimination of 6.2 kcal/mol of the MI analog must be due to indirect readout as its interactions with the protein are essentially identical to those with cognate DNA. It is well known from theoretical and thermodynamic studies that TA steps are among the least stable, and the differential energetic cost for base pair unstacking has been suspected to be an important factor in the indirect readout of EcoRY. As the central base pair becomes essentially unstacked but not solvated in the complex with EcoRV the energetic cost must be mostly enthalpic. The difference in binding energy between TA and MI (3.2 kcal/mol) should include all of this differential unstacking cost as a 50° bend amounts to the full loss of stacking energy. Stacking enthalpies differ up to 5 kcal/mol for the natural base pairs and are highest for CG and GC steps.

Differential unstacking cost could thus fully account for the reduced binding of the MI analog. However, there also is the possibility of a differential energy cost for major groove desolvation at the central two base pairs. We have already discussed the intrastrand hydrogen bonding/electrostatic contacts between the exocyclic 4-oxo (dT) and 6-amino (dA) groups at the TA bend which may partly compensate for the loss of major groove desolvation. In a CG step and likewise in the MI and CI analogs the positions of the exocyclic atoms are preserved within experimental error, but the amino and oxo groups are interchanged. The differential desolvation cost may thus be small in this case. However, for steps like AT or CT with unstacking enthalpies as small as that of TA other contributions to this indirect readout must exist to explain the similarly large total discrimination.

A differential cost of major groove desolvation may therefore contribute significantly to indirect readout for noncognate base steps where a favorable intrastrand donor/acceptor arrangement is not possible. The additional discrimination seen in the catalytic step cannot be assigned to differential unstacking energies. As indicated by the analysis of the different specific complexes the central DNA bend and the resulting compression in the major groove may need to be further increased towards the transition state. The anticipated structural changes in the recognition interface are small and nonbonded interactions probably change by at most a few tenths of one A. Discrimination may therefore be the result of few critical interactions that become more strongly repulsive at noncognate sequences.

General Aspects of Phosphodiester Hydrolysis

Like all restriction endonucleases, EcoRV cleaves DNA yielding 3′-OH and 5′-phosphate ends and the reaction can be considered as a phosphoryl transfer to water. Hydrolysis occurs with inversion of configuration at the phosphorus atom, suggesting an attack of a water molecule in line with the cleaved P-3′O bond. Two principal mechanisms, associative or dissociative, represent two extremes of possible intermediate mechanisms. Despite the lack of experimental evidence, a more associative-type mechanism is usually assumed to be operative in DNA cleavage by restriction endonucleases. This mechanism requires (1) a general base to generate the active hydroxide nucleophile for attack at the scissile phosphodiester group, (2) a Lewis acid stabilizing the extra negative charge at the pentacoordinated phosphorane transition state and (3) a general acid that stabilizes or protonates the alcoholate leaving group.

No detailed mechanism of DNA cleavage has thus far been elucidated for any restriction endonucleases but it appears that there is considerable variation in detail despite striking similarities in the active site residues and architecture. In the case of EcoRV, different hypotheses that will be discussed in section 3.3 have been developed and experimentally probed using mutant enzymes and modified substrates.

Steady State and Rapid Reaction Kinetics

Steady-state kinetics for EcoRV cleavage at a single site on a plasmid yields a k_{cat} of 0.015 s^{-1} (Halford and Goodall 1988) and a K_m of 0.5 nM. With small oligonucleotides k_{cat} is found to be about 50-fold faster (0.7 s^{-1}) and K_m values are typically 0.1-1 μM. As steady-state kinetics provides little information about intermediates in reaction pathways, rapid reaction analysis has been carried on a small 12 bp duplex and on plasmid DNA that contains a single site using stopped-flow fluorescence and quench-flow methods under single turnover reaction conditions. With both substrates, similar rate constants of 2-3 S^{-1} for the DNA cleavage step are measured and they appear to be determined by the chemical reactions (k_c). For the 12-bp duplex this rate is within an order of magnitude of k_{cat} which is thus partly determined by the cleavage step and partly by product dissociation. For the plasmid substrate, the open-circle nicked form is transiently formed and the rate constants for cleavage of the two strands are very similar.

The ratelimiting step for the complete catalytic cycle occurs after cleavage and must be slow product release as proposed earlier. If product release rather than product formation is rate limiting, the K_D of the enzyme-substrate complex is expected to be larger than K_m. In agreement with this, determination of associa-tion and dissociation rate constants yields a K_D of 10 nM for the plasmid substrate, 20-fold higher than the K_m. Intramolecular transfer from nonspecific sites to the recognition site, known to occur from competition experiments, was too fast to be measurable in these experiments. The mechanism of this transfer has been assumed to be linear diffusion by sliding on DNA, based on the observed cleavage rate enhancement with increasing length of the DNA as compared to a small reference substrate. This view has been challenged based on the measurement of processive cleavage at two sites separated by varied distances and using higher DNA/enzyme ratios to avoid having more than one enzyme bound to one substrate molecule.

A model that combines multiple dissociation/reassociations through three-dimensional space with one-dimensional diffusion over relatively short distances (~25 base pairs) has been proposed instead. This model explains the slower release from product with large DNA substrate equally well because dissociation/reassociation events can continue for a relatively long time before the protein eventually escapes the product DNA molecule that contains no more cleavage sites. Clearly, the two models are mechanistically quite different as linear diffusion proceeds without breaking all interactions. The about 200-fold difference between the rates of product release and product formation with plasmid DNA means that steady state kinetics is inappropriate to assess mechanistic effects of mutants. For example, a 2000-fold reduction of the cleavage rate can result in just a ten-fold reduction of k_{cat}. Such a mutant, Q69L, has indeed been found. Different divalent metal ions show differential effects on the kinetics and specificity of the EcoRV catalyzed reaction.

The relative activity (k_{cat}) using a plasmid DNA substrate has been found to vary in the following order Mg^{2+}>Co^{2+}>Mn^{2+}>>Cd^{2+}>Zn^{2+}>Ni^{2+}>>Ca^{2+}. With Ca^{2+} there is virtually no reaction and the reaction with Mg^{2+} is inhibited in its presence. Using a 12-bp oligonucleotide substrate, steady-state experiments showed that Co^{2+} has a very similar turnover rate (k_{cat}) as Mg^{2+} while that of Mn^{2+} is about tenfold reduced. The reaction with these three metal ions was also analyzed under single turnover conditions after rapid mixing. While Mg^{2+} and Co^{2+} yielded very similar cleavage rate constants around 1 s^{-1}, that for Mn^{2+} was found to be about five- to sevenfold faster. However, product release is nearly 30 times slower with Mn^{2+}. A very similar pattern is also observed with plasmid substrates. In the presence of Mn^{2+}, but not Co^{2+}, EcoRV looses much of its ability to discriminate between cognate and noncognate sequences.

The combination of increased phosphodiester hydrolysis rate and increased lifetime of the, enzyme-substrate complex may at least be partly responsible for this reduced specificity. In the case of EcoRV the rationalization of these metal specific effects is complicated by the fact that at least two metal ions appear to be involved in catalysis. Furthermore, the additional binding sites of soft divalent ions (Mn^{2+}, Co^{2+}) outside the active site may contribute to metal specific effects.

Mechanistic Hypotheses

The mechanistic studies of EcoRV have been largely driven by structure-based hypotheses. The identification of the essential catalytic residues, the understanding of their specific role and the determination of the number, location and coordination of the essential metal binding site(s) have been important issues. Ultimately one would like to have a chemically and stereo-chemically convincing model of the transition state and of the path leading from the productive ground state through the transition state to product. Structural information is critical to develop such detailed models but in most cases only low-energy, ground state structures are experimentally accessible. Unavoidably, this raises the question of how close these are to the unknown transition state. The structure of a high-energy intermediate close to the transition state has recently been determined for the phospho-group transfer reaction of phospoglucomutase. It shows a stretched pentacoordinate trigonal bypiramidal oxyphosphorane whose axial bond lengths of 2.0 A suggest a partly associative in-line displacement.

A similar geometry may be expected for the pentavalent phosphorane transition state of restriction endonucleases. For EcoRV, studies of steady-state, single turnover and rapid kinetics with mutant enzymes and modified substrates, studies of the effects of different metal ions and studies of the pH dependence of catalysis have provided us, together with the abundance of structural studies, with a wealth of data. Several mechanisms, differing in the number and location of the metal binding sites, in the nature of the general base and in the conformation of the DNA near the scissile group have been considered and experimentally explored. Although some aspects remain controversial, the more recent two or three metal ion mechanism proposed by the Perona group appears to us as the most plausible.

The Role of the Conserved, Catalytic-Motif Residues

The structurally conserved nature and essential role of D74, D90 and K92 has been recognized early and has been verified by mutagenesis. They are part of the catalytic sequence motif PD...(D/E)XK that occurs in many if not most Type II restriction endonucleases. A clear indication of the critical role of the two conserved acidic residues is that their mutants are equally inactive with Mg^{2}+ and Mn^{2+}. Based on its location on the 3′ side of the scissile phosphodiester group, the amine function of K92 appeared as a primary candidate for the general base.

It has been argued that a general base is not an absolute requirement for water activation, and that a hydroxide ion from the bulk solution could fulfill this role as in the case of the 3′-5′ exonuclease activity of DNA polymerase I. For EcoRV, this can almost certainly be ruled out on the basis of the pH dependence of the maximal rate of cleavage.

However, the pK_a of K92 would have to be lowered by perhaps 2-3 units to make it an efficient base at intracellular pH. This seems unreasonable at first sight in view of its proximity to three negatively charged groups (D90 and the two phosphodiester groups at P_0 and P_{+1}). However, the binding of divalent cations during catalysis may more than compensate for this. The lysine residue of the catalytic motif is not strictly conserved, and BamHI for example has an acidic residue at this position.

In the absence of decisive data, other candidates for the general base have been considered. One such hypothesis is based on modeling studies using the structures of the substrate complexes of EcoRV and EcoRI and has been substantiated by experimental evidence with modified oligonucleotide substrates. It suggests a sub-strate-assisted mechanism in which the P_{+1} phosphate (on the 3′ side of the scissile group) acts as the general base. A major problem with this mechanism is the low intrinsic pK_a of this group (<2). It seems more likely that this group is critically involved in generating the optimal geometry of the hydrogen bonding network that positions the functional groups and water molecules for metal binding and water activation.

The Role of Metal Ions and the Need of Conformational Changes in Different Mechanisms

It is now generally accepted that at least two metal ion sites are essential for catalysis by EcoRV. As described, the crystallographic metal binding studies have revealed three distinct sites, but at most two of them have been seen occupied together in the same structure. Evidence for more than one catalytically important binding site has also been derived from studies using mixtures of divalent cations and from stopped-flow fluorescence studies revealing two Mg^{2+} concentration dependent rates. As binding site I was unknown at that time, these were assigned to sites III and II respectively. It was, however not obvious why binding to these two sites in the triclinic crystal form did not cause DNA cleavage even under drastic conditions. Therefore, a larger conformational change that would be inhibited by crystal packing forces seemed indicated. The uncertainty about the need of a larger conformational change inspired a molecular dynamics (MD) simulation starting from the observed structure with two metals bound at sites III and II as observed in the triclinic crystal form.

A spontaneous transformation of the DNA conformation into a new state in which both nonbridging phosphodiester oxygens had become ligated to a metal ion was observed. The displacement of the scissile group was produced by a change in the sugar pucker of the thymidine deoxyribose at the central TpA step from C3′-endo to C2′-endo. The side chain of

E45 changed to an outer shell ligand of the site II metal and thus fulfilled two important roles in the postulated mechanism: first, it participates in the initial binding of the site II metal and secondly, it activates the attacking water after the conformational change. Such an important role appears, however, difficult to reconcile with the fact that E45 is not a conserved residue. The above mechanistic hypothesis has been further refined by looking at the pH dependence of the cleavage reaction under single-turnover conditions and by assigning an important role to another acidic residue, D36.

In the different structures, the carboxylate of D36 of one subunit is located 5-7 Å away from that of E45 in the other subunit. Mutations of D36 cause large cleavage rate reductions of up to 10^4-fold but show normal specific binding in the presence of Ca^{2+} suggestive of a role in catalysis. The pH dependence of the wild-type enzyme was studied with Mn^{2+} to guarantee full metal occupancy at all pH values up to pH 8.5. An asymptotic, increase of the cleavage rate reaching a plateau above pH 7.5 was best fitted with a model requiring the deprotonation of two ionizing groups with similar pK_a values of about 7. The close interaction between the carboxylates of D36 and E45 was postulated to raise the pK., of E45 and make it more effective in deprotonating the attacking water. This modification of the 'conformational change' mechanism did explain the observed pH profile and the large effects seen with D36 mutations but has a number of unsatisfactory aspects. Neither D36 nor E45 are conserved, and the two acidic groups are not in direct contact as to justify a large pK_a shift.

Furthermore, the essential roles of the conserved residues D90 and K92 remain insufficiently defined. Yet another modification of the 'conformational-change' mechanism was provoked by the observation that mutants of the Q-loop residue Q69 displayed strongly reduced cleavage rates. Crystal structure analysis of DNA complexes of the Q69E and Q69L mutants did not reveal large structural changes and failed to provide an obvious explanation for their strongly impaired catalytic activity. An MD simulation provided a structural link between the Q-loop conformation and the stabilization of the postulated pentavalent transition state and thus a possible explanation for the impaired catalytic rate of the Q69 mutants.

It must be emphasized though that these MD simulations are open to several kinds of criticism and, in our opinion, they can at best provide ideas on possible structural states. We think that the analysis of active and inactive conformations of EcoRV as discussed below provides an alternative and more plausible explanation for the rate reductions of the Q69 mutations.

Catalytically Active and Inactive Conformational States

The postulate of a larger conformational change during catalysis originated from the observation that no DNA cleavage was observed in the triclinic crystals despite binding of the metal cofactor. Four different crystal forms that contain specific complexes with substrate DNA have been reported. Only form I crystals, containing a decamer DNA duplex, show *in crystallo* cleavage activity upon soaking with Mg^{2+} and will be referred to as active. The detailed comparison of the protein and substrate DNA conformations in all four crystal lattices has revealed a very significant variation of the DNA-bending angle and a correlation of this variation with protein structural changes. Very importantly, however, the active (I) and inactive (II-IV) crystal forms cluster into two distinct states with respect to the B helix interface across the molecular twofold, as illustrated. At the quaternary structure level the two states differ to first approximation by (1) a rotation of the two DB-D's (together with the bound DNA) by about 10° about the dimer axis and (2) by a relative translocation of the two antiparallel B helices of about one turn.

The translocation is such that symmetry equivalent residues on this helix are 4-5 A further apart in the active state. As helix B extends out to the surface of the molecule it is not surprising that such an expansion can become forbidden in a crystal lattice. Apart from the change in the direct interactions between the symmetry related B helices one also observes a number of distinct structural differences between the two states. They mainly concern the N-terminal turn of helix B with residues D36, T37 and K38 that are located near the molecular twofold on the minor groove side of the DNA. The coupling of this change in protein structure to one in DNA conformation may be mediated by the contact of T37 with the ribose sugar of the thymine nucleotide of each ATC half-site. In the two states, T37 also interacts in different ways with the side chain of Q69 from the Q-loop of the symmetry related subunit and this contact appears important for intersubunit communication. A superposition of the active and inactive state structures based on the atomic positions of the central TpA step of one DNA strand reveals that the catalytic residues D90, D74 and K92 are very differently positioned with respect to the scissile phosphate.

The positional differences are in the range 1.5-2.5 Å and must severely affect metal binding. In particular, the ligating groups of site I are not properly positioned in the inactive state. A significant difference is also observed for the position and orientation of P_{+1}, whose chemical modification has large effects on catalysis and whose pro-ftp oxygen has been implicated in substrate-assisted catalysis. The restriction endonuclease Bgll, belonging to the EcoRV-related enzymes bond to P_{+2} is only observed in the inactive state conformation. The severe effect on catalysis observed for mutants of these three residues indicates that their interactions are critical for defining the active conformation.

From studies with heterodimers in which only one subunit carries a T37A, K38A or D36A mutation it was concluded that these mutations only affect the activity of their own subunit. These results were in contradiction to the role of D36 in the Halford mechanism where it exerts its effect on the active site of the other subunit. However, if the interaction between T37 and the thymidine sugar is critical for transmitting the conformational switch to the active site, this would represent an intrasubunit process consistent with the heterodimer results. Starting from the idea that all charged residues in the vicinity of the scissile phosphate could exert some effect on catalytic activity through influencing the active site electrostatics,

Perona and coworkers examined the role of eight charged residues within 12 A of the scissile phosphate in single-turnover experiments. Most interestingly they observed that the double mutant K38M/D36N, with both charges removed was 100 times less impaired in catalytic activity than the single mutant D36N. The authors sug-gest that the larger charge imbalance introduced by the single mutations may be responsible for their increased rate reductions. The observed effects are strong evidence against a direct role of D36 in catalysis as proposed by the Halford group. An important question concerns the relative stability of the two states.

With the exception of the decamer substrate, which makes fewer contacts with the protein and cannot induce the ordering of segment 221-229, the other complexes with undecamer and dodecamer DNA duplexes are observed in the inactive state. At least in the absence of catalytically competent divalent metnal ions this state must be significantly populated in solution and cannot be considered as a crystallization artifact. Crystals grown in the presence of Ca^{2+} ions also trap the inactive state and it must be assumed that this inhibitory metal ion preferentially stabilizes this state also in solution.

Factors that destabilize the active state relative to the inactive one may be the increased DNA bending and the less extended and less tightly packed B helix interface. On the other hand, increased DNA bending appears to correlate with an increase of the surface area buried between the Q-loop/Nt-aB segments and DNA, which may partly compensate for the destabilizing components. An important factor for tipping the balance in favor of the active conformation may be the binding of the catalytically competent metals. In view of these newer data, the rapid reaction analysis of the catalytic cycle of EcoRV that suggested the binding of two essential metal ions may now be reinterpreted. Addition of $MgCl_2$ to the preformed complex with the dodecamer cognate DNA fragment leads to tryptophan fluorescence enhancement on a 2-s time scale followed by an even slower fluorescence decay that was shown to reflect product formation. The initial cofactor dependent conformational change has a time constant of about 7 s^{-1} and may well represent the transition from the inactive to the active state. This pre-catalytic ternary complex then binds the second metal ion needed for catalysis. The metal binding studies in the inactive state would suggest that site III, possibly together with the structural site II, is first occupied and that site I is the one associated with catalysis.

However, when EcoRV binds to DNA in the presence of Mg^{2+} the initial enhancement is not seen suggesting that the inactive state is not an obligatory intermediate for the wild type enzyme. Interestingly, transient kinetics on Q69L indicated that its reaction involved a slow conformational change with a rate constant of 0.025 s^{-1}. The tight protein-DNA interface of the active conformation means that any perturbation introduced by a mutation or substrate modification cannot easily be relaxed without distorting the optimal stereochemistry for productive Mg^{2+} binding. The Mn^{2+} catalyzed reaction may be more tolerant to such perturbations because Mn^{2+} has a higher affinity for the binding site and is more tolerant to deviations from an optimal inner ligand arrangement. For the Mg^{2+}-catalyzed reaction of the Q69L mutant and of other such perturbations, the less tightly packed inactive state may be further stabilized with respect to the active one and may thus become an obligatory intermediate in the reaction.

The Three Metal Ion Mechanism

The mechanism elaborated by Perona and coworkers defines the active state as the pretransition state and therefore no longer requires a larger conformational change. A first important postulate is that a metal bound at site I plays an essential catalytic role. As previously discussed, this site is only seen in crystalline complexes of a mutant enzyme (T93A) or with poor or refractory DNA substrates. This illustrates that relatively small structural perturbations in the protein-DNA interface can have profound effects on metal binding.

One of the water ligands of site I bridges to the P_{+1} *pro*-S_p oxygen while the other is suitably positioned for in-line attack and is within hydrogen bonding distance to K92. A very similarly located metal binding site (occupied by Ca^{2+} or Mn^{2+} in crystal soaking studies) has been observed in all six restriction endonucleases for which metal complexes have been determined. Two variants of a transitionstate model invoking three distinct metal sites but differing in the role of K92 have been proposed. A purely structural role is assigned to site II, consistent with a nonessential role of E45. Quite different results from those obtained by the

Halford group have been reported with respect to the pH dependence of the reaction. Using Mg^{2+} and a wider pH range, a bell-shaped dependence of the cleavage rate with a relatively' sharp optimum at pH 8.5 was observed. This and the fact that a linear slope of log (k_c) against pH was observed has led to a different interpretation with respect to the identity of the titrated groups.

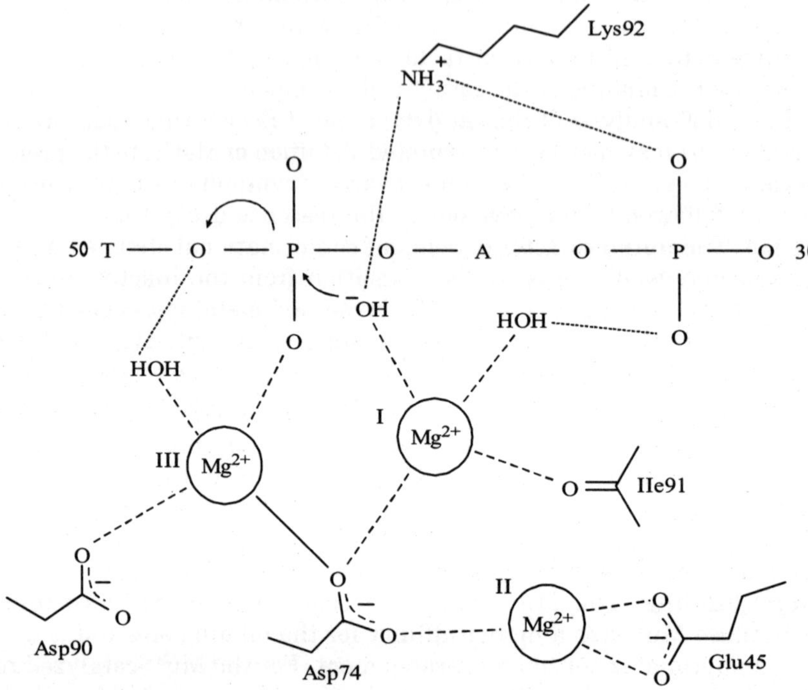

Fig. 11.3. Proposed state on the reaction pathway of the three metal ion DNA cleavage mechanism after water activation. An inner shell water of the site I metal ion has become activated for nucleophilic attack, most likely by K92 acting as the general base. The metal in site III stabilizes the increasing negative charge at the scissile group on the way to the transition state. A water ligand of the site III metal serves as the general acid to protonate the 3′-alcoholate upon dissociation. The role of the metal in site II is proposed to be structural. *Dashed lines* connect metal ions with inner sphere ligands (some additional water ligands are omitted). *Dotted lines* indicate hydrogen bonds. Schematic diagram adapted.

The curve is interpreted as resulting from the titration of an acidic and a basic group with the pK_a values differing by at most one unit from the observed pH optimum. The acidic function is postulated to be a site III bound water molecule that protonates the leaving alcoholate. A water molecule in the inner sphere of the site I metal is postulated to become activated for in-line attack. Mg^{2+}, Mn^{2}+ and Co^{2+} bound water molecules have pK_a values of 11.4, 10.5, and 9.7, respectively, for the first deprotonation of their hexaquo-complexes. It must be emphasized, however, that the pK_a of a specific water molecule bound to a metal coordinated by negatively charged ligands in a protein environment may be quite different. K92 is a natural candidate for the general base in this mechanism, as it makes a hydrogen bond to the attacking water molecule.

As discussed previously, its pK_a needs to be lowered, which appears structurally possible in view of the close site I metal ion. The site III and possibly site I metal provide the stabilization of the doubly charged phosphorane transition state, which is generally required for phosphoryl transfer reactions. Altogether, the three-metal ion mechanism is consistent with the majority of the experimental data, it is stereochemically convincing and, equally important, it shares essential features with the other restriction endonucleases.

Concluding Remarks

After its initial characterization, EcoRV endonuclease has been extensively investigated over the last twenty years through a combination of detailed mechanistic and structural studies by a number of research groups. We have gained insight into the time course of the different processes along the kinetic pathway from the initial association with DNA to target site location, DNA cleavage and product release. The specificity of the enzyme with respect to many naturally occurring or artificially introduced changes at the hexameric target sequence has been quantified and correlated with the extensive structural data. With the exception of EcoRI, other restriction endonucleases have not been investigated in such detail. Apart from some features of the hydrolytic mechanism that appear shared by all investigated Type II restriction endonucleases, the EcoRV appears unique in most of its detailed mechanistic aspects. Perhaps the most intriguing such aspect is how the catalysis is made highly sensitive to perturbations in the spatially remote recognition interface with cognate DNA.

In the case of EcoRV, and in contrast to EcoRI endonuclease, hardly any modifications within the tightly packed part of the interface are known to reduce binding but not the rate of cleavage. This indicates that strain relaxation in this part of the interface almost inevitably requires structural adaptations that propagate to and severely distort the active site. At the transition state of the reaction this distortion must be recorrected and the relaxed strain energy must be paid back. The structural basis of this general coupling mechanism may be different for each restriction endonuclease, and only for EcoRI and EcoRV have some important elements been identified.

At most DNA sites, EcoRV will be bound transiently in the nonspecific binding mode. To probe the DNA sequence at cognate and at noncognate sites the complex must pass through intermediate stages with only partial formation of the recognition interactions. The inactive conformation is deceptively similar to the active one and permits the formation of all base specific interactions. Its less tightly packed interface with DNA in the major groove is compatible with a range of bent DNA conformations. It is tempting to speculate that the inactive conformation is an obligatory intermediate for cleavage of large DNA substrates. Probing recognition in an inactive conformation could be a safeguard against undesired cleavage. Can we get even deeper insight or resolve some of the open questions in the near future?

It is certainly possible that similarly detailed studies with other restriction endonucleases will help to clarify some of the open questions. Sound molecular dynamics simulations for such a complex system on the necessary time scales of milliseconds will remain a dream for some time. On the experimental side, single molecule spectroscopy with suitably labeled probes appears to us as a possible new avenue that could yield some exciting new information on the relevance of the different structural states along the reaction path.

CHAPTER 12 Tetrameric Restriction Enzymes

Type II restriction endonucleases recognize specific DNA sequences, typically 4-8 bp in length, and cleave phosphodiester bonds in the presence of Mg^{2+}, within or close to their recognition sites. Around 3500 species, from variety of bacteria with nearly 240 differing specificities, have now been characterized. Most of the sequences recognized by Type II restriction endonucleases are palindromic, *i.e.,* possess a twofold rotational axis of symmetry. On the basis of this observation, Kelly and Smith proposed the first model for the interaction of these restriction enzymes with DNA. According to their model, recognition of the palindromic DNA sequence is achieved by two identical protein subunits related by a twofold axis of symmetry. Each subunit faces the same nucleotide sequence on the opposite DNA strand and contains one active site.

Symmetrical nicking of opposite DNA strands by both monomers within a homodimer generates the double-strand break. Hence, the symmetry of the recognition sequence implies the oligomeric state of the restriction enzyme. This simple strategy was subsequently confirmed by structural studies of Type II restriction endonuclease complexes with DNA. Despite individual structural peculiarities, differences in the recognition sites and cleavage patterns, these enzymes interacted with DNA as dimers, composed of two identical subunits.

Since the majority of the sequences recognized by restriction endonucleases are palindromic, it is generally thought that most Type II restriction endonucleases interact with DNA as homodimers. However, studies of SfiI restriction endonuclease by Steve Halford and coworkers provided first biochemical evidence that higher oligomeric forms of restriction endonucleases might be of functional importance. A tetramer has been reported to be the functionally active form of the SfiI restriction endonuclease that recognizes the interrupted palindromic sequence 5′-GGCCNNNN/NGGCC (cleavage point is indicated by /). Biochemical and kinetic studies revealed that SfiI interacts simultaneously with two copies of its recognition site before cleaving the DNA.

The oligomeric structure of SfiI and its mode of interaction with DNA were not dictated by its discontinuous recognition site. Further studies revealed that some restriction enzymes that recognize uninterrupted DNA sites, *e.g.,* Cfr10I, are tetramers like SfiI. The architecture of the tetrameric restriction enzymes and their complexes with DNA, however, remained unknown. In our efforts to elucidate structural and molecular mechanisms of the sequence discrimination by Type II restriction enzymes recognizing closely related DNA sites, we undertook the structure determination of Cfr10I, Bse634I and NgoMIV specific for the 5′-Pu/CCGGPy (CfrlOI/Bse634I) and 5′-G/CCGGC (NgoMIV) sequences. The isoschizomeric restriction endonucleases Cfr10I and Bse634I were originally isolated from the *Citrobacter freundii* and *Bacillus stearothermophilus* strains, respectively, and share ~30% of identical amino acids in their protein sequences.

Restriction endonuclease NgoMIV was originally identified in *Neisseria gonorrhoeae*. The crystal structures of Cfr10I and Bse634I were determined in the absence of DNA whereas the structure of NgoMIV was solved in the enzyme-product DNA complex. Crystallographic analysis revealed that Cfr10I, Bse634I, and NgoMIV are arranged as tetramers and provided us with the first structural models of the tetrameric restriction endonucleases.

Solution experiments performed in parallel demonstrated that the tetrameric architecture of CfrlOI, Bse634I and NgoMIV is of functional importance. Here, we summarize results obtained by structural and functional studies of tetrameric restriction enzymes CfrlOI/Bse634I and NgoMIV which recognize closely related nucleotide sequences.

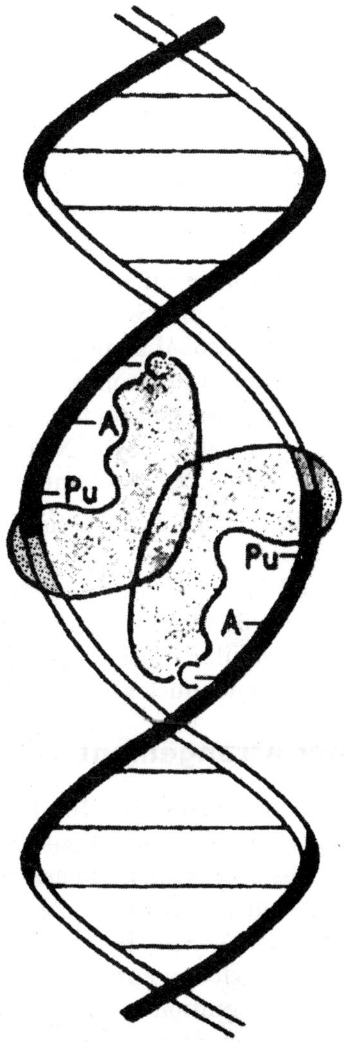

Fig. 12.1 Model of nucleotide sequence recognition by endonuclease R according to Kelly and Smith (1970). Two identical protein subunits related by a twofold axis of symmetry interact with CAPu sequences located on the opposite strands of the DNA duplex.

STRUCTURAL ANATOMY

Monomer Structure

The polypeptide chain of the Cfr10I, Bse634I and NgoMIV restriction endonucleases is folded into a compact α/β structure. A twisted fivestranded mixed β-sheet sandwiched by α-helices makes up the core of the globule and is conserved within all three proteins. Spatially conserved pairs of helices α7 and α8 of Cfr10I, α6 and α7 of Bse634I, and α7 and α8 of NgoMIV, respectively, flank the convex side of the β-sheet.

Structurally equivalent α3 and α9 helices of Cfr10I, α3 and α8 of Bse634I, and α3 and α9 of NgoMIV, respectively, pack on the concave side. Both Cfr10I/Bse634I and NgoMIV have an

N-terminal substructure outside the conserved common core. In addition, NgoMIV possesses a C-terminal substructure formed by $\alpha10$ and $\alpha11$ helices, and $\beta6$-strand, that is missing in Cfr10I/Bse634I. The structural analysis of Bse634I suggests that the N-terminal substructure forms a domain that is connected to a C-terminal part by a hinge located between residues 70 and 90 in helix $\alpha3$. The N-terminal domain of Bse634I is comprised of two β-strands and two α-helices, whereas the C-terminal domain bears a conserved α/β core. Cfr10I exhibits similar domain organization that is also supported by limited proteolysis experiments.

Indeed the N-terminal domain of Cfr10I that is structurally equivalent to that of Bse634I could be cleaved off by trypsin at K56 residue keeping the C-terminal subdomain intact (Skirgailiene A. and Siksnys V., unpublished data). The isoschisomeric restriction endonucleases Cfr10I and Bse634I share ~30 % of sequence identity. The crystal structures of both proteins are very similar and conserved structural cores overlap with RMSD of 1.0 Å for all atoms. To achieve superimposition of the N-terminal domains with a similar RMSD value, however, a nearly 10° rotation of the N-terminal domain of CfrlOI around the axis passing through the middle of the $\alpha3$ helix was necessary.

Thus, structural analysis suggests that the N-terminal subdomain of CfrlOI/Bse634I is connected to the conserved core by a relatively flexible hinge. The N-terminal substructure of NgoMIV is smaller and predominantly comprised of two α-helices. The general topology of Cfr10I, Bse634I, and NgoMIV monomers is similar to that of dimeric restriction enzymes. Hence, a similar conserved building block is used to construct both dimeric and tetrameric restriction endonucleases.

Dimer Arrangement

Two monomers of Cfr10I, Bse634I, and NgoMIV associate to build a primary dimer similar to that of the *bonafide* dimeric restriction enzymes, like EcoRI, that cleave hexanucleotide sequences producing four base pair staggered 5′-overhanging ends. In order to achieve such a cleavage pattern, restriction endonucleases should position their catalytic sites ~18-19 Å apart along the DNA axis. This constraint most likely is achieved through the conserved dimerization mode of restriction enzymes exhibiting the same cleavage pattern despite of the differences in their recognition sites. For example, EcoRV-like enzymes that recognize hexanucleotide sequences and cleave leaving blunt ends, exhibit a different dimerization mode.

A common structural core and the active site structure, however, are conserved between EcoRI- and EcoRV-like enzymes. The similarities of dimer organization between Cfr10I, Bse634I, and NgoMIV and restriction enzymes producing four base-pair staggered 5′-overhanging ends further support the idea that the cleavage pattern rather than recognition sequence plays a key role in the determination of the overall structure of the dimer.

Three conserved noncontiguous structural elements of the NgoMIV, Bse634I, and Cfr10I contribute to the dimer interface. First, amino acid residues located on the N-terminal parts of the conserved helices $\alpha7$, $\alpha6$, and $\alpha7$ of NgoMIV, Bse634I, and Cfr10I, respectively, make a set of van der Waals and hydrogen bond contacts at the interface between the two monomers.

Noteworthy, the N-termini of the latter α-helices are directed towards the major groove and carry aspartate and arginine residues involved in DNA base contacts (see below). Hence, the dimerization and DNA recognition interfaces are intertwined. The second segment of the dimer interface is comprised of the amino acid residues located on the loops and α-helical stretches bridging the first two conserved β-strands (β3-β4 of Bse634I/Cfr10I and β1-β2-strands of NgoMIV, respectively). The third conserved module of the intersubunit interface spans across the C-terminal parts of the conserved helices α8, α7, and α8 and subsequent loops of NgoMIV, Bse634I, and Cfr10I, respectively. In NgoMIV, additional contacts at the dimer interface eminate from amino acid residues located at the C-terminus of the α10 helix which is missing in the Cfr10I and Bse634I. Structure-based sequence alignment demonstrates that most of the 30% of sequence identities between the Cfr10I and Bse634I correspond to the amino acid residues involved in the organization of the active site and DNA recognition and dimerization interfaces.

Analysis of the inter-subunit interface of the typical dimeric restriction enzymes (*e.g.*, EcoRI, MunI,BamHI) producing four-base-pair staggered 5′-overhanging ends indicates that amino acid residues involved in dimer formation are located predominantly on the same conserved structural elements described for the CfrlOI, Bse634I, and NgoMIV. A U-shaped cleft large enough to accommodate a B-DNA molecule is formed between two monomers of Cfr10I/Bse634I. In the NgoMIV a similar cleft is indeed occupied by the DNA.

Tetramer Organization

Tetrameric restriction enzymes Cfr10l, Bse634I, and NgoMIV are arranged as dimers of primary dimers positioned back-to-back to each other with DNA binding clefts facing opposite directions. The mutual orientation of the primary dimers, however, differs between Bse634I/ Cfr10I and NgoMIV. In the NgoMIV tetramer the concavely shaped backsides of the dimers are slightly rotated around their local twofold axis, therefore the two bound DNA duplexes are positioned in an X-like fashion with their helical axes enclosing an angle of 60°. The monomers in the Cfr10l and dimers of Bse634I crystal are related by a twofold crystallographic axis and form a symmetric tetramer with a similar back-to-back arrangement of dimers. Two primary dimers become positioned at 90° angle in respect to each other.

The differences in the mutual orientation of the two primary dimers of NgoMIV and Bse634I/Cfr10I are due to the distinct dimer-dimer interfaces. While the mode of the dimerization is mostly conserved between NgoMIV and Bse634I/Cfr10I, much less of the structural conservation is found at the dimer-dimer interfaces. In the Bse634I/Cfr10I tetramers, structural elements involved in the dimerization also contribute to the interface between the two primary dimers.

For example, amino acid residues located on the N-termini of the structurally conserved α7/α6 helices of Bse634I/Cfr10I, respectively, make intersubunit contacts in the dimer, while residues located on the C-termini of the same helices are involved in tetramerization. Of note is that some of the residues positioned on these helices contribute both to the dimerization and tetramerization interfaces. A single subunit of Cfr10I or Bse634I makes contacts to both subunits of the neighboring primary dimer. Totally, each subunit makes contacts with all three other protein subunits within the tetramer. The tetrameric assembly of NgoMIV is fixed by side-by-side contacts (between subunits A/C and B/D) and cross-contacts (between subunits A × D and B × C) of primary dimers, respectively.

Cross-contacts are made exclusively by the helices a4 and cc5 which also contribute to the dimerization interface. Similar contacts are conserved at the tetramerization interface of Bse634I/Cfr10I. The most extensive contacts in the NgoMIV tetramer, between subunits A/C and B/D, are made by the tetramerization loop (residue 147 to 176 including helix $\alpha 6$) that spans across the neighboring monomer. The residues located on the $\alpha 6$ helix contacts the C-terminus of the $\alpha 3$ helix while other residues positioned on the loop make an intricate set of contacts to the residues with the N- and C-terminal residues of NgoMIV. Additional contacts at the A/C interface come from amino acid residues located on the C-termini of helices $\alpha 7$ and $\alpha 8$ and downstream regions.

Similar contacts are made by structurally equivalent elements of Cfr10I/Bse634I. In total, the contacts between subunits at the tetramerization interface comprise eight saltbridges and numerous hydrogen bonds and van der Waals interactions. The tetramerization loop that contributes major contacts to the interdimer interface of NgoMIV, is replaced by a short loop between two structurally equivalent p-strands in the dimeric restriction enzymes EcoRI, BamHI, and Muni. Interestingly, both Cfr10I and Bse634I tetramers possess loops (residues 140-182 and 152 to 190, respectively) that are topologically similar to the tetramerization loop of NgoMIV. However, in contrast to NgoMIV, loops bridging $\beta 4$-$\beta 5$-strands of CfrlOI/Bse634I are not involved in tetramerization but rather contact the N-terminal subdomain of the same subunit. The dual role performed by the topologically conserved loops in Cfr10I/Bse634I and NgoMIV remotely resembles a domain swapping mechanism for forming oligomeric proteins from their monomers.

Indeed, in the Cfr10I/Bse634I the structurally conserved loop makes contacts to the N-terminal domain of the same monomer, while in NgoMIV it contacts the N-terminal domain of the neighboring subunit. The structures of Bse634I/Cfr1OI were solved in the apo-form, while the structure of the NgoMIV in complex with product DNA. One might speculate that DNA binding by Bse634I/Cfr1OI might induce conformational changes and replace intermolecular contacts of the conserved loop by intersubunit contacts similar to those of NgoMIV.

The requirement for the stringent conservation of the intra- and intersubunit loop contacts, however, might put a major constraint on the loop swapping in Bse634I/Cfr10I complexes with DNA. For example, fifty hydrogen bonds are made across the intersubunit interface by the residues located on the tetramerization loop of NgoMIV. It is more likely that topologically equivalent loops in Bse634I/ Gfr10I and NgoMIV were adopted to perform different structural roles. Accessible surface area calculations indicate that in Cfr10I/Bse634I the largest accessible surface area of ~1350 and ~1550 Å^2, respectively, is buried at the primary dimer interface.

Much less of the monomer protein surface is buried in contacts with both subunits of the neighboring primary dimer. In NgoMIV, the contribution of individual intersubunit contacts to the total surface accessible area buried differs notably. The cross-contacts between monomers A × D comprise only around 400 Å^2. The contacts between subunits AB of the primary dimer hide ~1000 Å^2, while the largest surface area of nearly 2900 Å^2 is buried in side-by-side contacts between subunits A/C and B/D of the two primary dimers, respectively. Thus, the surface of each monomer shielded during tetramerization almost three times exceeds the surface buried at the dimer interface. The tetramerization interface of CfrlOI/Bse634I is much smaller in comparison to that of NgoMIV. Of note is that a single alanine replacement of the W220 residue of Cfr10l or structurally equivalent W228 residue in Bse634I located at the C-terminus of the conserved α7/α6-helices, respectively, disrupts the tetramerization interface generating a protein dimer.

The tryptophan W220/W228 residue at the interface between two primary dimers most likely represents a hot spot that makes a major contribution to the binding energy. If the loop would be involved in the stabilization of the tetramerization interface between the two primary dimers of Cfr10I.

Table 12.1. Protein interface parameters of the tetrameric restriction endonucleases

Protein interface parameter	***Cfl0I***			***Bse634I***			***NgoMIV***		
	A/B	***A/C***	***A/D***	***A/B***	***A/C***	***A/D***	***A/B***	***A/C***	***A/D***
Interface ASA	1347	491	682	1544	411	504	1002	2886	409
Interface residue segments	3	3	2	4	2	2	4	8	2
Polar atoms in interface (%)	25.6	30.2	30.7	32.1	33.2	34.2	36.9	38.1	33.8
Nonpolar atoms in interface (%)	74.4	69.8	69.3	67.8	66.8	65.7	63.1	61.8	66.1
Hydrogen bonds	8	4	0	10	2	1	4	50	2

Bse634I similarly to NgoMIV, the disruption of the tetramers would be much less likely.

ACTIVE SITES

Type II restriction endonucleases except of *Bfil* (Sapranauskas et al. 2000) require cofactor Mg^{2+} for the catalysis to occur. The presence of two Mg^{2+} ions nearby the cleaved phosphate in the crystal structure of enzyme-product complex allowed us to identify unambiguously the catalytic/metal binding site of NgoMIV. There are two Mg^{2+} ions in the active center. Both Mg^{2+} ions exhibit octahedral coordination. The 02P of the 5′-phosphate, carboxylate of D140 and the acetate molecule (from the reservoir of the crystallization buffer) bridge both Mg^{2+}-ions.

The remaining three ligands of the Mg^{2+} ion A are the backbone oxygen of C186 and water molecules 1 and 2. Water molecules 3,4, and 5 complete the octahedral coordination of Mg^{2+} ion B. The ligand oxygen atoms are within 2.2-2.3 Å from Mg^{2+} ion A, and within 2.1– 2.4 Å from Mg^{2+} ion B. All Mg^{2+}-coordinated waters are fixed by either protein and/or DNA residues. Waters 1 and 2 are within the hydrogen bonding distance from NZ of K187,O1P of the adjacent Cytl and OE2 of E70 carboxylate, respectively. Water molecules 3 and 4 are hydrogen-bonded to the carboxylate group of E201 while water 5 is in contact with released O3' hydroxyl group of Gua3 (3.3 Å), the carboxylate of D140 (3.0 Å) and O5' oxygen of the 5'-phosphate (3.2 Å).

The location of the metal ions of NgoMIV resembles the location of the metal ions at the active site of the BamHI restriction enzyme and the 3'-5' exonuclease domain of DNA polymerase I. Indeed, Mg^{2+} ions at the active site of NgoMIV and Mn^{2+} ions at the active site of BamHI can be superimposed in their enzyme-product complexes. Structural comparisons reveal the conserved spatial location of the metal chelating residues of NgoMIV, Cfr10I, and Bse634I suggesting similar architecture of the catalytic/metal binding sites. Therefore, it is very likely that Cfr10I and Bse634I also coordinate two metal ions at their active sites. The metal ion B Cfr10I, Bse634I, and NgoMIV would be chelated by acidic residues of the conserved sequence motif $PDX_{46-55}KX_{13}E$. Aspartate residues belonging to the latter sequence motif and conserved glutamates E70/E71/E80 located on the a3-helix would contribute to the binding site of the metal ion A.

Site-directed mutagenesis of the putative catalytic residues of Cfr10I supports their active site function. Orthodox dimeric restriction endonucleases contain a single (EcoRI, Bglll) or two metal ions (BamHI, PvuII, EcoRV) positioned at their active sites. One of the two-metal binding sites is homologous to the single site observed in EcoRI and BgIII. Two acidic residues and a lysine defined by a weakly conserved sequence motif PDX_{10-30}(D/E)XK are located at the ends of p-strands and form the first catalytic/metal binding site that is conserved within most restriction enzymes.

Structural comparisons between the active sites of Cfr10I/Bse634I/NgoMIV and those of orthodox restriction enzymes reveal that the spatial location of the metal-chelating residues is conserved, however, the order and position of the active site residues within the polypeptide sequence is different. This results in a sequence divergence in the second parts (shown in bold) of the catalytic motifs $PDX_{46-55}KX_{13}E$ and $PDX_{19\cdot30}$(E/D)XK characteristic for Cfr10I/ Bse634I/NgoMIV and canonical restriction enzymes, respectively. In EcoRI an acidic El 11 residue from the EAK part of the conserved active site motif $PDX_{19}EAK$ is positioned at the end of the β-strand. In NgoMIV/Bse634I/Cfr10I an equivalent position is occupied either by glycine or serine residue.

The conserved glutamate residue located at the N-terminus of the α7-helices of NgoMIV and Cfr10I, however, spatially substitutes the acidic residue located at the end of β-strand of EcoRI and most other canonical restriction enzymes. Site-directed mutagenesis confirmed the functional significance of the E204 residue in Cfr10I catalysis. Thus, the spatial arrangement of the active sites of NgoMIV/Bse634I/ Cfr10I is similar to those of EcoRI or EcoRV, however, the order and position of the active site residues is not conserved. Such plasticity of the active sites has been reported for a number of different enzymes. The primary importance of structure rather than sequence conservation of catalytic/metal binding residues is further supported by a residue swapping experiment.

Double mutation S188E/E204S in Cfr10I transposed the E204 residue from the α-helix to the end of the β-strand restoring a canonical active site structure. The Cfr10I mutant S188E/ E204S with a reengineered metal binding site retained significant catialytic activity. The sequence motif $PDX_{46-55}KX_{13}E$ first found in the NgoMIV/Bse634I/ fiCfr10I, however, does not seem to be limited only to the tetrameric restriction lenzymes and might be more common than it was initially thought.

Indeed, a siamilar catalytic motif was identified in the protein sequences of 11 restriction enzymes (and three putative restriction enzymes) which belong to jibe different subtypes (Type II, Type HE, and Type IIP) but all have CCNGG, ffiCWGG or CCGG in their respective recognition sequences. Mutational analysis of SsoII, Eel 18 kl and PspGI (Pingoud, pers. comm.) confirmed the functional significance of the putative active site residues. Hence, the catalytic sequence motif rst revealed for the tetrameric restriction enzymes most likely represents a ifferent solution for coordination of metal ions at the active site of restriction enzymes. According to the crystal structure D140 from $PDX_{46}KX_{13}E$ motif and E70 located on the α-helix constitute a second metal ion binding site in NgoMIV llDeibert et al. Both E70 and D140 residues of NgoMIV are conserved pflthin CfrlOI/Bse634I.

Multiple sequence alignment reveals that E70 of NgoMIV is also conserved within a group of 11 restriction enzymes discussed above. Site-directed mutagenesis of the latter conserved residue supports its active site function in Cfr10I, SsoII, EcllSkI and PspGI (Pingoud, personal communication). The structurally equivalent glutamate residues E70/E71/E80 of NgoMIV/ CfrlOI/Bse634I are involved in coordination of the

NGOMIV RESTRICTION ENDONUCLEASE

The crystal structure of the NgoMIV restriction endonuclease in complex with product DNA provides a detailed picture of the protein-DNA recognition interface. Two 10 bp oligonucleotide duplexes containing the recognition sequence 5′-GCCGGC-3′ are bound to the NgoMIV tetramer arranged as a dimer of dimers. Two primary dimers are stacked back-to-back, placing the duplexes on the opposite sides of the tetramer. Each duplex is contacted by just two protein subunits of the primary dimer.

A single subunit of NgoMIV makes a set of hydrogen bonds in the major groove to the bases of the first 5′-GCC half site of the recognition sequence and a single minor groove contact to the C base of the second GGC-3′ half site. Thus, major groove contacts between the primary dimer of NgoMIV and palindromic recognition sequence follow the model suggested by Kelly and Smith. The recognition of the specific sequence 5′-GCCGGC-3′ by NgoMIV is achieved through a number of direct contacts between protein side chains and DNA bases.

The structural elements of NgoMIV involved in the sequence recognition form two separate modules that are comprised by different segments of the protein chain. Since both half sites of the recognition sequence make an identical set of contacts to the protein residues, only interactions with the 5′-GCC half-site are discussed. Two central C:G base pairs are contacted by residues located on a short contiguous stretch of amino acids with the sequence 191RSDR just upstream and at the N-terminus of the α7 helix.

Arginines R191 and R194 donate bidentate hydrogen bonds to the guanines while carboxylate oxygens of D193 bridge exocyclic amino groups of the neighboring cytosines. Thus, three residues R191, D193, and R194 of a single subunit unambiguously specify inner and middle C:G nucleotides within the first half-site of the NgoMIV recognition site 5′-GCC. Equivalent residues located on the symmetry related α7 helix make the same set of contacts to the central C:G base pairs of the second GGC-3′ half-site of the recognition sequence. Besides their key role in the inner C:G recognition, D193 and R194 residues also contribute to the dimerization interface making two salt-bridges with R194 and D193 residues of neighboring subunit, respectively. Hence DNA recognition and dimerization interfaces in NgoMIV are intertwined. The recognition of the outer G:C base pair by NgoMIV is achieved by a recognition module comprised of three noncontiguous structural elements: loop preceding helix α2, helix α8, and helix α3 of the neighboring subunit.

The major groove contacts to the outer G:C base pair arise from the D34 and R227 residues located on loops upstream of helix $\alpha2$, and helix $\alpha8$, respectively. D34 makes a hydrogen-bond to the exocyclic amino group of Cyt3′. The side chain of D34 is also buttressed by a salt bridge to R227 and a hydrogen bond to S36. The NH2 atom of R227 makes a direct hydrogen bond to the O6 oxygen of Gua3, while the NE of R227 in companion with the main chain nitrogen of SI92 sandwiches a water molecule, which is hydrogen bonded to the N7 nitrogen of Gua3. The amino group of Q63, located at the N-terminus of the $\alpha3$ helix of the neighboring subunit, donates a hydrogen bond to the oxygen atom of Cyt3′ in the minor groove.

The side chain conformation of Q63 residue is probably fixed through the interaction of its side chain carbonyl oxygen atom with the backbone nitrogen atoms of Q63 and G62 residues. Thus both subunits of a primary dimer contribute to the recognition of the outer G:C pair of one half-site. Interestingly, there is one direct protein-base contact outside of the 6-bp recognition sequence which may explain the possible C:G preference for the bases flanking the NgoMIV recognition sequence. S36 of each subunit A, B and C makes a single hydrogen bond to Cyt4 outside of the recognition site of the DNA strands E, F, and H, respectively (not shown).

The conformations of the S36 residues however slightly differ between subunits. Therefore S36 of monomer D makes a water-mediated contact to the backbone phosphate of the complementary Cyt3′ in strand H instead of direct contacts to the C base. Furthermore the backbone oxygens of S36 in subunits B and C are involved in water mediated contacts to the outermost guanine. The SgrAI restriction endonuclease recognizes 5′-CR/CCGGYG-3′ sequence that overlaps completely with the recognition sequence of Bse634I/Cfr10I. Protein sequence comparisons of SgrAI and Cfr10I/Bse634I restriction endonucleases reveal that key catalytic and sequence recognition residues are conserved suggesting that recognition of the common 5′-R/CCGGY-3′ sequence is probably achieved by similar mechanisms. It would be interesting to see, how recognition of the extra C:G base pair is achieved by the SgrAI restriction endonuclease.

BSE634I/CFR10I RESTRICTION ENDONUCLEASES

Besides the common tetrameric architecture, NgoMIV and Bse634I/Cfr10I also share similarities in their recognition sites. Their recognition sequences, G/CCGGC and Pu/CCGGPy of NgoMIV and Bse634I/Cfr10I, respectively, differ only at the outer base pair flanking the common CCGG core. The crystal, structures of both Bse634I and Cfr10I were solved in the absence of the DNA. The superimposition of the active sites and common structural cores of NgoMIV and Bse634I/Cfr10I, however, allowed us to predict the putative DNA binding interface of Bse634I/Cfr10I. The N-terminus of the oc6 helix of Bse634I overlaps with the helix $\alpha7$ of NgoMIV involved in the sequence-specific contacts to the inner C:G base pairs.

Moreover, Bse634I residues R202, D204, and R205 (and equivalent residues R194, D196, and R197 in Cfr10I) superimpose with the R191, D193, and R194 residues of NgoMIV that make discriminating contacts to the inner C:G base pairs. The conserved sequence motif R(P/S)DR corresponds to the structurally equivalent residues of CfrlOI, Bse634I and NgoMIV, suggesting that these enzymes use the same residues and structural mechanisms to interact with the central CCGG bases. The recognition sequence 5′-Pu/CCGGPy-3′ of Bse634I/Cfr10I is less strin-gent in comparison to that of NgoMIV.

Indeed, Bse634I/Cfr10I can tolerate two alternative purine nucleotides (A or G) at the first position of its recognition site while eliminating pyrimidines. The 5′-GCCGGC-3′ site of NgoMIV therefore overlaps only with one of the possible recognition sequences of, Bse634I/Cfr10I. Does the structural comparison of NgoMIV and Bse634I/Cfr10I allow us to predict how the recognition of the Pu:Py base pair is achieved by Bse634I/Cfr 101? The stringent discrimination of the outer G:C base pair by NgoMIV is achieved predominantly by three different residues. R227 located on the 018 helix makes one direct (to the 06 oxygen) and one water-mediated (to the N7 nitrogen) hydrogen bonds to the outer Gua base. Such a hydrogen bond pattern is specific for the Gua base.

Structural superposition of the Bse634I and NgoMIV indicates that structurally equivalent helix $\alpha 7$ of Bse634I is shifted away from the DNA in respect of the $\alpha 8$ helix of NgoMIV. The position of the helix $\alpha 7$ of Bse634I therefore prevents direct interactions of amino acid residues with outer Ade/Gua bases of the recognition site. The amino group of Q63 located at the N-terminus of $\alpha 3$ helix of NgoMIV donates a hydrogen bond to the oxygen atom of the outer Cyt base in the minor groove.

Noteworthy, in Bse634I N73 (or K64 in the case of Cfr10I) is a structural equivalent of Q63 in NgoMIV and in principle can make a hydrogen bond to the oxygen atom of the Thy or Cyt bases in the minor groove. Such a minor groove contact alone in the case of Bse634l/ Cfr10I, however, is not sufficient to discriminate unambiguously the A:T and G:C base pairs against C:G and T:A base pairs, respectively.

The D34 of NgoMIV located at the N-terminal domain makes a hydrogen bond to the amino group of the outer Cyt base in the major groove. Superim-position of the crystal structures of Bse634I and NgoMIV reveals significant differences in the conformations of the N-terminal parts of both proteins (residues 1-89 in Bse634I and 1-79 in NgoMIV, respectively). Whereas the conserved structural cores of Bse634I and NgoMIV overlap, the N-terminal domain of Bse634I is rotated about 90° in respect to its equivalent in NgoMIV. Due to the different conformations of the N-terminal sub-domain the loop bridging $\beta 1$- and $\beta 2$-strands in Bse634I becomes positioned close to the outer G:C base pair at the opposite recognition half-site than in the case of NgoMIV. Amino acid residues/backbone atoms situated in this loop or flanking regions might contact outer Pu:Py base pair of the PuCCG-GPy sequence.

The crystal structure analysis of the BsoBI restriction endonuclease, specific for the C/PyCGPuG sequence, revealed that recognition of the degenerate base pair occurred in the major groove predominantly through the hydrogen bond of K81 to the N7 of the purine base (Ade in the crystal) and the water molecule sandwiched between the N6 of Ade and side chains of K81 and D246. The hydrogen bond of the lysine amino group to the N7 nitrogen atom is purine specific. In the crystal structure of Hindi, specific for the GTPyPuAC sequence, a similar inner Gua N7 interaction with the amino group of N141 residue was the only hydrogen bond observed at the center C:G base pair (as seen in the crystal), consistent with the degenerate recognition of any Py-Pu dinucleotide at this position. Similar interactions with the N7 nitrogen atom of the outer purine base of the Bse634I/Cfr10I sequence in principle would allow to discriminate Gua/Ade bases against Cyt/Thy, consistent with the less stringent specificity of Bse634I/Cfr10I in comparison to the NgoMIV restriction enzyme.

Formation of the protein-DNA complexes, however, often evokes conformational rearrangements. One cannot exclude the possibility that DNA binding might induce conformational changes of the N-terminal domain of Bse634I leading to a different set of contacts with DNA. Therefore a proposal regarding the detailed mechanism of the external base pair recognition could only be made after the elucidation of the crystal structure of Bse634I(Cfr10I)-DNA complex.

Structural comparisons between NgoMIV and Bse634I/Cfr10I, however, suggest that recognition mechanisms of the common CCGG tetranucleotide are similar while the mechanism of outer base pair discrimination seems to be different.

A similar interaction pattern at the protein-DNA recognition interface was shown previously for the EcoRI and Muni restriction enzymes, specific for G/AATTC and C/AATTG sequences, respectively. The recognition of the AATT tetranucleotide common to both EcoRI and Muni recognition sites is achieved by a similar structural mechanism. In the case of EcoRI and Muni a short stretch of conserved residues (GNAXER) located on the α-helices structurally equivalent to the helix α7 of NgoMIV, are involved in similar hydrogen bonding and van der Waals interactions with AATT bases. The mechanism of the discrimination of the outer base pair, however differed between EcoRI and Muni. Structural studies of EcoRI/ Muni and NgoMIV/ Bse634I/Cfr10I, however, demonstrate that enzymes recognizing overlapping nucleotide sequences can employ conserved mechanisms to interact with overlapping parts of their recognition sites. However, it seems not to be a general rule. The comparison of structural mechanisms of sequence recognition by Bglll (A/GATCT) and BamHI (G/GATCC) indicate that both proteins display considerably different protein-DNA contacts at the common GATC sequence within their recognition sites. It is worth to note that structural mechanism of the CCGG sequence recognition employed by NgoMIV/Bse634I/Cfr10I is probably shared by a wide group of restriction enzymes recognizing CCGG, CCNGG, or CCWGG in different sequence contexts. Moreover, these enzymes belong to different subtypes, viz. orthodox Type II enzymes (like SsoII or EclISKI), Type HE enzymes (like EcoRII) and Type IIP enzymes (like NgoMIV). It was suggested (Pingoud et al. 2002) that these subfamilies share a not too distant common ancestor, which presumably was a homodimeric enzyme.

By acquisition of an extra domain a Type HE enzyme like EcoRII was derived. Alternatively, by development of a new interface for protein-protein interaction enabled tetramerization giving rise to representatives of Type IIP enzymes like CfrlOI and NgoMIV. Two lines of evidence come support this hypothesis. First, limited proteolysis experiments of EcoRII revealed that an extra N-terminal domain can be removed yielding a dimeric restriction enzyme that lost an absolute requirement for the second recognition site. Second, the alanine replacement of a single tryptophane residue located at the tetramerization interface of Cfr10I and Bse634I restriction enzymes generates dimeric restriction enzymes that in the case of Bse634I W228A mutant retains a significant catalytic activity. These experiments. Drovide first experimental evidence supporting evolutionary relationships between different subtypes of restriction enzymes.

FUNCTIONAL SIGNIFICANCE

The crystallographic evidence reveals that NgoMIV tetramer is composed of two dimers with a separate DNA binding site in each. Crystal structures of Bse634I and Cfr10I show similar architecture. Therefore, tetrameric restriction enzymes might interact with two recognition sites. Processing at these sites in principle may be independent or cooperative. Cleavage

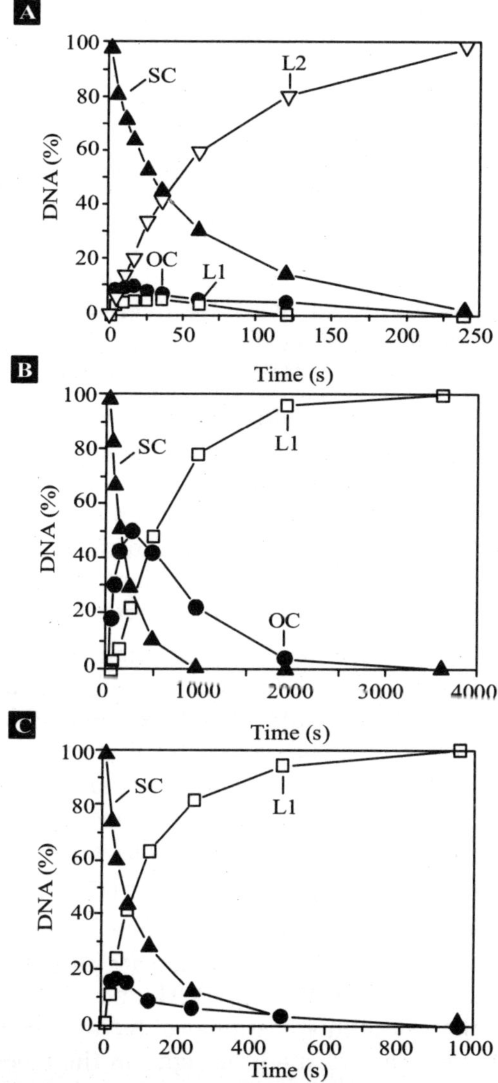

Fig. 12.2 Cleavage of plasmid DNA by NgoMIV a Cleavage of supercoiled plasmid pUCGC2 bearing two recognition sites of NgoMIV. Reaction mixtures contained 1.15 nM of pUCGC2 (2.3 nM of target sequence), 50 nM of NgoMIV (as a tetramer), 10 mM $(CH_3COO)_2Mg$, 33 mM Tris-acetate (pH 8.0, 25 °C) and 66 mM CH_3COOK at 25°C. Amounts of supercoiled *(filled triangles)*, open circle *(filled circles)*, linear DNA with a single double-stranded break *(open squares)* and linear DNA with two double-stranded breaks *(inverted triangles)* forms are shown, b Cleavage of supercoiled plasmid pUCGCl bearing a single recognition site of NgoMIV. Reaction mixtures contained 2.3 nM pUCGC1, 50 nM of NgoMIV (as tetramer) and other components as in a. Amounts of supercoiled *(filled triangles)*, open circle *(filled circles)* and linear *(open squares)* forms of pUCGCl are shown, c Cleavage of supercoiled plasmid pUCGCl bearing a single recognition site of NgoMIV in the presence of cognate oligodeoxynucleotide. Reaction mixtures contained 2.3 nM pUCGCl, 200 nM of specific self-complementary 12 bp duplex 5′-GAG GCC GGC CTC-3′, 50 nM of NgoMIV (as tetramer) and other components as in a. Amounts of supercoiled (*filled triangles*), open circle (*filled circles*) and linear (*open squares*) forms of pUCGCl are shown.

patterns of plasmids containing a single or two recognition sites provide a general test whether a restriction enzyme acts independently on the individual site or acts simultaneously at two copies of the recognition site. Representative data for NgoMIV cleavage of plasmids containing a single or two recognition sites are shown in; Cfr10I and Bse634I exhibit qualitatively similar cleavage patterns.

Kinetic studies revealed that tetrameric enzymes NgoMIV, Bse634I, and Cfr10I all cleave plasmid DNA containing two recognition sites much faster than a single site plasmid. Moreover, most of the supercoiled two-site DNA is directly converted into the final reaction product – linear DNA with two double-stranded breaks and only a small fraction of open-circle DNA and linear DNA cut at one site is formed. We have demonstrated that Cfr10I interacts simultaneously with both recognition sites located *in cis* through DNA looping. We suggest that a similar model could be applied for NgoMIV and Bse634I.

Hence, tetrameric restriction enzymes NgoMIV, Bse634I, and Cfr10I simultaneously bind two copies of their recognition sequence and rapidly cleave all four phosphodiester bonds at both sites during the lifetime of the enzyme-DNA complex. The cleavage rate of a plasmid containing a single recognition site by tetrameric restriction enzymes NgoMIV, Bse634I, and Cfr10I was much slower than that of two-site DNA. Addition of the oligodeoxynucleotide containing recognition sequence, however, increased the cleavage rate of the supercoiled form of pUCGCl and significantly reduced the amount of the open circle DNA intermediate.

In contrast, oligodeoxynucleotide lacking the recognition sequence had no effect on the cleavage rate of plasmid containing a single site (data not shown). We suggest that the activation of one-site plasmid DNA cleavage by cognate oligodeoxynucleotides is consistent with a tetrameric architecture of NgoMIV, Bse634I, and Cfr10I. In the presence of the specific oligodeoxynucleotide one of the dimers of the NgoMIV tetramer interacts with recognition site on the plasmid while the other one interacts with a specific oligodeoxynucleotide and such interactions *in trans* increase the cleavage rate of pUCGCl.

Transactivation by cognate oligonucleotides of plasmid DNA cleavage has been reported previously for tetrameric restriction enzyme Sfil. Hence, cleavage studies of plasmids containing a single and two recognition sites demonstrate that two recognition sites are required for effective DNA cleavage by NgoMIV, Bse634I, and Cfr10I and are consistent with a tetrameric architecture. Further mechanistic studies of tetrameric restriction enzymes and putative dimeric mutants are necessary to shed the light on the mechanism of cross-talking interactions between dimers.

CHAPTER 13

IIE Restriction Endonucleases

Restriction endonucleases (REases) and DNA methyltransferases (MTases) are encoded by chromosomal, plasmid, or viral genes. In eubacteria as well as in archaea, they often form biologically active DNA restriction-modification (R-M) systems. The first R-M systems, discovered by reversible growth reduction of bacterial viruses, were found to be encoded by chromosomal genes in *Escherichia coli* K-12, *Escherichia coli* B, and *Salmonella typhimurium*. Besides the prophage PI and the related plasmid p15, also naturally occurring drug resistance plasmids ("R factors") of the *fi*⁻ (fertility inhibition-minus) type have been shown to restrict and modify infecting bacteriophage or trans-conjugated plasmid DNAs in vivo. The first R-factor controlled R-M systems, called EcoRI and EcoRII today, have been defined by those phenotypic properties.

Endonucleolysi and cytosine methylation, respectively, have been proposed as the molecular mechanisms of EcoRII-specific restriction and modification. EcoRII was among the very first R-M enzymes for which the DNA substrate site was identified. The REase EcoRII recognizes the sequence 5′-CC(A/T)GG which exhibits a twofold rotational symmetry with a dyad axis corresponding to the central A-T pair. EcoRII cleaves the phosphodiester bond at the 5′ end of the first cytosine of the unmethylated sequence. The resulting DNA cleavage products exhibit cohesive ends with overhangs of five bases. By electron microscopy, Boyer et al. (1973) observed that EcoRII-generated cleavage products of SV40 DNA tend to reassociate at room temperature, which could be explained by the length and high CG-content of the "sticky" ends. Such re-association of cleaved EcoRII sites seemed to support experimental data that intracellular growth of bacteriophage λ is more efficiently restricted by EcoRI (efficiency of plating, eop=10^{-4}) than by EcoRII (eop=10^{-2}) even though there are 14 times as many EcoRII as EcoRI substrate sites per single genome. The EcoRII MTase converts the internal cytosine to 5-methylcytosine and thus, protects the DNA recognition sequence against endonucleolytic digestion.

The chromosomally encoded cytosine methylase of *Escherichia coli* K-12, M.EcoKDcm, exhibits the same sequence specificity and crossprotects DNA substrates towards EcoRII digestion. Genes coding for the EcoRII REase and MTase were cloned and their nucleotide sequence was determined. The two genes are convergently transcribed from separate promoters on opposite DNA strands. The open reading frame of the EcoRII REase codes for a protein with a molecular mass of 44 and 45.6 kDa, respectively, and that of the MTase gene codes for a protein of about 54.6 kDa. Investigations on the in vivo resistance of DNA towards the EcoRII R-M system have only been carried out to a limited extent.

The single-stranded DNA phages, fd and M13, were found not to be subject of restriction in cells carrying EcoRII-encoding plasmids (Arber 1966) but they are in vivo sub-strates for EcoRII methylation (Hattman 1973). Vovis et al. (1975) noted that complete digestion of the replicative form (RF) DNA of the closely related phage fl (2 sites per 6,407 bp) by EcoRII could not be obtained, and Hattman et al. (1979) reported that phage phiX174 RF DNA (two EcoRII sites per 5386 bp) could be completely digested only in the presence of unmodified heterologous DNA. The authors did not follow up this phenomenon.

The General Problem

Testing *E. coli* strains expressing different R-M systems regarding their ability to control growth of the virulent bacteriophages T7 and T3, it was found that these viruses are not restricted by EcoRII-producing cells though both phage genomes contain EcoRII-specific recognition sites. In contrast to the isoschizomeric REases BstNI and Mval, EcoRII can not even cleave the purified DNA genomes in vitro. The reason for this cleavage resistance can not be DNA methylation, because at least T3 DNA is not methylated at all due to the activity of a virus-encoded S-adenosyl-L-methionine (SAM)-hydrolase. Noticeable was the fact of strong avoidance of the pentameric EcoRII recognition sequence 5′-CC(A/T)GG in the virus genomes; T7 and T3 contain this nucleotide sequence only two and three times, respectively.

However, on the basis of the occurrence of its subsequences, 5'-CC(A/T)G, 5′-C(A/T)GG, and 5′-C(A/T)G, the EcoRII site is suspected to occur 56 times at least in the T7 genome demonstrating a strong counterselection against this particular pentameric sequence in the evolution of the virus. The resistance of the rarely occurring EcoRII-specific recognition sites in the genomes of the closely related phages T3 and T7 against cleavage by EcoRII led to the question whether this phenomenon is due to the effect of neighboring bases.

Upon cloning of an EcoRII recognition site including the 5′- and 3′- adjacent bases out of the T7 genome into the vector pUC 19, which is itself cleavable by EcoRII, the T7-derived recognition site became susceptible to EcoRII cleavage. It was generalized from this result that the site is not refractory per se and base sequences immediately around the EcoRII recogni-tion site are not the primary reason for cleavage resistance. Independently, it was observed that the recognition sites in T7 and T3 phage genomes become susceptible to EcoRII cleavage in the presence of DNAs, which contain an abundance of EcoRII-sensitive sites, *e.g.*, pBR322 plasmid or phage lambda DNA.

Studies using fragments of pBR322 DNA containing different numbers of EcoRII sites showed that the susceptibility to EcoRII cleavage is proportional to the number of sites in the individual fragment. An isolated fragment containing only one site was demonstrated to be highly resistant to cleavage. In the presence of increasing amounts of pBR322 DNA and a constant amount of T3 DNA, cleavage only went to completion when the molar ratio of recognition sites in pBR322:T3 DNAs was at least 2:1. Lower pBR322 concentrations in the reaction mixture did not suffice. The direct correlation between the amount of activator DNA and the extent of T3 DNA cleavage is indicative of a stoichiometric relationship of the DNA molecules involved in EcoRII cleavage activation.

It was postulated that EcoRII is the prototype of REases that require the presence of at least two simultaneously bound DNA sites for their activity a feature that so far was only known for enzymes involved in DNA transposition, recombination or replication. Today, REases with similar reaction mechanisms and substrate requirements as EcoRII form a group of their own–named Type IIE – within the family of Type II restriction enzymes.

ACTIVATION OF TYPE IIE REases

The inability of EcoRII to cleave non-methylated recognition sites in certain DNA molecules is due to an unusual reaction mechanism which is characterized by the simultaneous interaction of the protein with two DNA sites. What are the structural and functional prerequisites of an activator DNA molecule? Synthetic oligonucleotide duplexes of only 14-bp length with a single recognition site for EcoRII are sufficient to activate the cleavage of T3 DNA as well as the other high-molecular-weight DNAs mentioned above. In these reactions, EcoRII cleaves the short activator molecules themselves. A single-stranded oligonucleotide does not activate the cleavage of T3 DNA by EcoRII. Activation of EcoRII depends on the concentration of the oligonucleotide duplexes.

The complete cleavage of T3 DNA requires a molar ratio of recognition sites on the oligonucleotide duplexes to recognition sites on T3 DNA of 140:1. This high molar excess of the oligonucleotide duplexes compared to pBR322 plasmid DNA (2:1) is probably due to differences in the stability of the formed enzyme-substrate complexes that correlate with the length of the DNA. The fact that the 14-bp oligonucleotide duplexes are cleaved themselves in the reaction implies that EcoRII can simultaneously bring together two of these molecules in an enzyme-substrate complex. This was the first example of a REase that can bind two copies of its' DNA recognition sequence *in trans.* Conrad and Topal (1989) described two classes of DNA recognition sites for Nael, a Type HE REase isolated from *Nocardia aerocolonigenes*– sites susceptible and sites resistant to cleavage. They showed by kinetic analyses that Nael is activated by addition of DNA containing cleavable sites to cleave resistant sites.

The activation relies on an increase of the rate of DNA cleavage but did not alter substrate binding. In addition, they observed that in the presence of 1 mM spermidine, resistant Nael sites become cleavable although digestion of cleavable DNA is inhibited. Besides EcoRII, Nael is the second Type HE REase that has been comprehensively investigated.

BY CLEAVAGE PRODUCTS AND BY NON-CLEAVABLE OLIGONUCLEOTIDE DUPLEXES

To understand the activation mechanism of Type HE REases, the question had to be addressed whether the activator DNA molecules necessarily have to be cleavable. The six EcoRII recognition sites of pBR322 plasmid DNA, which can activate the cleavage of a priori resistant EcoRII DNA recognition sites in phage T3 DNA, were destroyed by EcoRII digestion. From the reaction sample, the enzyme was removed by phenol extraction and ethanol precipitation and the resulting pBR322 DNA fragments were incubated with T3 DNA and

EcoRII again. It became evident that the EcoRII-derived pBR322 cleavage products stimulate T3 cleavage. Remarkably, no enzyme activation was obtained using pBR322 cleavage products derived by digestion with the isoschizomers BstNI or Mval.

The data demonstrate that only the authentic EcoRII DNA cleavage products with their characteristic five-base 5′overhangs can activate EcoRII. BstNI and Mval do recognize the same DNA recognition sequence as EcoRII, but release DNA fragments with only a single-base 5′ overhang. Moreover, even oligonucleotides mimicking EcoRII cleavage products are efficient activators as long as they provide the 5′ phosphate group.

Another unconventional reaction mechanism] that is similar to the EcoRII enzyme activation by cleavage products has recently been reported for the REase SgrAI. Moreover, in contrast to the sensitivity of EcoRII to the 5′ phosphate group, oligonucleotide duplexes displaying intact recognition sequences but with a modified base (*e.g.*, 6-methyl-2′-deoxyadenosine, N^4-methyl-2′-deoxycytidine, I-2′-deoxyinosine) are still capable of activating the enzyme without being cleaved themselves. Interestingly, though confirming that EcoRII does not cleave oligonucleotide duplexes containing 6-methyl-2′-deoxyadenosine in the recognition sequence.

Petrauskene et al. reported that DNA molecules with this modified base in place of the central adenine are hydrolyzed in the presence of a canonical substrate. For Nael, it was demonstrated that oligonucleotide duplexes, which contain a Nael recognition sequence with an uncleavable phosphorothioate at the cleavage position, are similarly potent Nael-activators as the respective oligonucleotide duplex with a scissile phosphodiester bond. By investigating the cleavage of oligonucleotide duplexes of different lengths by EcoRII, an indirect correlation between their length and their cleavability by the enzyme was found. Oligonucleotide duplexes with a length of 14,30, and 71 bp, respectively, are cleaved with decreasing efficiency. Though EcoRII has a two orders of magnitude higher affinity for a 30-bp substrate compared to a 14-bp substrate, the former substrate is hydrolyzed more slowly.

The isoschizomeric REase Mval cleaves DNA substrates independent of their length. Moreover, EcoRII does not cleave DNA fragments of around 500-600 bp with only a single EcoRII recognition sequence, which were generated by pBR322 digestion with different REases. If there are two or more recognition sites on a DNA fragment, the distance between them plays a decisive role. The intersite distance must not exceed about 1000 bp. Furthermore, EcoRII simultaneously interacts with both strands of the substrate, but independently cuts them in a single binding event. When compared to the isoschizomeric Mval REase, EcoRII is much more sensitive towards structural modifications in the DNA recognition sequence. These data prove that it is not the cleavage of the activator molecule which activates Type HE REases, because cleavage products as well as uncleavable substrate analogues are good activators.

Although the activator DNA has not to be cleaved itself, it has to fulfil particular requirements: (1) EcoRII has to bind and to recognize the activator DNA, (2) for an effective intermolecular enzyme reaction (interaction of EcoRII with two recognition sites located *in trans*) the activator molecule has to be small enough to sterically allow the formation of the active enzyme-DNA complex, and (3) for an intramolecular reaction of EcoRII (interaction with two recognition sites located *in cis*), the distance between two recognition sites must be within a certain limit. One can summarize these experimental details in the following model: in the active enzyme-substrate complex, Type HE REases interact with two copies of their recognition sequence.

One of them serves as an allosteric activator and does not have to be cleaved. The allosteric site (activator) and the site that has to be cleaved can be located on the same or on separate DNA molecule(s). The interaction of Type HE REases with DNA sites on different DNA molecules depends on their length and concentration, whereas the interaction within the same DNA molecule is limited by the distance between two DNA sites.

THE ENZYMES' REACTION MECHANISM

Cooperative Interaction with Two Recognition Sites

The suggested two-site reaction model of Type HE REases implies some kind of substrate cooperativity. Studying the DNA cleavage rate of numerous REases on various substrates, Oiler et al. (1991) differentiated between resistant, slow, and cleavable substrate sites. Cleavage of the former two kinds of DNA sites could be significantly enhanced by the addition of cleavable DNA or spermidine. Based on the effect of the activator DNA on V_{max} and K_m of the reaction, they established two different action mechanisms. The REases Nael and BspMI give hyperbolic substrate saturation curves with varying V_{max} and constant K_m Cleavage was not significantly enhanced by increasing substrate concentrations.

In contrast, substrate cleavage by the enzymes Hpall, Narl, and SacII showed sigmoidal dependence of the initial rate on the substrate concentration. This indicates that the substrate binding at the active site was cooperative. With increasing concentration of the respective activator DNA, K_m decreased, whereas V_{max} remained constant. Based on these results, it has been concluded that all five enzymes possess independent binding sites for the activator and the substrate DNA, but follow different positive allosteric enzyme activation mechanisms. Moreover, it has been proven that the two DNA-binding sites of Nael are nonidentical and that they recognize different families of sequences flanking the recognition sequence.

By measuring the relative binding affinities of Nael for a 14-bp oligonucleotide duplex, which contained the Nael recognition sequence 5'GCCGGC flanked by various sequences, Yang and Topal (1992) demonstrated that the activator binding site of Nael prefers GC-rich flanking sequences, whereas the substrate binding site of Nael most efficiently binds AT-rich flanking sequences. Though detailed studies on how different neighboring bases influence the cleavage efficiency of EcoRII recognition sites have not been carried out, it is unlikely that EcoRII can strictly discriminate between activator and substrate molecules. Recognition site density, distance between recognition sites within a DNA molecule, and DNA concentration determine the catalytic activity of EcoRII. Gabbara and Bhagwat (1992) showed that, at low concentrations, short 14-bp activator oligonucleotide duplexes themselves are poorly cleaved by EcoRII.

The reaction showed positive cooperativity, and at high concentrations oligonucleotide duplexes are likewise activators and good substrates. The turnover number for EcoRII was determined to be about 700 double-strand scissions per minute per dimer, a number that is significantly higher than that of other REases. Moreover, these authors showed that the amount of enzyme-substrate complexes is not the limiting factor for product formation at substrate concentrations at which poor yields of the product are obtained. The enzyme-substrate complexes are stable, even in the absence of Mg^{2+} ions. Cleavage of longer synthetic oligonucleotide duplexes (30 and 111 bp) with a single EcoRII recognition site confirmed the sigmoidal dependence of DNA cleavage efficiency on DNA site concentration. In contrast, a 111-bp oligonucleotide duplex with two EcoRII recognition sites was still cleaved significantly at the lowest DNA concentrations tested this is due to the intramolecular cooperation of EcoRII with the two recognition sites.

Stoichiometry of the Active Enzyme-Substrate Complexes

EcoRII and Nael preferentially exist as dimeric molecules in solution, Nevertheless, the cooperative interaction of Type HE REases with two DNA recognition sites can be achieved either by one dimer alone or by binding of one dimer per DNA recognition site and subsequent formation of an active tetrameric protein-DNA complex. Both EcoRII and Nael bind to their DNA recognition sequence in a symmetrical manner. Using a 259-bp fragment with a single Nael recognition site, Baxter and Topal showed by DNasel digestion that Nael protects a DNA sequence of 24 bp, which is symmetrically located around the 6-bp recognition sequence.

The DNA region protected by EcoRII has been estimated to be 20-22 bp based on the dissociation kinetics of the protein bound to synthetic concatemeric DNA substrates (Vinogradova et al. 1990). DNasel footprinting analyses with a DNA substrate of 282-bp length containing two EcoRII recognition sites 82 bp apart reduced the DNA region specifically contacted by EcoRII to 16-18 bp. The DNasel footprinting patterns were symmetic in the CCAGG as well as in the CCTGG strands of both recognition sites. However, EcoRII did not protect the DNA between the two recognition sites against DNasel. In Nael cleavage assays, activator DNA stimulated either DNA nicking or DNA cleavage depending on Nael concentration.

The transition from nicking to cleavage activity depends on the increasing Nael concentrations and was interpreted to be due to two different structures of Nael. The first structure is a Nael monomer that is able to cleave one strand of a recognition site and the second structure is a dimer able to cleave both strands of a recognition site. Both, the nicking and the cleavage activity were postulated to require bound activator DNA. In gel electrophoretic mobility shift experiments, Nael formed two different complexes with DNA -monomer-DNA and dimer-DNA complexes. It was stated that both, the Nael monomer and the Nael dimer possess two DNA-binding sites.

Nevertheless, the completely active enzyme-substrate complex, which cleaves both strands of the recognition sequence, can only be formed by the Nael dimer and when the activator binding site is occupied. To determine the stoichiometry of the active EcoRII-substrate complex, Petrauskene et al. carried out single turnover experiments using increasing enzyme concentrations. The reaction rates increased with increasing enzyme concentration until the ratio of dimeric EcoRII molecules and DNA recognition sites of 1:2. Excess of EcoRII over DNA recognition sites inhibited the enzyme activity. This data indicated that the active EcoRII-DNA complex is formed when two subunits of the enzyme interact with two DNA recognition sites.

Reuter et al. confirmed that the cooperative interaction with two recognition sites can be achieved by one EcoRII dimer. These authors demonstrated that the enzyme inhibition of

EcoRII at high enzyme concentrations is reversible and that it is not due to high EcoRII concentration per se; a balanced increase of EcoRII recognition sites to the optimal molar ratio of one dimer per two recognition sites restored DNA cleavage to the normal extent. Moreover, the molecular weight of the EcoRII-DNA complex eluted from a gel filtration column corresponded to a dimeric enzyme structure bound to two DNA substrate sites. In contrast, studying EcoRII binding to various synthetic DNA duplexes, which contained either the modified or the canonical EcoRII recognition sequence, with native gel electrophoresis, Karpova et al. claimed that the active complex is most likely formed by one monomeric EcoRII subunit and one DNA recognition site.

This model implies that after occupation of both DNA-binding sites on the EcoRII dimer, the complex dissociates into two monomer-DNA complexes that are both catalytically active in the presence of Mg^{2+}. After product release the monomeric EcoRII subunits are supposed to form again dimeric enzyme molecules. Indeed, Mucke et al. demonstrated that a monomeric EcoRII subunit binds to the EcoRII recognition sequence. Single amino acid replacement of Val^{258} by Asn yielded an EcoRII mutant that was unaffected in substrate affinity and in its DNaseI footprinting pattern when compared to EcoRII wild type, but exhibited a profound decrease in cooperative DNA binding and cleavage activity.

Analysis of the molecular mass of mutant V258N showed a high percentage of protein monomers in solution. The dissociation constant of mutant V258N confirmed a 350-fold decrease in the enzyme's dimerization capability. The electrophoretic mobility of the mutant enzyme-DNA complex was significantly higher than that of the wild-type EcoRII, and both can easily be distinguished in gel shift assays. Taking into account that EcoRII forms stable dimers with a K_D value for dimerization of 2.89 nM and that comparably fast-migrating enzyme-DNA complexes as in the case of the monomeric V258N mutant do not occur in the case of the wild-type EcoRII, the data contradict the model proposed by Karpova et al.

Communication Between Remote DNA Recognition Sites in a DNA Molecule

In principle, different enzymatic processes can be discussed for the interaction of protein molecules with distant, *in cis* located DNA elements – ATP-dependent DNA translocation and ATP-independent DNA sliding and loop-ing mechanisms. Because the enzyme activity of Type HE REases does not depend on ATP, DNA translocation seems unlikely. Topal et al. demonstrated by electron microscopy that in the presence of Ca^{2+} or Mg^{2+} ions, NaeI simultaneously binds to multiple recognition sites in pBR322 DNA to form multiloop structures with NaeI bound to the loop's base. The maximum number of loops formed with a common base suggests four binding sites per enzyme molecule.

NaeI binds also to other than the cognate DNA sites. These complexes probably represent enzyme-DNA intermediates trapped during the process of scanning through the DNA for specific binding sites. In contrast, Milsom et al. did not detect NaeI-mediated DNA loops using plasmids with two NaeI recognition sites interspersed with two *res* sites for sitespecific recombination by the Tn21 resolvase in the presence of Ca^{2+}. This assay evaluates the degree of inhibition of the resolvase from the extent to which the *res* sites were sequestered into separate loops by the REase.

Recombination by the resolvase can only occur after the loop formed by the REase has dissociated. Because the resolvase inhibition test can only detect DNA loop structures if the rate of loop disassembly is slower than the rate of recombination by the resolvase, it can be assumed that DNA loops formed by Nael had very short life-times under the used conditions. Several biochemical data provided the first hints that EcoRII also uses DNA looping to bridge the distance between two recognition sites *in cis* a Lac represser molecule as a "molecular barrier" bound between two cooperating EcoRII recognition sites, which were located 191 bp apart did not inhibit REase activity. This indicated that EcoRII communicates with two DNA sites by bending or looping of the intervening DNA stretch.

In contrast, the bound Lac represser blocks the essential cooperative interaction of the Type III REase EcoP15 with two inversely oriented DNA recognition sites. EcoP15-DNA complexes cooperate via an ATP-dependent DNA translocation mechanism. Bound proteins and irregular DNA structures also hinder other REases, *e.g.,* EcoRI that moves along the DNA helix by linear diffusion. Comparative cleavage of linear DNA substrates with differently spaced interacting EcoRII sites revealed an inverse correlation between site distance and cleavage efficiency starting from more than 80% at a distance of 10 bp to 6% at a distance of 952 bp. EcoRII probably bridges distances of more than 10 bp by distortions introduced into the enzyme-substrate complex, as by DNA bending or looping. Furthermore, this demonstrated that the probability of one DNA site to be occupied by an EcoRII dimer finding the second DNA site decreases with increasing distance.

At least at the optimal distance of one helical turn, the EcoRII cleavage is independent of the orientation of the two recognition sites in the DNA double strand. Comparable quantitative data on the correlation between site distance and cleavage efficiency as well as on optimal and limiting site distances are not available for Nael. Transmission electron microscopy provided evidence that EcoRII induces loop formation after binding of two EcoRII recognition sites 191 bp apart on linear DNA molecules in the presence and in the absence of Mg^2+ ions. The asymmetric location of the two EcoRII sites in the DNA mol-ecule allowed to distinguish both DNA ends. In contrast, binding of EcoRII to a single-site substrate did not induce DNA loops. DNA loop formation specifically occurred between the two recognition sites.

When comparing the frequency of loop formation of the dimeric wild-type EcoRII (20%) and the monomeric mutant enzyme V258N (3%), it was observed that induction of loop formation by EcoRII directly depends on an intact dimeric protein structure. These results again illustrate that the complex of the EcoRII dimer and two copies of the DNA recognition sequence represents the active form which is required for DNA hydrolysis.

SEQUENCE-SPECIFIC DNA RECOGNITION

By employing random mutagenesis of the Nael-encoding gene with N-methyl-N′-nitrosoguanidine and selecting bacteria by their ability to survive in the presence of mutagenized Nael and in the absence of a protecting methylase, Holtz and Topal (1994) isolated cleavage-deficient Nael mutant enzymes. Analysis of the enzymatic properties of different classes of mutants assigned amino acids involved in catalysis and DNA recognition and predicted the existence of two independent protein domains; an N-terminal substrate-binding domain that also contains the catalytic center and a C-terminal domain that isi only involved in DNA binding, probably the activator-binding domain. To answer the question of how EcoRII achieves sequence-specific DNA binding Reuter et al. used matrix-bound peptide scans that represent, the complete EcoRII amino acid sequence as overlapping peptides.

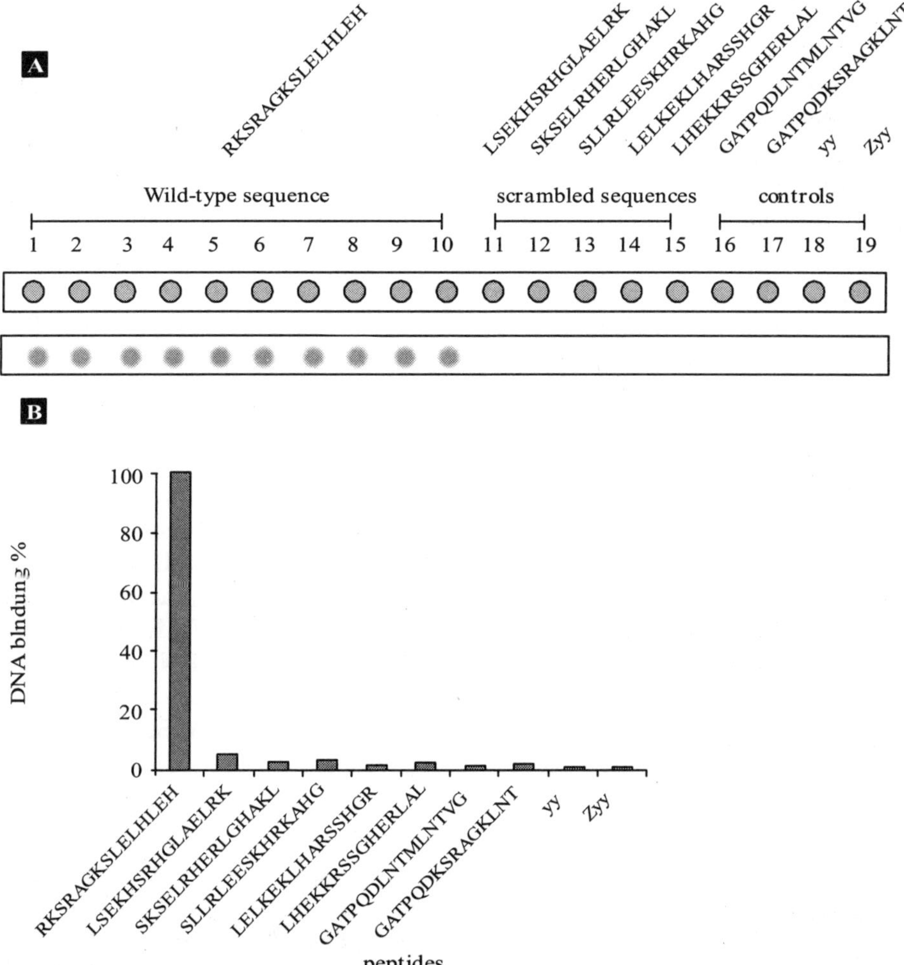

Fig. 13.1. DNA binding to synthetic EcoRII peptides comprising the 15 C-terminal amino acids of DNA binding region II arranged as wild-type or as scrambled sequence, a, upper part: Diagram of a cellulose strip with membrane-bound peptides and their sequences. *Spots 1-10* Amino acids 262-276 of the EcoRII wild-type sequence; *11-15* randomly scrambled sequences of amino acids 262-276 of EcoRII; *16* HIV-1 capsid epitope peptide; 17 HIV-1 capsid epitope peptide with the EcoRII consensus sequence KxRxxK; *18* b-alanine that is used as linker between peptide and membrane; *19* acetylated b-alanine residues; *lower part* scan of a cellulose strip after incubation with radioactively labeled specific oligonucleotide duplex in the presence of a twofold molar excess of unspecific oligonucleotide duplex over bound peptides. b Quantitation of bound radioactivity per peptide spot was performed as described in Reuter et al. The average value of bound radioactivity to the ten wild-type sequences was calculated and defined as 100%.

Two separate DNA-binding regions emerged; amino acids 88-102 (binding site I, RHFGKTRNEKRITRW) and amino acids 256-273 (binding site II.NSVSNR-, RKSRAGKSLELH), which share the consensus motif KXRXXK. Substitution analogs were obtained by exchanging every residue of both binding sites for' all other amino acids. Results obtained with these substitution analogs demonstrated that the replacement of the basic amino acids within the consensus motif strongly affected DNA binding of the respective peptides. EcoRII mutant enzymes generated by replacing the consensus lysine residues for alanine or glutamic acid in one or both DNA-binding sites showed different effects-mutations in DNA-binding site I attenuated DNA binding, whereas corresponding mutations in DNA-binding site II inhibited DNA cleavage without changing the DNA dissociation constant when compared to the EcoRII wild-type.

Apparently, the binding site is important for DNA recognition by EcoRII, because introducing alanine instead of glutamic acid at position 96, nearly abolished DNA binding of EcoRII. The motivation for this E96A mutant came from a search for putative catalytically relevant amino acids. Based on the catalytic consensus motif P(D/E)X_n (D/E)X(K/R), E96 seemed to be part of a potential catalytic site.

To investigate whether the specific amino acid sequence or the overall amino acid composition of DNA-binding sites I and II is important for the DNA binding, "scrambled" peptides of both DNA-binding sites have been used. The data obtained indicated that in the C-terminal part of binding I region II (NH_2-RKSRAGKSLELHLEH-COOH) the specific amino acid sequence mediates specific DNA binding. In contrast to the original peptide sequence, peptides with an altered amino acid sequence did not bind DNA EcoRII-specifically. DNA binding was not affected by "scrambling" the amino acids in the DNA-binding site I of EcoRII. In 1993, Kosykh and coworkers described an amino acid sequence homology between the target recognition domains of three DNA MTases (M.EcoRII, M.Dcm, M.SPR) and the REase EcoRII. Each of the enzymes recognizes the IDNA sequence 5'CCWGG.

Interestingly, EcoRII DNA-binding site II, which was detected by synthetic peptide scans, is part of a homology region. Moreover, sequence alignments identified analogous motifs to EcoRII DNA-binding site II in various REases with terminal G:C or C:G base pairs in their recognition sequences. Particularly, the great similarity of binding site II of EcoRII (/CCWGG) to a sequence in the EcoRII-homoisoschizomeric thermophilic restriction enzyme PspGI, to SsoII (/CCNGG) and its considerable resemblance to a ScrFI sequence (CC/NGG) lent independent support to a role of this site in sequence-specific DNA recognition. More comprehensive sequence and structure-based alignments have predicted a third DNA-binding motif and a potential active site in the C-terminal part of EcoRII between amino acids Pro298 and Glu337. Beside the basic lysine residues, these data also suggested a critical role for several arginine residues in the earlier described EcoRII binding sites I (RHFGKTRNEKRITRW) and II (NSVSNRRKSRAGKSLELH).

The generated EcoRII enzyme mutants confirmed the effect of the lysine substitutions (1) replacing the single arginine residues R88, R94, R98, and R101 by the neutral alanine in DNA-binding site I decreased the DNA-binding capability of the enzyme mutants R94A, R98A, and R101 A, and indirectly their catalytic activity; (2) though amino acid substitutions at positions R262 and R265 in the DNA-binding site II did not affect the DNA binding, they yielded inactive enzyme mutants. To further support the finding that the identified DNA-binding regions of EcoRII contribute to a different extent to DNA recognition and cleavage, data of a few other available EcoRII mutant enzymes were introduced into.

DOMAIN ORGANIZATION

Proteolytic cleavage of proteins provides information regarding protein folding units. Stably folded protein domains are expected to be better protected against proteolytic cleavage than linker or loop regions that connect protein domains. Nael and EcoRII were subjected to limited proteolysis to map their domain organization, and the domains thus identified were purified after cloning of the respective coding DNA sequences. The trypsin and chymotrypsin digestion patterns were determined in the presence and in the absence of specific DNA. Both enzymes revealed two protease-resistant domains involving approximately the N- and C-terminal halves of the proteins. The experimental data showed that Nael consists at least of one N-terminal domain (amino acids 1-120) and one C-terminal domain (amino acids 151-317). Both domains are linked by a protease-sensitive linker (amino acids 121-150). EcoRII consists of an N-terminal domain (amino acids 1-192) and a C-terminal domain (amino acids 173-404). Amino acids 173-192 probably represent a connecting linker region.

DNA binding mainly protected the C-terminal domain of Nael and the N-terminal domain of EcoRII, respectively, against proteolysis. This indicates, that both domains present dominant DNA-binding sites. Photocrosslinking of EcoRII to an oligonucleotide duplex containing a single recognition site with 5-iododeoxcytidine substitutions for each C and 5-iododeoxyuridine substitutions for A, G, or T, strongly supported the predominant contribution of the N-terminal domain to DNA binding. Interestingly, EcoRII crosslinked specifically to the 5′C of the 5′CCAGG strand, but not to the 5′C of the 5′CCTGG strand yielding an asymmetrical cross-linking pattern.

Mutational analyses of the electronrich amino acids of the photocross-linked peptide in EcoRII revealed Tyr41 as the photocross-linking amino acid. Characterization of the purified protein domains demonstrated that only the C-terminal domain of Nael and the N-terminal domain of EcoRII specifically bind DNA. The apparent dissociation constants differed only by one order of magnitude from those of the complete enzymes. Mutational analysis of the N-terminal domain of Nael implied that this domain mediates catalysis. Nevertheless, the purified N-terminal domain of Nael showed neither DNA binding nor catalytic activity. Although the N-terminus of EcoRII specifically bound to the DNA, it was not able to cleave DNA substrates. In contrast, the EcoRII C-terminal domain cleaved DNA specifically and much more efficiently than wild-type EcoRII.

The fact that EcoRII-C specifically cleaved DNA implies that it also can specifically bind to DNA. Based on these data, it was concluded that the C-terminal EcoRII domain (termed EcoRII-C) contains all structural and functional components of an orthodox Type IIP REase that cleaves DNA at single recognition sites. Indeed, this truncated form of EcoRII no longer needs the simultaneous interaction with two copies of the recognition sequence and cleaves DNA substrates inde-pendent of the number and the distance of the DNA recognition sites in vitro and in vivo. DNA of phage T3 that resists cleavage by EcoRII wild-type enzyme is completely cleaved by EcoRII-C as efficiently as by the iso-schizomer BstNI. These data obtained in vitro have been confirmed in vivo by plating phage T3 on *E. coli* cells that express EcoRII-C. In contrast to EcoRII wild-type, EcoRII-C restricts growth of phage, T3 by one and a half (1.5) orders of magnitude. Remarkably, the restriction of phage λ by EcoRII-C is less efficient as by the EcoRII wild-type.

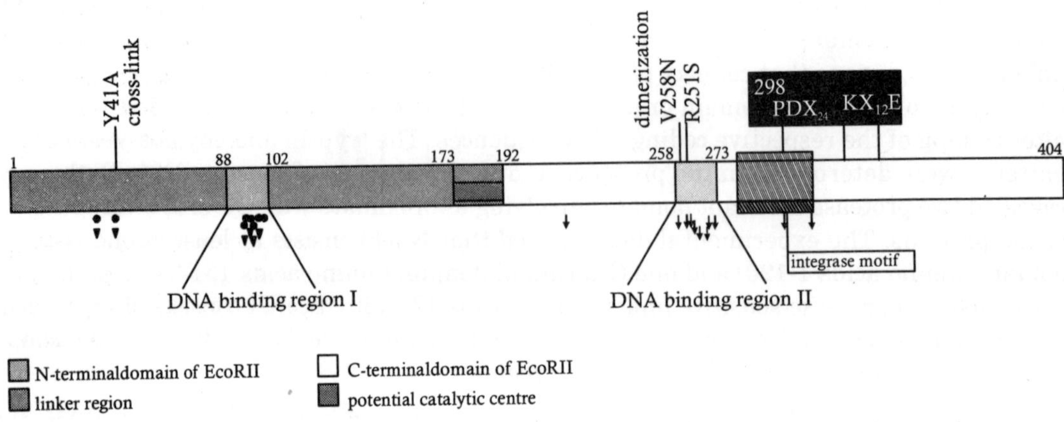

Fig. 13.2. Overview of the EcoRII domain organization, its functional important regions find amino acids. The bar corresponds to the 404 amino acids of the EcoRII monomer. The black box in the C-terminal domain shows the predicted active site. Y41 was photochemically cross-linked to the 5'C of the t'CCAGG strand of the EcoRII recognition sequence, V258 and R261 apparently are revolved in EcoRII dimerization. EcoRII mutants that are compromised in DNA binding and catalysis are indicated as dots and arrows, respectively. Mutants at several positions in the N-terminal domain show predominantly impaired DNA binding, but often also linpaired catalysis. Because the latter is probably a secondary effect it is marked by arrow heads.

Analytical ultracentrifugation experiments confirmed that EcoRII-C , mainly forms dimers in solution. The dimeric structure allows specific cleavage of double-stranded DNA. The Nael N-terminal domain was shown to form various aggregates in solution. In contrast, the N-terminal domain of EcoRII is monomeric in solution as is the C-terminal domain of Nael. Studies using synthetic peptide scans to determine protein-protein contacts of EcoRII further support that mainly interactions in the C-terminal part of the protein stabilize the EcoRII dimmer, However, these data do not exclude that the N-terminal domains of EcoRII dimerize on DNA binding.

Table 13.1. Efficiency of plating of unmodi-fied phages T3 and X on *E. coli* cells expressing EcoRII wild-type or EcoRII-C

E.coli strain	T3	λ
JM109	1	1
JM109(EcoRII)	1	10^{-3}
JM109(EcoRII-C)	5×10^{-2}	5×10^{-2} D

In brief, NaeI and EcoRII show a similar two-domain organization with a linker structure that connects both domains. The C-terminal domain of NaeI appears to correspond functionally to the N-terminal domain of EcoRII in that both domains exhibit high DNA-binding affinity. The N-terminal domain of NaeI and the C-terminal domain of EcoRII apparently harbor the endonuclease catalytic sites, even though in contrast to EcoRII-C, the purified N-terminus of NaeI had no enzymatic activity. As an example, the EcoRII monomer is shown schematically together with the available amino acid substitutions and the corresponding loss of function supporting the predicted two-domain organization.

MODULAR ARCHITECTURE

The crystal structure of NaeI was solved to 2.3 Å and showed that NaeI is a dimeric molecule with two domains per monomer. Each domain contains one potential DNA recognition motif. The N-terminal domain folds like a Type II REase, A-exonuclease, and the DNA repair enzymes MutH and Vsr, and therefore was named Endo domain. The C-terminal domain contains the catabolite activator protein DNA-binding motif, a helix-turn-helix motif, which is present in a variety of DNA-binding proteins including topoisomerases. Because this structural homology matches the finding that replacement of Leu43 by lysine changed the NaeI endonuclease activity to a topoisomerase/recombinase activity, the C-terminal domain was referred to as Topo domain. The Topo domain consists of a two-layer β-sandwich adjacent to four helices. In contrast to the well-ordered Endo domain, the Topo domain is loosely packed and appears to be of considerable flexibility. Both domains pack against one another by van der Waals interactions and hydrogen bonds.

The assignment of the Endo domain to substrate binding implies that the Topo domain is responsible for the activator binding. Compared to the domains that were predicted from proteolysis studies, the domain location was revised to the following amino acid residues: Endo domain 1-162, hinge loop 163-171, and Topo domain 172-313. The structure of NaeI bound to a 17-bp cognate DNA substrate, which was determined to 2.5 Å, demonstrated that each of the two domains of NaeI recognizes one DNA duplex. DNA recognition induces structural rearrangements. The binding site of the Topo domain was narrowed and that of the Endo domain was widened to encircle and bend DNA by 45° for cleavage.

The interface between the two NaeI monomers was completely restructured. The structure of the NaeI-DNA complex was the first example that two copies of the same DNA sequence are bound by two different amino acid sequences and two different structural motifs in one polypeptide chain. The structure of the DNA duplex is different depending on where it is bound; the DNA bound to the Endo domain shows B-DNA conformation, but with increased pitch and number of bases per helical turn. Furthermore, the DNA is bent approximately by 45° around the central CG base pairs of the recognition sequence. In contrast, the DNA bound to the Topo domain activator site is not bent and shows B-conformation with the expected values for the pitch and number of bases per helical turn. Moreover, it was shown that the Endo domain recognizes GCCGGC mainly through hydrogen bonds in the major and in the minor groove.

In contrast, the Topo domains only contact the DNA via the major groove. In the complex with DNA, the ten hydrogen bonds that mediate the NaeI dimerization in the unligated form are replaced by five new hydrogen bonds. In addition, most of the van der Waals interactions between the Nael jnonomers also rearrange. The Nael dimer-DNA complex is symmetric, in contrast to the asymmetric monomer arrangement in the unligated Nael dimer. Structure-based alignments detected similarities of Nael with eight REases (EcoRV, PvuII, Bgll, BamHI, EcoRI, Muni, Cfr10I, FokI), with DNA repair endonucleases Vsr (very short patch repair) and MutH (mismatch repair), and with 5′-3′ A-exonuclease (recombination and repair). These structural and topological similarities between REases and other DNA processing enzymes probably imply that DNA processing enzymes evolved from a common ancestor and share a similar catalytic mechanism. For the EcoRII wild-type enzyme, two different crystals have been described that were obtained under varying experimental conditions.

In the absence of Mg^{2+}, cubic EcoRII crystals were obtained which diffracted to about 4.0 Å. In another attempt using His_6-tagged EcoRII, monoclinic crystals appeared in the presence of Mg^{2+} and diffracted to 2.8 Å. Both EcoRII wild-type crystals revealed two EcoRII monomers per asymmetric crystal unit. This is consistent with the fact that EcoRII is also a dimer in solution. The asymmetric unit of the crystals of a single amino acid substitution mutant of EcoRII, R88A, also contained two EcoRII monomers. The enzymatic properties of this mutant did not differ significantly from that of the EcoRII wild-type. The crystals were easier to grow and diffracted to 2.1 Å. Their structure was now determined. The overall EcoRII structure consists of 18 α-helices and 13 β-strands and can be divided into two domains linked through a hinge loop. These are the N-terminal domain (residues 1-172), a linker, and a C-terminal domain.

The N-terminal domain is built up from an eight-stranded β-sheet. The C-terminal domain has an endonuclease-like fold with a central five-stranded mixed β-sheet surrounded on both sides by α-helices. Its topology is similar to that of other Type II endonuclease structures with a common structural core. The search for structural homologues did not identify any known protein domain that is similar to the N-terminal domain of EcoRII and thus indicated that the N-domain has a novel fold. The EcoRII N-terminal domain corresponds to the functional effector binding domain identified by limited proteolysis. It has a prominent cleft near its interface with the C-terminal domain. This cleft harbors numerous amino acids residues that were shown to be important for EcoRII DNA-binding earlier by different biochemical methods, *e.g.*, Tyr41 that cross-linked to the 5′ of the CCAGG strand of the EcoRII recognition sequence is right inside the cleft.

In addition, the positively charged DNA-binding site identified by membrane-bound peptide repertoire is also part of the cleft. Furthermore, several amino acid substitutions located in the cleft yielded mutant enzymes that were severely impaired in DNA binding (H36A, Y41A, K92E/K97E, R94A, E96A, R98A, and R101A). Therefore, it is concluded that this cleft is involved in binding the effector DNA. The most similar structures to the C-terminal domain of EcoRII identified from structure databases are the REases NgoMIV and Cfr10I followed by BsoBI, FokI, and Muni. This homology implies structural conservation and similar catalytic mechanisms among these endonucleases despite very low sequence identities.

Structure-based alignments of the EcoRII C-terminal domain with NgoMIV and Cfr10 supported earlier predictions of the active center of EcoRII and putative catalytic residues E271, D299, K324, and E337. Based on the described structural alignment, Zhou et al. modelled a DNA duplex containing the EcoRII recognition sequence into the putative active center of EcoRII and showed that the catalytic site is spatially blocked by parts of the N-terminal domain and thus, is inaccessible for the substrate DNA. In this inactive state, specific DNA can only bind to the N-terminal effector domain. The active site only becomes accessible after the effector domain has moved away. This probably will be accomplished by effector DNA binding induced conformational rearrangements.

REACTION MECHANISM OF TYPE IIE REases

The homodimeric Type HE REases have to bind two copies of the recognition site to cleave the DNA. Therefore, the actual recognition sequence is 5′GCCGGC-$(N)_X$-GCCGGC for Nael and 5′CCWGG-$(N)_X$-CCWGG for EcoRII. Nael and EcoRII form loops with DNA molecules that possess two or more recognition sites in order to bring them together in the active complex. Though interaction with two recognition sites preferably occurs *in cis* (Schleif 1992), both REases can also interact with two recognition sites on different DNA molecules *in trans*. The cleavage efficiency *in trans* depends on DNA recognition site concentration and is inversely correlated to the length of the interacting DNA molecules. The results of cleavage experiments with Nael and EcoRII using plasmid DNAs with one and two recognition sites, respectively, emphasize that both enzymes require a second DNA recognition site to cleave efficiently.

Nael and EcoRII cleave supercoiled plasmid DNA with two recognition sites faster than plasmids with a single site (G. Tamulaitis, M.R., M.M., V. Siksnys, in prep.). Furthermore, Nael cleaves catenanes that consist of two plasmid rings each with a single recognition site faster than plasmids with single sites. At least under the experimental conditions tested, EcoRII and Nael mainly convert a two-site plasmid into the full-length linear DNA that results from cleavage at only one of the two bound recognition sites (G. Tamulaitis, M.R., M.M., V. Siksnys, in prep.). Cleavage at only one out of two recognition sites and the reduced cleavage efficiency of Type IIE REases with one-site substrates in general compared to orthodox Type IIP enzymes can be explained by the two-domain structure found for the Type IIE enzymes investigated so far. Nael and EcoRII consist of two stably folded domains flexibly connected by a linker structure. These two domains are an endonuclease-like domain and a DNA-binding domain that both specifically interact with DNA.

The DNA-binding domain binds one of the DNA recognition sites, the activator site. This binding activates the endonuclease-like domain in turn to bind and to cleave the second DNA recognition site, the substrate site. The essential positive cooperativity observed for DNA cleavage by EcoRII or Nael is due to this two-site DNA-binding process. For Nael, the crystal structure revealed a structural motif characteristic for the DNA-binding of the catabolite activator (CAP) protein of *E. coli* in the DNA-binding domain. In the case of EcoRII, the DNA-binding domain shows a new fold.

Because the DNA-binding dissociation constants for the separately purified DNA-binding domains of Nael and EcoRII differ only marginally from those of the complete enzymes, and because the isolated catalytic domains do not bind (Nael) or bind only weakly (EcoRII) to DNA, it is supposed that in the complete enzymes the initial DNA-binding event is that to the activator domain. Structural comparison of the unligated and ligated Nael showed that in the absence of DNA, the DNA-binding cleft at the catalytic site is too narrow to accommodate DNA. Only after DNA binding to the Topo domain, the DNA-binding cleft of the Endo domain widens by several A and allows DNA substrate binding. The situation in the EcoRII structure even seems more complex. In the absence of DNA, EcoRII forms an inactive dimer structure. In the inactive structure the effector domain site is accessible for DNA, but the catalytic site is spatially blocked by the effector domains. It is assumed that in the presence of DNA, DNA binds first to the effector site.

Effector binding induces domain rearrangements by moving the two effector domains away from the catalytic domains and by bringing the catalytic domains closer to each other to form the active dimer with the substrate DNA bound. This could mean that in the biological context, EcoRII is regulated by a typical autoinhibition mechanism. This kind of tight "on-site" repression of enzymatic activity was described for numerous transcription factors as well as for proteins involved in signal transduction in eukaryotic cells, but this would be the first example observed among REases.

This autoinhibition mechanism matches the observation that in limited proteolysis, initial DNA binding protects the N-terminal domains of EcoRII against proteases, whereas the C-terminal domains are accessible to proteolytic digestion. In contrast, in the absence of DNA the C-terminal domains are more resistant towards protease digestion, possibly because of their spatial protection by the N-terminal domains. Moreover, it was shown that the N-terminal domain of EcoRII (EcoRIl-N) specifically binds to the DNA recognition site but does not cleave it, while the C-terminal domain of EcoRII (EcoRII-C) alone acts as an orthodox Type IIP endonuclease, which specifically cleaves also at single 5′CCWGG sites. Apparently, cutting off the N-terminal domain relieved the autoinhibition of EcoRII and activated the catalytic C-terminal domain.

Together with the modular EcoRII structure as a prerequisite, these experimental data perfectly match the definition of autoinhibition; activity, enhancement of a particular protein domain in the absence of some other region of the protein indicates autoinhibition. Because autoinhibitory domains usually negatively regulate the function of another domain, a repression *in trans* can be expected. Cleaving pBR322 DNA with EcoRII-C is inhibited by increasing amounts of EcoRII-N from a molar ratio of EcoRII-N:EcoRII-C of 1:1 upward. DNA cleavage by EcoRII-C is completely abolished at a 20-fold molar excess of EcoRII-N over EcoRII-C. The same inhibitory effect can be achieved by leaving the protein amount in the sample constant, but successively substituting EcoRII-N for EcoRII-C. However, these experiments can not differentiate whether cleavage by EcoRII-C is inhibited by EcoRII-N due to inter-domain interac-tions or by competitive DNA-binding to EcoRII recognition sites on the DNA. The structure-derived EcoRII activation model is different from that of Nael whose effector-binding domains (Topo domains) have neither sequence nor structural homology with those of EcoRII and are physically distant from the substrate binding cleft.

In summary, the complex of the dimeric Type HE REases with two DNA recognition sites is the active form of the enzymes. Only the DNA recognition site bound to the endonuclease–like domains of the dimer serves as substrate that is cleaved.

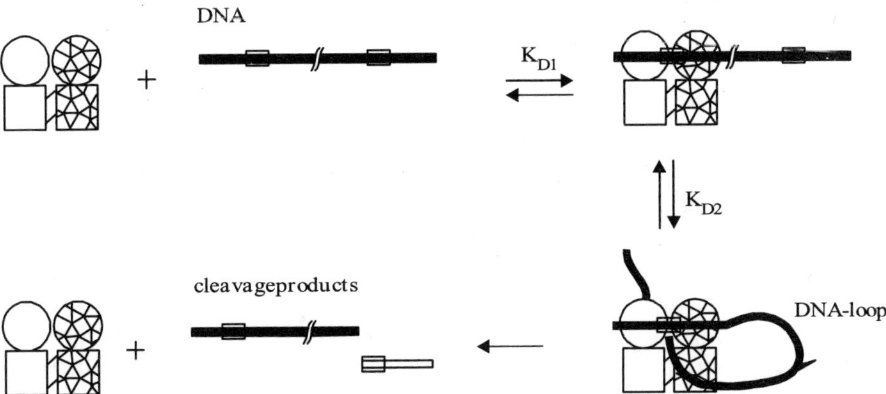

Fig. 13.3. Type HE restriction endonuclease interaction with DNA. The two domains of a monomer are symbolized by a *circle* and a *rectangle*, respectively. *White* Monomer A, *black grid* monomer B of the homodimer. *Circles* DNA-binding domain; *rectangle* endonuclease domain. The *lines* between the *rectangles* symbolize dimerization contacts. *Black lines with boxes* DNA with specific DNA recognition sites. DNA first binds to the DNA-binding domains of the dimer. In a second event, the DNA-binding site in the endonuclease-domains becomes occupied. The enzyme forms a DNA loop with DNA containing two recognition sites. The DNA recognition site bound to the endonuclease-domains will be cleaved

EVOLUTIONARY LINKS BETWEEN REASES, TOPOISOMERASES AND DNA RECOMBINASES

Although both enzymes, Nael and EcoRII, possess an additional DNA-binding domain, EcoRII was the first REase that after dissection into functional domains revealed a still intact and even more active endonuclease unit that is regulated in the complete enzyme by an autoinhibition mechanism. Based on these results, we like to suggest that during evolution the C-terminal domain of EcoRII acquired an additional DNA-binding domain and thus evolved a new protein function. Sequence homology studies among different subtypes of REases have also suggested this acquisition of an additional DNA-binding domain for EcoRII. The newly acquired function for EcoRII is the simultaneous interaction with two identical DNA recognition sites and, moreover, allows to control the "off" and the "on" state of the EcoRII enzyme activity, probably to prevent inappropriate function.

Activity will only be achieved in the presence of an activating signal, namely by binding of DNA to the effector domain. In the activated EcoRII the autoinhibitory N-terminal domain fulfils a new function - mainly to achieve DNA binding of the enzyme. The essential interaction of Type HE enzymes with two recognition sites to form the catalytically active complex could be considered as a safety mechanism to avoid accidental cleavage of single unmethylated recognition sites in the host genome.

Unmethylated sites can occur after incomplete DNA methylation or DNA repair. Furthermore, the interaction of Nael and EcoRII with their bipartite complex recognition sequences 5'GCCGGC-$(N)_X$-GCCGGC and 5'CCWGG-$(N)_X$-CCWGG, respectively, also provides the functional benefit of increased DNA specificity. The specificity of Nael measured as k_{cat}/K_M *for* a noncognate recognition site was determined to be 10^{11}-fold lower than that for a cognate recognition site. This accuracy of Nael cleavage is approximately 10^4-fold higher than that of restriction enzymes that require only one site. The requirement for two DNA sites relates Type IIE REases to enzymes that participate in cellular processes such as DNA recombination and transposition in a wider sense.

Evidence for a connection between EcoRII and these enzymes came from conserved amino acids between EcoRII and the integrase family of site-specific recombinases. However, it was neither possible to isolate a covalent intermediate between EcoRII and DNA, the formation of which is a hallmark of integrase proteins nor to demonstrate EcoRII-mediated DNA recombination so far. Moreover, an evolutionary link between REases, topoisomerases, and recombinases has been suggested. The Nael amino acid sequence uncovered similarity to the active site of human DNA ligase I, except that for Leu43 in Nael a lysine is essential for the ligase activity. The replacement of Leu43 in Nael by lysine produced a variant with Nael-typical DNA topoisomerase activity, which includes DNA cleavage, strand passage, and religation. The topoisomerase activity of mutant NaeI-L43K is Nael-sequence dependent.

Intercalating topoisomerase inhibitors completely inhibited DNA cleavage by NaeI-L43K, but only marginally inhibited Nael wild-type. This single amino acid substitution significantly changed DNA recognition by Nael. In contrast to Nael, which recognizes double-stranded DNA, NaeI-L43K preferably interacts with single-stranded and mismatched DNA regions. Because NaeI-L43K changes the linking number of a single negatively supercoiled topoisomer of pBR322 by units of one and decatenated nicked DNA circles, it behaves like a prokaryotic Type I topoisomerase. Recently, it has been shown that a positive charge at position Leu43 of Nael dramatically changes the Nael protein fold. The changes affect both, the Endo and the Topo domains. The discovery of a covalent Nael-DNA intermediate provided further direct evidence for an evolutionary link between Type HE REases and the recombinase protein family.

The amino acid residue involved in the covalent contact to DNA has not yet been identified. Interestingly, endonuclease-like structures have been found in DNA transposases, recombinases and DNA repair enzymes in prokaryota, eukaryota, and archaea, *e.g.*, in the X-exonuclease, the MutH protein, the Vsr (very short patch repair) endonuclease, and a transposase subunit coded by Tn7. Furthermore, sequence and structural alignments predicted evolutionary relationships between Type II REases and eukaryotic non-LTR retrotransposable elements, the Hjc resolvase, a recombination enzyme from archaea, the endonuclease VII, a repair enzyme of bacteriophage T4, and homing endonucleases.

The structural homologies could mean that these DNA-processing enzymes have evolved from a very distant ancestral nuclease fold, and that the function of this hypothetic nuclease was adapted to different fundamental cellular processes. This hypothesis is supported by an evolutionary tree of the endonuclease-like superfamily. Based on this phylogenetic relationship, it has been proposed that the nonspecific DNA cleavage domain of the Type US REase FokI is evolutionarily older than other endonucleases of the same branch. FokI might have acquired a separate domain for specific DNA binding. Because of the numerous hints for an evolutionary relationship between REases and DNA-processing enzymes, we suppose that acquiring an additional DNA-binding domain could be the first step on the evolutionary way to protein functions beyond phosphodiester bond cleavage.

NATURAL PROTEINS WITH NEW FUNCTIONS

The described structural homology between protein folds apparently emphasizes the evolutionary link between enzymes involved in site-specific recombination and transposition on the one hand, and REases on the other – a link that might rely on the functions of R-M systems in nature. Besides the commonly accepted explanation that R-M systems counteract virus infections in bacterial cells, two other hypotheses have been discussed to explain the devel-opment and maintenance of the impressively high number and diversity of R-M systems during evolution. Genetic stability for an individual or for a population at short term is as important as, genetic variability at long term.

The endonucleolytic and methylating activities of DNA R-M systems likewise can accomplish both demands. Arber referred to R-M genes as "evolution genes" and termed them together with repair genes "frequence modulators" of genetic variation; on the one hand, these gene products restrict the uptake of foreign genetic material into bacterial cells to a tolerable extent. On the other hand, they produce a pool of potential recombinogenic DNA fragments that can be incorporated into the bacterial genome. Comparison of the accumulated bacterial genomic sequences, their GC-content, and codon usage give evidence for the mobility of R-M genes within and between different genomes that often causes rearrangements of genomic sequences.

Further evidence for the frequent horizontal transfer of DNA R-M genes includes (1) that they are often flanked by insertion elements, (2) that DNA R-M genes themselves are often located on transposable elements and (3) that they were found in the vicinity of sequences encoding DNA transposases and recombinases. R-M genes were much more often found by genomic approaches than expected, though they are not always expressed. *Helicobacter pylori* strains, for instance, which can colonize the human stomach and are associated with the onset of stomach ulcer and cancer, contain 14 or 15 putative R-M systems corresponding to approximately 1% of the genome. In this regard, it seems exceptionally surprising that only one phage has been described so far that infects *Helicobacter pylori* strains. Otherwise, *Helicobacter* is naturally competent to accept the uptake of foreign DNA.

Though interstrain-specific horizontal gene transfer is apparently a major source of genetic diversity in *Helicobacter*, the numerous R-M systems might indeed function as a barrier to control the DNA exchange. Moreover, in the genomes of *Neisseria gonorrhoeae* strains twenty or more putative R-M systems were identified. It is hard to believe that all these R-M systems serve as protectors against foreign DNA. Thus, it can be expected that right in these human pathogenic bacterial species, R-M systems influence further biological features, *e.g.*, bacterial pathogenicity. Recently, an interrelation between the expression of the R-M system NmeSI and the hypervirulence of a *Neisseria meningitidis* strain has been described. Another hypothesis postulates R-M systems to be selfish genetic elements like viruses or transposons. R-M systems ensure their maintainance within a population at the expense of their host cells, in that they kill cells that have lost the R-M genes. This observation was described for Type II restriction enzymes.

In contrast, loss of Type I R-M systems does not seem to be lethal for the host cell. Rocha et al. analyzed to what extent small palindromic sequences occur in viral and bacterial genomes. The comparison demonstrated that in most cases bacteria much more counterselected palindromic sequences in their genomes than viruses. This result might support an invasive character of R-M systems and the resulting evolutionary pressure to the bacterial genomes to avoid palindromic sequences for their integration. Taking into account that R-M genes presumably live like nomades and that the encoded enzymes can integrate DNA fragments into the host's genome, the shuffling of protein domains seems to be an efficient evolutionary strategy to generate proteins with new functions.

There is not only evidence that genes have been passed between prokaryotes, but also from vertebrates into bacteria. Comprehensive investigations on several protein superfamilies led to the conclusion that nature has re-engineered the same protein fold over and over to provide the manifold protein functions that are needed. In either case, Type HE REases benefit from their additional DNA-binding domains in that they acquired the capability to interact simultaneously with two recognition sites. This is at the same time a striking feature of enzymes participating in cellular site-specific recombination and transposition processes. Achieving the capacity to transpose DNA could be regarded as a great evolutionary advantage for Type HE REases because that way they could accomplish horizontal distribution of their encoding genes, like transposable elements.

MAPPING EPIGENETIC DNA MODIFICATIONS

Besides the prototypes EcoRII and Nael, several other REases originating from different microorganisms were found to require the interaction with two recognition sites in the DNA for their endonucleolytic activity. REases AtuBI, BspMI, Cfr9I, Hpall, Narl, SacII, SauBMKI, Eco57I, and Ksp632I can be stimulated by addition of synthetic oligonucleotide duplexes containing the specific recognition sequence to cleave all copies of the recognition sequence in a DNA molecule. In the case of BspMI, addition of cleavable sites *in trans* led to an increase of V_{max}, and in the case of Hpall, Narl, and SacII to an increase of K_m, respectively. It is conceivable that the enzymatic activity of these enzymes also significantly depends on the recognition site frequency and the distance between recognition sites.

For the tetrameric Type US restriction endonuclease BspMI, it was reported that it acts concurrently with two DNA sites and cuts both DNA strands at both sites. Nevertheless, detailed investigations on the other enzymes have to be done. Inasmuch as these REases, as EcoRII and Nael, are commonly used as tools in molecular biological research, the described method of enzyme activation is valuable to provide complete DNA digests. Exploiting the experiences with stimulation of EcoRII and Nael by modified oligonucleotide duplexes, this method was commercialized for Nael with an oligonucleotide duplex containing a modified Nael recognition sequence that is not cleaved itself and offered as Turbo Nael. Genome mapping, determination of restriction fragment length polymorphisms, and DNA methylation pattern analyses rely on the complete cleavage of all unmethylated specific recognition sites in a DNA molecule.

Using some of the described REases for these procedures may lead to misinterpretations of the DNA cleavage patterns, *e.g.,* in DNA methylation analyses whenj uncleaved DNA recognition sites are stringently associated with an interfering DNA methylation. It was shown that such potentially misinterpreted results can appear during detection of Dcm methylation (5'Cm5CA/TGG) in bacterial DNA with the isoschizomeric REase pair EcoRII and BstNI as well as the detection of eukaryotic 5′-C5 mCGG methylation with the pair Hpall and Mspl. Methylation-sensitive REasesi EcoRII and Hpall belong to the enzymes that only incompletely cleave DNA depending on the frequency and on the distance of the recognition sites and thus, they become sources of errors in determining DNA methylation patterns. This experimental problem was solved by adding site-specific oligonucleotide duplexes to the EcoRII and Hpall restriction digests, respectively.

The DNA activator molecules stimulated DNA cleavage at all unmethylated recognition sites leaving the resistance of methylated DNA recognition sites unaffected. Furthermore, the method of REase stimulation was adapted for cleavage of agarose-embedded high-molecular DNA. Thus, activationi of methylation-sensitive REases Hpall and EcoRII by cognate site-containing oligonucleotide duplexes improves the reliability of CpG and CpNpG cytosine methylation detection in small as well as in large DNA molecules. This approach is of some importance for the correct determination of epigenetic signals in current genomics.

FINAL REMARKS

During the last years several other Type II REases with different oligomeric structures and reaction mechanisms, thus belonging to different subtypes of REases, were reported to require two DNA recognition sites for exertion of catalytic activity. Moreover, the functional dependence on two copies of the recognition sequence in the substrate DNA is also shared by REases of Types I, III, and IV, albeit the different enzymes follow different molecular mechanisms to interact with the DNA.